Power Systems

Abhisek Ukil

Intelligent Systems and Signal Processing in Power Engineering

Abhisek Ukil
Author

# Intelligent Systems and Signal Processing in Power Engineering

With 239 Figures and 36 Tables

*Author*
Dr. Abhisek Ukil
ABB Corporate Research
Segelhofstrasse 1K
CH-5405 Baden-Daetwill
Switzerland
abhisek.ukil@ch.abb.com
abhiukil@yahoo.com

ISSN 1612-1287
ISBN 978-3-540-73169-6 Springer Berlin Heidelberg New York

Library of Congress Control Number: 2007929722

Matlab® is a registered trademark of Mathworks Inc.

In short, no guaraentees, whatsoever, are given for the example computer programs provided in this book. They are intended for demonstration purpose only. The author or the publisher would not be responsible for any consequences or damages of any sort from the usage of the programs or any other relevant ideas from the book.



Springer is a part of Springer Science+Business Media
springer.com


Typesetting: Integra Software Services Pvt. Ltd., India

Cover Design: deblik, Berlin

Printed on acid-free paper SPIN: 11884910 60/3180/Integra 5 4 3 2 1 0

*Dedicated to all my teachers who enabled me to write this book, and my family & friends for supporting me all along.*

# Preface

Power engineering is truly one of the main pillars of the electricity-driven modern civilization. And over the years, power engineering has also been a multidisciplinary field in terms of numerous applications of different subjects. This ranges from linear algebra, electronics, signal processing to artificial intelligence including recent trends like bio-inspired computation, lateral computing and the like. Considering the reasons behind this, in one hand, we have vast variety of application sub-domains in power engineering itself; on the other hand, problems in these sub-domains are complex and nonlinear, requiring other complementary techniques/fields to solve them.

Therefore, there is always the need of bridging these different fields and the power engineering. We often encounter the problem of distributed and scattered nature of these different disciplines while working in various sub-domains of power engineering, trying to apply some other useful techniques for some specific problem. Oftentimes, these are not direct field of work for many power engineers/researchers, but we got to use them. This book is urged by that practical need.

As the name suggests, the book looks into two major fields (without undermining others!) used in modern power systems. These are the intelligent systems and the signal processing. These broad fields include many topics. Some of the common and useful topics are addressed in this book.

In this book, the intelligent systems section comprises of fuzzy logic, neural network and support vector machine. Fuzzy logic, driven by practical humanoid knowledge incorporation, has been a powerful technique in solving many nonlinear, complex problems, particularly in the field of control engineering. Neural network, on the other hand, is inspired by the biological neuronal assemblies that enable the animal kingdom (including us!) to perform complex tasks in everyday life. Support vector machine is a relatively newer field in machine learning (neural network also falls in this category) domain. It augments the robustness of machine learning scenario with some new concepts and techniques. Although there are many more extensions of the concept of machine learning and intelligent systems, we confine ourselves to these three topics in this book. We look at some theories on them without assuming much particular background. Following the theoretical basics, we study their applications in various problems in power engineering, like, load forecasting, phase balancing, disturbance analysis and so on. Purpose of these, so called, application studies in power engineering is to demonstrate how we can utilize the theoretical concepts. Finally, some research information are included, showing utilizations of these fields in various power systems domains as a starting point for further futuristic studies/research.

In the second part, we look into the signal processing which is another universal field. Whenever and oftentimes we encounter signals, we need to process them! Power engineering and its enormous sub-fields are no exceptions, providing us with ample voltage, current, active/reactive power signals, and so forth. Therefore, we look in this section about the basics of the system theory, followed by fundamentals of different signal processing transforms with examples. After that, we look into the digital signal processing basics including the sampling technique and the digital filters which are the ultimate (signal) processing tools. Similar to the intelligent systems part, here also the theoretical basics are substantiated by some of the applications in power engineering. These applications are of two types: full application studies explained like in-depth case-studies, and semi-developed application ideas with scope for further extension. This also ends up with pointers to further research information.

As a whole, the book looks into the fields of intelligent systems and signal processing from theoretical background and their application examples in power systems altogether. It has been kind of hard to balance the theoretical aspects as each of these fields are vast in itself. However, efforts have been made to cover the essential topics. Specific in-depth further studies are pointed to the dedicated subject intensive resources for interested readers. Application studies are chosen with as much real implications as possible.

Finally, the book is a small effort to bridge and put together three great fields as a composite resource: intelligent systems, signal processing and power engineering. I hope this book will be helpful to undergraduate/graduate students, researchers and engineers, trying to solve power engineering problems using intelligent systems and signal processing, or seeking applications of intelligent systems and signal processing in power engineering.

April, 2007 Abhisek Ukil

# Contents

# Chapter 1
# Introduction

## 1.1 About the Book

Power engineering is an ever-growing, important, multi-dimensional field for electrical engineering students and the associated industry people. And with the increasing applications of the intelligent systems, signal processing techniques, power engineering has truly become a multi-disciplinary field. Modern applications of intelligent systems, e.g., fuzzy logic, neural network, support vector machines, etc and signal processing, like, the Fourier transform-based digital filters, etc are being applied more and more in various sub-domains of the vast power engineering. These include harmonic analysis, load-forecasting, load-balancing, load-profiling, disturbance analysis, fault classification, energy management, energy efficient operation and so on.

However, it is often difficult to find a resource off the shelf which can altogether provide basic understanding of the various important intelligent systems and signal processing technologies along with possible applications in the power engineering. Power engineering students, researchers and industry people often have to search exhaustively for the different scattered specialized literatures to work on some interdisciplinary applications. This book is exactly aimed at that. It is intended to provide concise and basic theoretical foundations in the various intelligent systems and signal processing technologies, along with detailed discussions of different applications of them under one cover in a modular fashion. Exclusive subject-intensive references are provided for further specialized studies. Modern, up to date applications as thorough case studies alongside many a prospective project idea would nurture the current research trends and inspire future exhaustive, inter-disciplinary research works involving intelligent systems, signal processing and power engineering.

The subject overview of the book is depicted in Fig. 1.1.

## 1.2 Prospective Audience

The book is primarily for the graduate and the undergraduate students of electrical engineering, power engineering and the related fields. This book is not a textbook

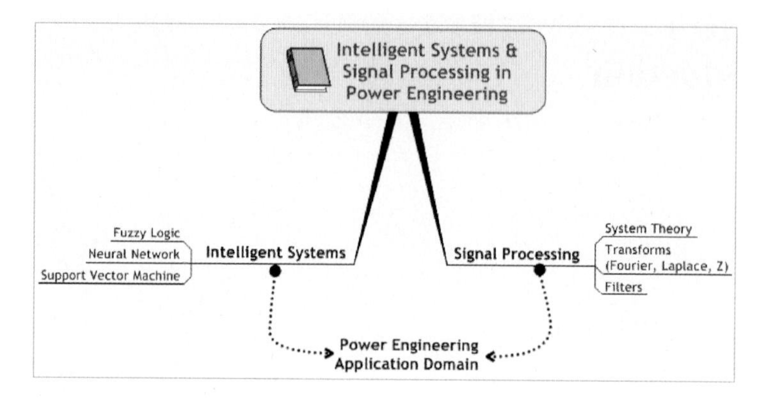


**Fig. 1.1** Subject overview of the book

for power engineering, rather it is mainly oriented towards inter-disciplinary applications in power engineering, demonstrating how to apply intelligent systems and signal processing techniques in power engineering applications. However, ample power engineering specific and application-oriented references are provided in order to follow up particular further research objective.

Also, power engineering researchers and industry people would be interested at the dispositions of the different multi-domain present and future trends of research in power engineering. Basic theoretical discussions are followed up with ample pointers to the up to date specialized references, which should be a starting/supporting point for multi-disciplinary projects involving the intelligent systems, signal processing and power engineering.

Basic understanding of power engineering concepts, matrix computation, complex numbers are in general assumed. Nevertheless, requisite reference books/resources are suggested for particular topics in parallel to the application discussions on those topics. Some application studies include example application/simulation computer codes using Matlab®. For this, basic understanding and availability of the Matlab® software (http://www.mathworks.com) to the reader is assumed. However, this is not a must.

## 1.3 Organization of the Book

### *1.3.1 Book Chapters*

The book is broadly divided into two parts, part I dealing with the intelligent systems and part II, the signal processing. Both part I and II would be modular in nature, treating each specific topics with in depth theory, ample practical applications, future directions and references. Here, 'modular' means mostly mutually exclusive, self-contained chapters. However, some chapters share some points, like the neural network and the support vector machine share some common grounds on machine learning.

The book starts with this introductory chapter explaining the scope and the layout of the book. This is followed by part I: intelligent systems and part II: signal processing. Part I contains three chapters: Chap. 2 on fuzzy logic, Chap. 3 on neural network, Chap. 4 on support vector machine. Part II contains Chap. 5 on signal processing. The chapters are shown in Fig. 1.2.

### *1.3.2 Chapter Structure*

As mentioned above, the chapters are mostly modular in nature. Each chapter is divided into four main sections. These are:

1. Section I: Theory
2. Section II: Application Study
3. Section III: Objective Projects
4. Section IV: Information Section.

For each chapter, Sect. I, i.e., the theory section starts with brief background description, history, then the detailed theoretical discussions accompanied by examples and reference. The theoretical section would be followed and substantiated by Sect. II: application study. This contains few full-length detailed application examples (like case-study, assignment work) in the power engineering domain. Each example will have its own reference section. In many cases, example application/simulation computer codes (using Matlab®) are provided. However, the reader should check the compatibility of his/her Matlab® version as some functionality might differ. This will be followed by several semi-developed project ideas (in line with current and future trends of research in power engineering) along with the

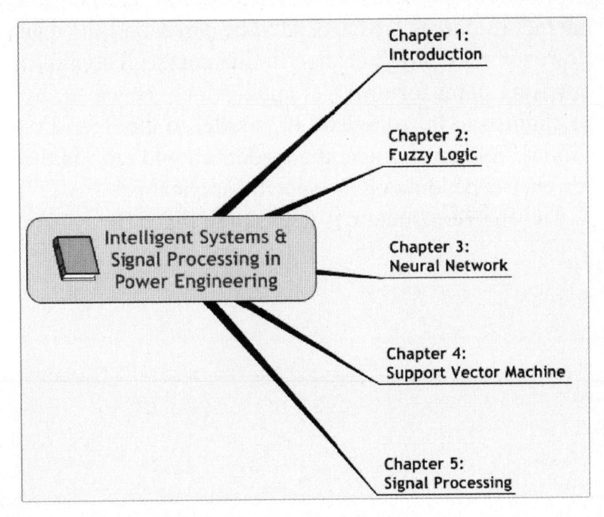


**Fig. 1.2** Chapters of the book

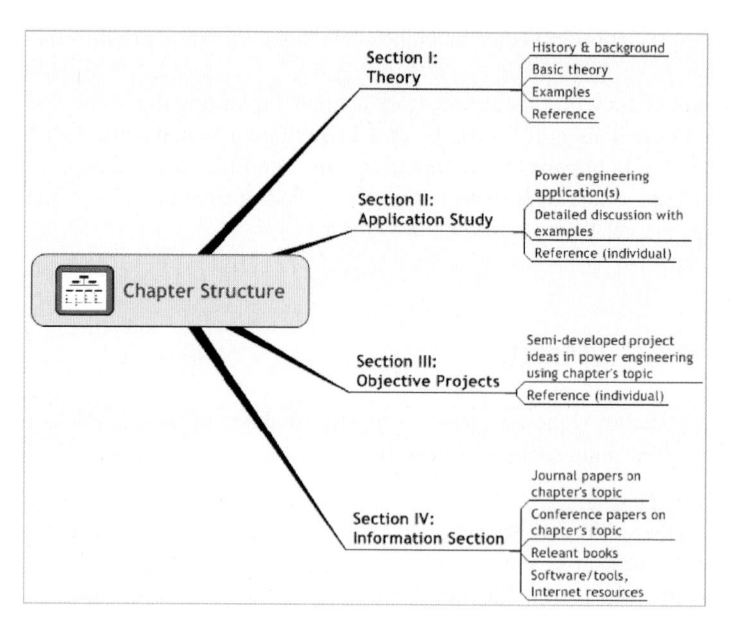


**Fig. 1.3** Chapter structure of the book

references in Sect. III: objective projects. In Sect. III, also each project idea will contain its own reference section at the end of its discussion. Next, there will be Sect. IV: information section about up to date ongoing research worldwide utilizing the specific chapter topic, e.g., Sect. IV of Chap. 2 would provide research information on the applications of the fuzzy logic in various power engineering subfields. For the specific chapter-topic, this section will comprise of ongoing research works worldwide along with exclusive reference pointers in terms of journal papers, conference papers, books, tutorial/technical notes, and possible software/tool information. Internet websites are cited where possible. It is to be noted that the Internet resources are checked to be correct till the date of publication of the book. However, they might change in due course. The aim of the Sect. IV is to provide a starting point for different applications. However, by no means, this section can be claimed to be complete. In parallel to the Sect. IV which should be viewed as a model reference guide, the readers should rely on their subject-specific literature searches depending on the specific application.

The chapter structure is depicted in Fig. 1.3.

# Chapter 2
# Fuzzy Logic

## Section I: Theory

## 2.1 Introduction

The word fuzzy invokes an image of uncertain vagueness. Associating such a term with logic might seem nonsense. But as a matter of fact, sometimes traditional logical theories fall short in practical scenario. Practical knowledge sometimes proves more fruitful than theoretical equations. Fuzzy logic was initiated with this view of emphasizing apparently vague practical knowledge into the real life solutions.

Fuzzy logic evolved as a key technology for developing the knowledge-based systems in the control engineering to incorporate practical knowledge for designing controllers. But gradually it progressed as a powerful technique for other fields as well.

Fuzzy logic tries to balance the question of precision. The question that fuzzy logic tries to solve is, should a rough practical answer be more effective than complex precision. Lotfi Zadeh, regarded as the creator of fuzzy logic, remarked on this: "As complexity rises, precise statements lose meaning and meaningful statements lose precision." The aim of fuzzy logic control is to model the human experience and the human decision-making behavior. Translating this statement into real situation means, given an input data space, we put this into a fuzzy black box system which maps it to the desired output space. This fuzzy black box system is a heuristic and modular way for defining the nonlinear process. As mentioned before, fuzzy logic evolved around the control engineering which traditionally uses linear PID (proportional-integral-differential) control around the setpoint. In contrast, fuzzy logic-based technique delinearizes the control away from the setpoint by describing the desired control with the situation and action rules.

### *2.1.1 History and Background*

Historically, fuzzy logic was created by Lotfi Zadeh in the 1960s (Zadeh 1965). Initially, fuzzy logic was aimed at control engineering. Fuzzy logic was designed to represent the knowledge using a linguistic or verbal form, but at the same time to be operationally powerful so that computers can be used. Fuzzy logic is a nonlinear technique to map the input to the output of the real life control objects which are nonlinear, in general. Historically, fuzzy logic was developed as knowledge-based system for control operation. A general definition can be stated (Driankov et al. 1993): "A Knowledge Based System (KBS) for closed-loop control is a control system which enhances the performance, reliability, and robustness of control by incorporating knowledge which can not be accommodated in the analytic model upon which the design of a control algorithm is based, and that is usually taken care of by manual modes of operation, or by other safety and ancillary logic mechanisms." A fuzzy control system is a KBS, implementing expertise of a human operator or process engineer, which does not lend itself to be easily expressed in PID parameters or differential equations but rather in situation–action rules.

### *2.1.2 Applications*

Since its initiation, fuzzy logic has been widely successful in the industry and different fields of study. This has been nicely summarized by Sugeno (Sugeno 1985). Few are mentioned below.

- Control Engineering
  - Supervision of the closed-loop operation, thus complementing and extending the conventional control algorithm
  - Directly realizing the closed-loop operation, thus completely replacing the conventional control algorithm
- Consumer Products
  - Washing machine
  - Camera, Camcorder
  - Microwave ovens
- Industrial Controller
  - Process controller
  - Medical instrumentation
  - Decision-support system
- Power Engineering
  - Power system analysis
  - Power system control
  - Optimal operation
  - Load profiling
- Pattern recognition
  - Neuro-fuzzy applications

  - Image recognition
  - Speech recognition
- Financial sector
  - Market analysis
- Mechanical
  - Process monitoring
  - Quality control.

### 2.1.3 Pros and Cons

- Pros
  - Fuzzy logic is conceptually easy to understand.
  - It is based on natural language which is more realistic compared to the equation-based modeling.
  - It is effective when dealing with poorly defined operations or imprecise data where traditional methods might not be effective, even not applicable at all.
  - Nonlinear functions of arbitrary complexity can be effectively and quickly modeled.
  - It combines experience of experts with the conventional control which enhances the overall operability.
  - Implementation of expert knowledge (like, if the situation is such and such, I should do so and so) also improves the degree of automation by reducing the human intervention.
  - It is flexible in terms of technique, data, application domain, yet a robust nonlinear technique.
  - It is generally faster and cheaper than conventional methods. This reduces the development and maintenance time.
- Cons
  - Performance of a fuzzy regulator is sometimes difficult to analyze due to nonlinear effects.
  - Fuzzy logic heavily depends on real life experience which is the critical factor for the success of such controller. So, lack of experience or desired correctness can hamper the operation.
  - Gaining proper experience can be time dependent. Hence it is difficult to apply fuzzy logic directly into a new field of which little is known. In comparison, well-established field like process control which has a long standing record or history, can easily and effectively adopt fuzzy logic.
  - Sometimes, it is difficult to replace a whole lot of existing system with real experience-based fuzzy system. Instead, fuzzy logic, in these situations, should be used as secondary or supporting element.

## 2.2 Fuzzy Logic

### 2.2.1 Linguistic Approach

In fuzzy logic, natural language-based linguistic notions are usually used for the knowledge representation. These linguistic terms are generally meaning-independent of the particular application domain. Moreover, our everyday experiences get reflected in the linguistic terms. Hence, the linguistic approach incorporates apparently vague terms and notions which are not only quantitative but qualitative as well. In principle, conventional logic-based knowledge representation does not include vague qualitative terms which cannot be measured. Here lies the major difference as well as the strength of fuzzy logic. Some typical linguistic terms are:

- *PL* – positive large
- *PM* – positive medium
- *PS* – positive small
- *ZE* – zero
- *NS* – negative small
- *NM* – negative medium
- *NL* – negative large.

A linguistic variable (Zadeh 1975) is a natural or artificial language-based variable whose values are expressed in words or sentences. A linguistic variable can be expressed using the following framework.

$$\text{Linguistic Variable Framework } \langle X, R_X, U_X, F_X \rangle, \tag{2.1}$$

where,

$X$ is the symbolic name of the linguistic variable (e.g., height, temperature, error, change of error, etc).

$R_X$ is the reference set of linguistic values that $X$ can take. It represents the property of $X$. For example, for the linguistic variable *height* $H$ we could have

$$R_H = \{tall, average, short\}.$$

For the linguistic variable *error* $E$ we get from the example shown earlier,

$$R_E = \{PL, PM, PS, ZE, NS, NM, NL\}.$$

$U_X$ is called the *universe of discourse* (Zadeh 1975) which is the actual physical domain over which the linguistic variable $X$ takes its quantitative values. In case of the height $H$ example, it could be something like [180 cm, 150 cm, 120 cm], for error $E$ it can use a normalized domain.

$F_X$ is the semantic function (Zadeh 1989) which provides a meaning (interpretation) of a linguistic value in terms of qualitative elements of $U$.

$$F_X{:}R_X \rightarrow \tilde{R}_X, \tag{2.2}$$

where $\tilde{R}_X$ is a notation for fuzzy set (Zadeh 1965, Driankov et al. 1993) defined on the universe of discourse $U$, i.e.,

$$\tilde{R}_X = \sum_U \mu_{R_X}(x)/x \quad \text{in case of discrete } U, \tag{2.3}$$

$$\tilde{R}_X = \int_U \mu_{R_X}(x)/x \quad \text{in case of continuous } U. \tag{2.4}$$

$\mu$ is called the *characteristic* function. We show an example below. For the linguistic variable $H$ denoting height, we represent it using the framework $\langle H, R_H, h, F_H \rangle$, where,
$R_H = \{T, A, S\}$, $T$–tall, $A$–average, $S$–short. $h = [200\,\text{cm}, 100\,\text{cm}]$, and
$F_H{:}R_H \rightarrow \tilde{R}_H$ (fuzzy sets are generally represented with $\tilde{}$).

$$\tilde{T} = \tilde{R}_{H1} = \int_{100}^{200} R(x;\, 200, 180, 170)/x, \tag{2.5}$$

$$\tilde{A} = \tilde{R}_{H2} = \int_{100}^{200} R(x;\, 170, 150, 130)/x, \tag{2.6}$$

$$\tilde{S} = \tilde{R}_{H3} = \int_{100}^{200} R(x;\, 130, 120, 100)/x. \tag{2.7}$$

### 2.2.2 Set Theory

Linguistic variables represent the knowledge, but we need a technique to manipulate the linguistic variables. That technique is the fuzzy set theory. However, fuzzy set theory has similarities and dissimilarities than the standard logic and the set theory. So, it will be beneficial to discuss about the classical set theory before getting into the fuzzy set theory.

A set is a collection of any elements, e.g., a set of fruits, set of sports, set of numbers etc. In comparison with the fuzzy sets, classical sets are often referred as *crisp* sets in fuzzy logic.

#### 2.2.2.1 Terminology

We can define the following basic terminologies for the set theory.

- **Elements** $x$: Elements of a set, $A = \{x_1, x_2, \ldots, x_n\}$.

- **Predicate** $P$: Predicate $P(x)$ means that every element $x$ of the set has the property $P{:}A = \{x|P(x)\}$.
- **Characteristic function** $\mu$: It is used (like fuzzy sets) to define the sets. For example, $\mu_A{:}X \to \{0, 1\}$,

$$\mu_A(x) = \begin{cases} 1 \text{ when } x \in A \\ 0 \text{ when } x \notin A \end{cases}.$$

- **Universal-set** $\varepsilon$: Set of all elements considered (equivalent to the universe of discourse in fuzzy logic).
- **Empty (null)-set** $\phi$: A set without any elements.
- **Sub-set**: Set $A$ is a sub-set of set $B$ if all the elements of set $A$ are within (contained in) set $B$. It is represented as, $A \subset B$. For example, if $B$ represents the set of all fruits and $A$ represents the set of apples, then $A$ is a sub-set of $B$.
- **Separate** or **non-overlapping set**: Sets which do not have any common elements (not properties) are called separate or non-overlapping sets. For example, sets of apples and sets of oranges are separate sets.
- **Venn diagram**: Sets can be effectively represented by the venn diagram. The fruit–apple example is shown using the venn diagram in Fig. 2.1.

#### 2.2.2.2 Operations in Set Theory

In this section, we will see certain operations on the sets. It will be effective if we discuss alongside examples. So, we define, the Universal-set $\varepsilon$ as the first ten positive integers, $\varepsilon = \{1, 2, 3, 4, 5, 6, 7, 8, 9, 10\}$, and then, we define three sets: $A = \{1, 2\}$, $B = \{2, 3\}$, $C = \{3, 4, 5\}$.

- **Complement:** Complement of set $A$ (indicated by a superscript $^C$) is defined as Universal – set. $A^C = \{Universal\text{–}set A\} = \{3, 4, 5, 6, 7, 8, 9, 10\}$.
- **Union**[1]**:** $A \cup B = \{1, 2, 3\}$, $A \cup B \cup C = \{1, 2, 3, 4, 5\}$. Figure. 2.2 shows the union operation.
- **Intersection**[2]**:** $A \cap B = \{2\}$, $B \cap C = \{3\}$, $A \cap C = \{\} = \phi$ (empty set). The intersection operation is indicated by the shaded area in Fig. 2.3.

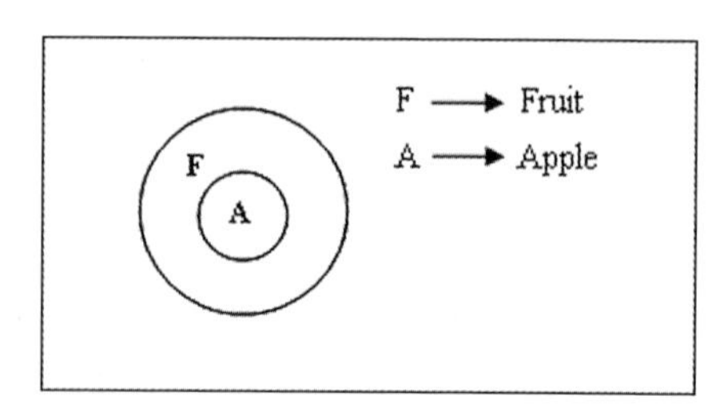


**Fig. 2.1** Venn diagram

[1] Combine all elements, consider common elements only once.

[2] Only consider the common elements.

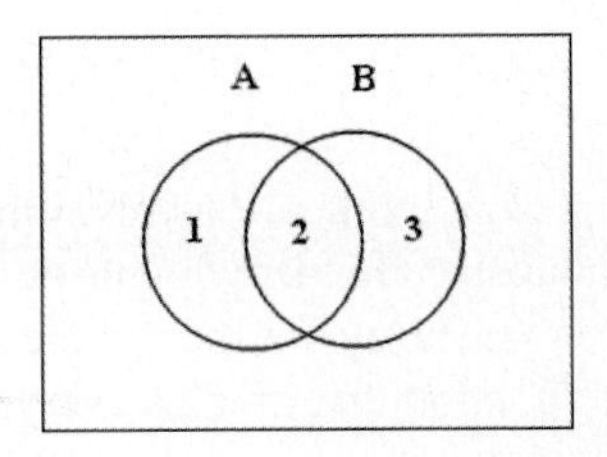


Fig. 2.2 Union operation

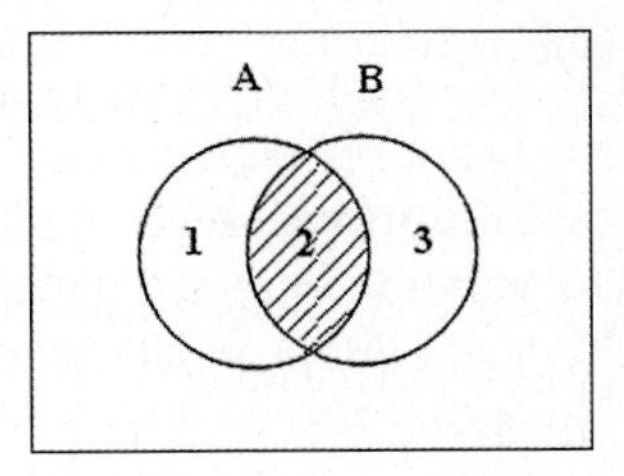


Fig. 2.3 Intersection operation

- **De-Morgan's Theorem:** $(A \cup B)^C = A^C \cap B^C$,

$$(A \cap B)^C = A^C \cup B^C.$$

Using the above example,

$$A \cup B = \{1, 2, 3\}, (A \cup B)^C = \{4, 5, 6, 7, 8, 9, 10\},$$
$$A \cap B = \{2\}, (A \cap B)^C = \{1, 3, 4, 5, 6, 7, 8, 9, 10\}.$$
$$A^C = \{3, 4, 5, 6, 7, 8, 9, 10\}, B^C = \{1, 4, 5, 6, 7, 8, 9, 10\},$$
$$A^C \cap B^C = \{4, 5, 6, 7, 8, 9, 10\}, A^C \cup B^C = \{1, 3, 4, 5, 6, 7, 8, 9, 10\}.$$

So, $(A \cup B)^C = A^C \cap B^C$ and $(A \cap B)^C = A^C \cup B^C$.

#### 2.2.2.3 $t$- and $s$-Norm

- Triangular norm or $t$-norm: Denotes a class of functions that can represent the intersection operation.
- Triangular co-norm or $s$-norm: Denotes a class of functions that can represent the union operation.

Any function $t$: $[0, 1] \times [0, 1] \rightarrow [0, 1]$ is called a $t$-norm if it satisfies the following four conditions.

1. Boundary conditions: $t(0, 0) = 1, t(x, 1) = x$,
2. Commutativity: $t(x, y) = t(y, x)$,
3. Monotonicity: If $x \leq \xi$ and $y \leq \upsilon$ then $t(x, y) \leq t(\xi, \upsilon)$,
4. Associativity: $t(t(x, y), z) = t(x, t(y, z))$.

### 2.2.3 *Fuzzy Set Theory*

Fuzzy set theory involves manipulation of the fuzzy linguistic variables. Fuzzy sets involve the fuzzy linguistic variables, as per the framework shown in (2.1).

In fuzzy set theory, the characteristic function is generalized to a membership function that assigns every element $x \in U$ a value from the interval [0,1] instead of the two-element set {0,1}. The membership function $\mu_F$ of a fuzzy set $F$ is a function

$$\mu_F : U \to [0, 1]. \tag{2.8}$$

So, every element $x$ from the universe of discourse $U$ has a membership degree $\mu_F(x) \in [0, 1]$. Fuzzy set $F$ is completely determined by the set of tuples (Zadeh 1965)

$$F = \{(x, \mu_F(x)) \,|x \in U\}. \tag{2.9}$$

In (2.9), on the left hand side, $F$ is called the *fuzzy set*; on the right hand side, the term $(x, \mu_F(x))$ is called the *membership function* and the term $x \in U$ is called the *universe* or *universe of discourse*.

The fuzzy set $F$ for discrete $U$ is described by

$$F = \mu_F(x_1)/x_1 + \mu_F(x_2)/x_2 + \ldots + \mu_F(x_n)/x_n = \sum_{i=1}^{n} \mu_F(x_i)/x_i\,, \tag{2.10}$$

and for continuous $U$ by

$$F = \int_U \mu_F(x)/x. \tag{2.11}$$

#### 2.2.3.1 Properties of Fuzzy Sets

1. The *support* of a fuzzy set $A$

$$S(A) = \{x \in X|\mu_A(x) > 0\}. \tag{2.12}$$

2. The *width*[3] of a fuzzy set $A$

$$width(A) = \max(S(A)) - \min(S(A)). \tag{2.13}$$

[3] It is possible to have *left* and *right width* for asymmetrical functions.

3. The *nucleus*[4] of fuzzy set $A$

$$nucleus(A) = \{x \in X | \mu_A(x) = 1\}. \quad (2.14)$$

#### 2.2.3.2 Operations on Fuzzy Sets

1. Two fuzzy sets $A$ and $B$ are equal if

$$\forall x \in X : \mu_A(x) = \mu_B(x). \quad (2.15)$$

2. $A$ is a subset of $B$ $(A \subseteq B)$ if

$$\forall x \in X : \mu_A(x) \leq \mu_B(x). \quad (2.16)$$

3. *Complement* of $A$ denoted as $A'$,

$$\forall x \in X : \mu_{A'}(x) = 1 - \mu_A(x). \quad (2.17)$$

4. *Intersection* of $A$ and $B$ denoted as $A \cap B$ (logical operator *AND*)

$$\forall x \in X : \mu_{A \cap B}(x) = \min(\mu_A(x), \mu_B(x)). \quad (2.18)$$

5. *Union* of $A$ and $B$ denoted as $A \cup B$ (logical operator *OR*)

$$\forall x \in X : \mu_{A \cup B}(x) = \max(\mu_A(x), \mu_B(x)). \quad (2.19)$$

6. *Convexity:* a fuzzy set $A$ is convex if for any $\lambda$ in [0, 1],

$$\forall x \in X : \mu_A(\lambda x_1 + (1 - \lambda)x_2) \geq \min(\mu_A(x_1), \mu_A(x_2)). \quad (2.20)$$

#### 2.2.3.3 Membership Functions

In fuzzy control, usually four kinds of fuzzy sets are considered, *increasing*, *decreasing* and two kinds of *approximating* functions, *triangular* and *trapezoidal.* Membership functions for these fuzzy sets are described below.

1. Increasing function

$$\Gamma(x; \alpha, \beta) = \begin{cases} 0 & x < \alpha \\ \dfrac{x - \alpha}{\beta - \alpha} & \alpha \leq x \leq \beta. \\ 1 & x > \beta \end{cases}. \quad (2.21)$$

   This is shown in Fig. 2.4.

[4] If there is only one point with membership degree equal to 1, then this point is called the *peak value* of $A$.

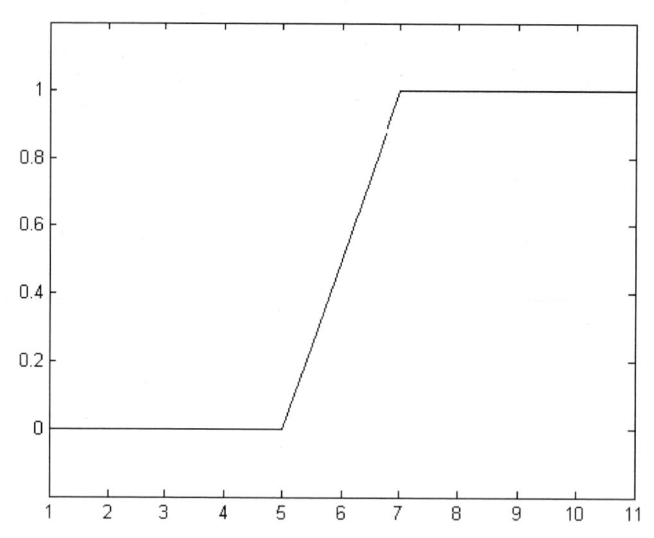


**Fig. 2.4** Increasing function

2. Decreasing function

$$L(x;\alpha,\beta)=\begin{cases}1 & x<\alpha\\ \dfrac{\beta-x}{\beta-\alpha} & \alpha\le x\le\beta\\ 0 & x>\beta\end{cases}\,. \tag{2.22}$$

This is shown in Fig. 2.5.

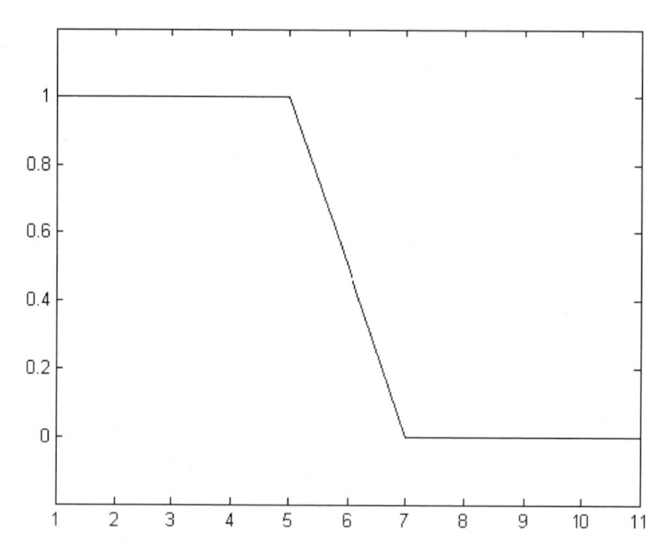


**Fig. 2.5** Decreasing function

3. Triangular approximating function

$$\Lambda(x;\alpha,\beta,\gamma)=\begin{cases}0 & x<\alpha\\ \dfrac{x-\alpha}{\beta-\alpha} & \alpha\le x<\beta\\ \dfrac{\gamma-x}{\gamma-\beta} & \beta\le x\le\gamma\\ 0 & x>\gamma\end{cases}. \tag{2.23}$$

This is shown in Fig. 2.6.

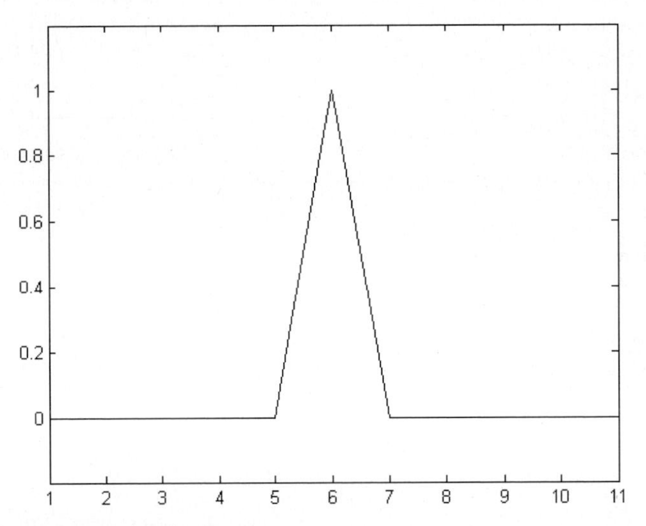


**Fig. 2.6** Triangular approximating function

4. Trapezoidal approximating function

$$\Pi(x;\alpha,\beta,\gamma,\delta)=\begin{cases}0 & x<\alpha\\ \dfrac{x-\alpha}{\beta-\alpha} & \alpha\le x<\beta\\ 1 & \beta\le x\le\gamma\\ \dfrac{\delta-x}{\delta-\gamma} & \gamma<x\le\delta\\ 0 & x>\delta\end{cases}. \tag{2.24}$$

This is shown in Fig. 2.7.

5. Sigmoidal approximating function
A smooth variant of the $\Gamma$ function is the sigmoidal $S$ function,

$$S(x)=\frac{1}{1+e^{-cx}},\quad \text{where}\quad 0<c\le 1. \tag{2.25}$$

The constant $c$ determines the shape of the sigmoid function. For simplification, we can consider $c=1$. Higher values of $c$ bring the shape of the sigmoid closer

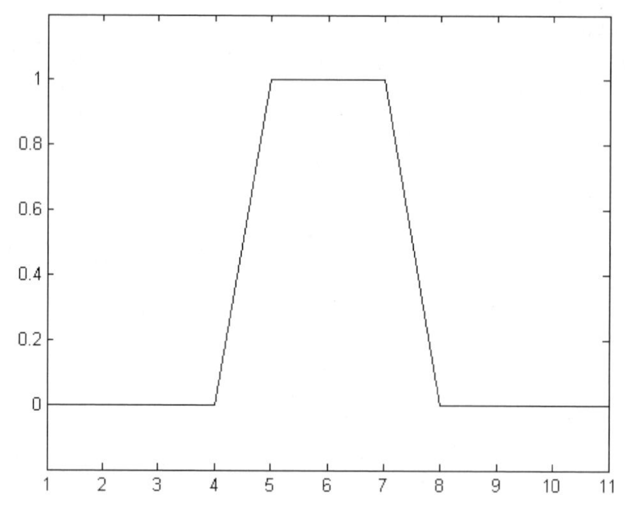


**Fig. 2.7** Trapezoidal approximating function

to that of the step function. And in the limit $c \to \infty$, the sigmoid converges to a step function at the origin. Figure 2.8 shows the sigmoid function for the different values of $c$.

6. Bell-shaped approximating function
   A smooth variant of $\Lambda$ function is the bell-shaped $\pi$ function,

$$\pi(x) = \exp\left(-x^2/\sigma^2\right) \tag{2.26}$$

where $\sigma$ is a real parameter (variance). This is shown in Fig. 2.9

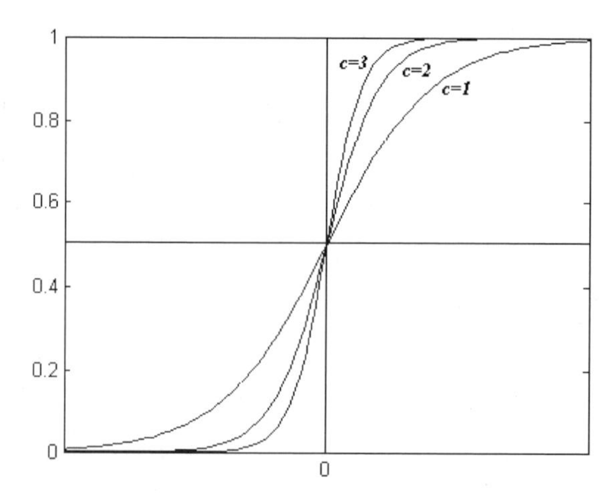


**Fig. 2.8** Sigmoidal function

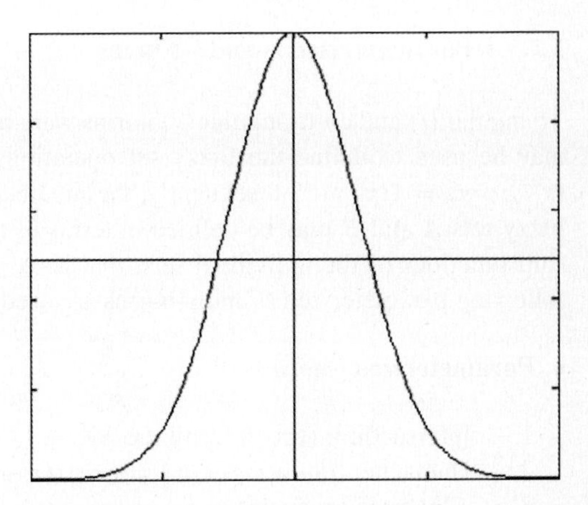

Fig. 2.9 Bell-shaped function

However, the sigmoidal and the bell-shaped functions have relatively low practical uses in fuzzy control. Other membership function terminologies like core, crossover points, $\alpha$-cut, support are depicted in Fig. 2.10.

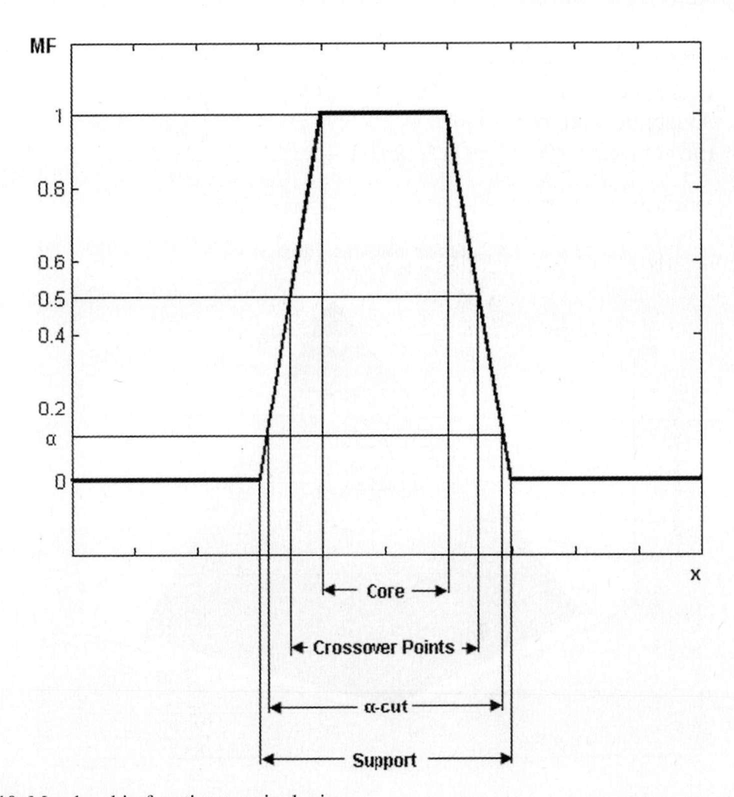


Fig. 2.10 Membership function terminologies

#### 2.2.3.4 Parameterized t- and s-Norms

Triangular ($t$) and co-triangular ($s$) norms were discussed in Sect. 2.2.2.2. A $t$-norm may be used to define the fuzzy set operations and of two fuzzy values, where $(x \ and \ y) = t(x, y)$. Subsequently, the membership function of the union of two fuzzy sets $A$ and $B$ may be defined in terms of the $t$-norm applied to the membership functions of the individual sets, that is, $\mu_{(A \cup B)}(x) = t\,(\mu_A(x), \mu_B(x))$. The following parameterized $t$- and $s$-norms are used to describe fuzzy sets.

- **Parameterized *t-norms***

  1. Intersection: $t(x, y) = \min(x, y)$.
  2. Hamacher product: $t(x, y) = (xy)/(x + y - xy)$.
  3. Algebraic product: $t(x, y) = xy$.
  4. Einstein product: $t(x, y) = (xy)/(1 + (1 - x)(1 - y))$.
  5. Bounded difference: $t(x, y) = \max\,(0, x + y - 1)$.
  6. Drastic product: $t(x, y) = \begin{cases} \min(x, y) & if \quad x = 1 \ or \ y = 1 \\ 0 & otherwise \ if \ x, y < 1 \end{cases}$.

  These are shown in Figs. 2.11–2.16.
- **Parameterized *s-norms***

  1. Union: $t(x, y) = \max(x, y)$.
  2. Hamacher sum: $t(x, y) = (x + y - 2xy)/(1 - xy)$.
  3. Algebraic sum: $t(x, y) = x + y - xy$.
  4. Einstein sum: $t(x, y) = (x + y)/(1 + xy)$.

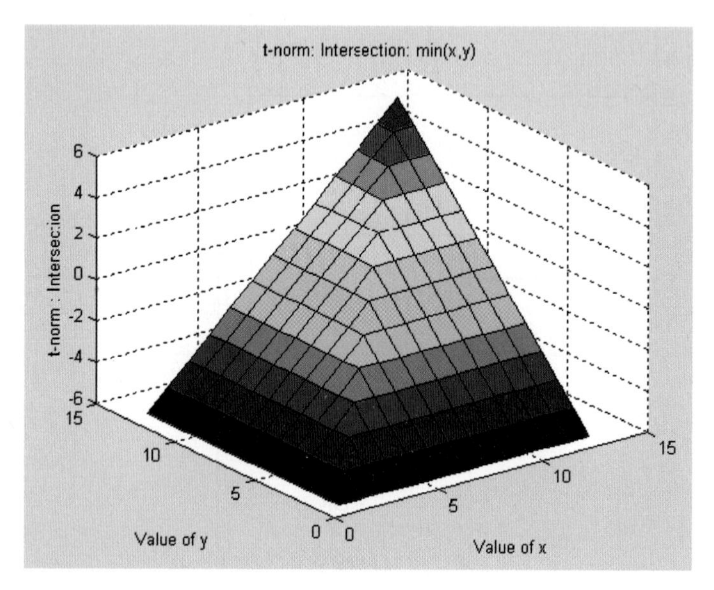


**Fig. 2.11** $t$-norm: intersection

5. Bounded sum: $t(x, y) = \min(1, x + y)$.
6. Drastic sum: $t(x, y) = \begin{cases} \max(x, y) & if\ x = 0\ or\ y = 0 \\ 1 & otherwise\ if\ x, y > 0 \end{cases}$.

These are shown in Figs. 2.17–2.22.

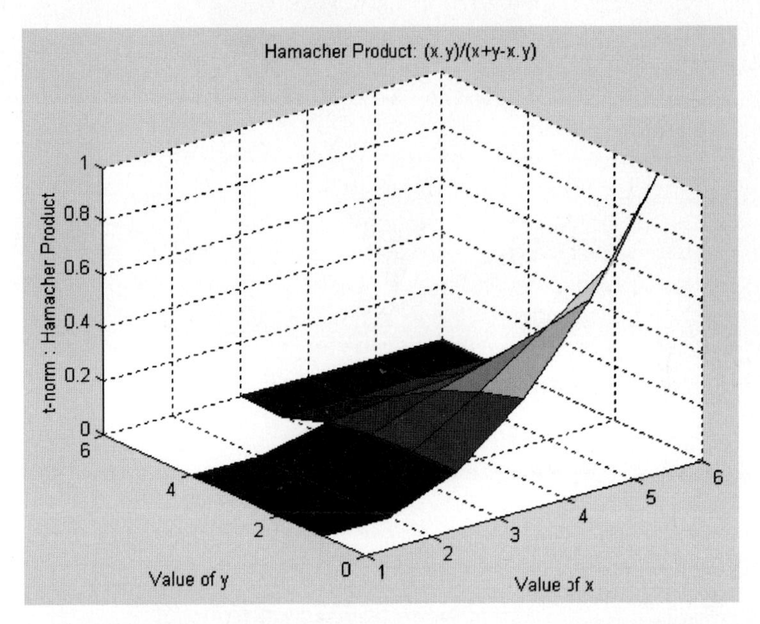


Fig. 2.12 $t$-norm: Hamacher product

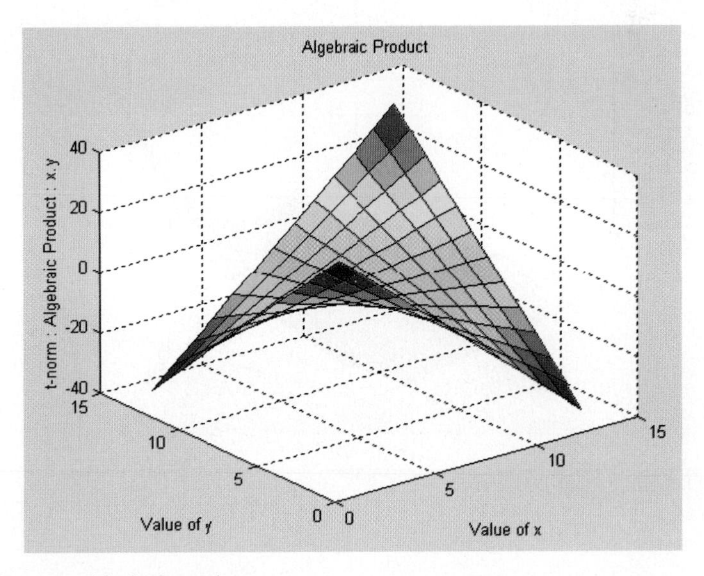


Fig. 2.13 $t$-norm: Algebraic product

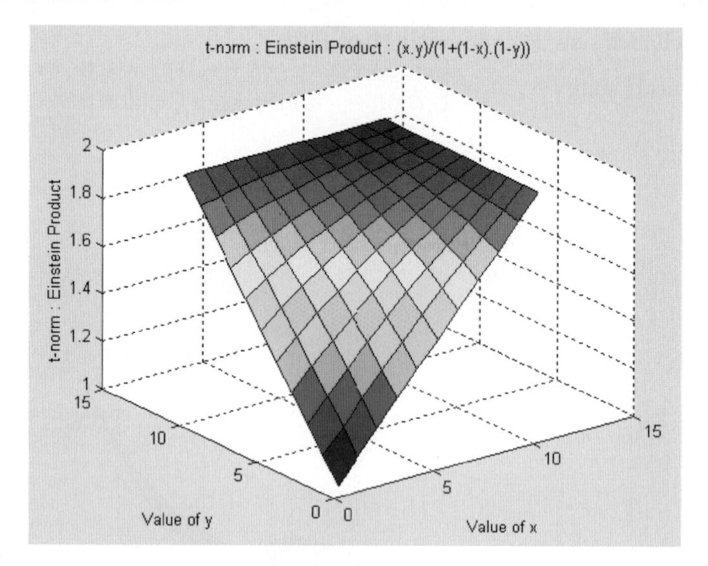


**Fig. 2.14** $t$-norm: Einstein product

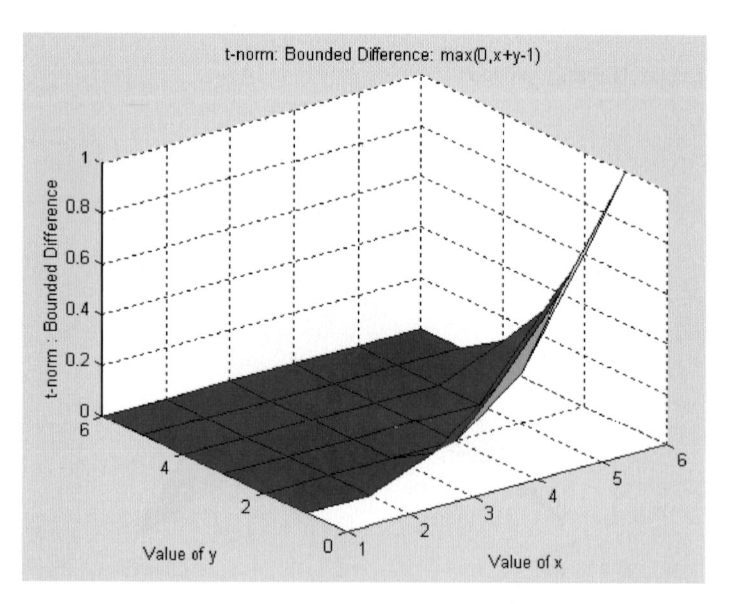


**Fig. 2.15** $t$-norm: bounded difference

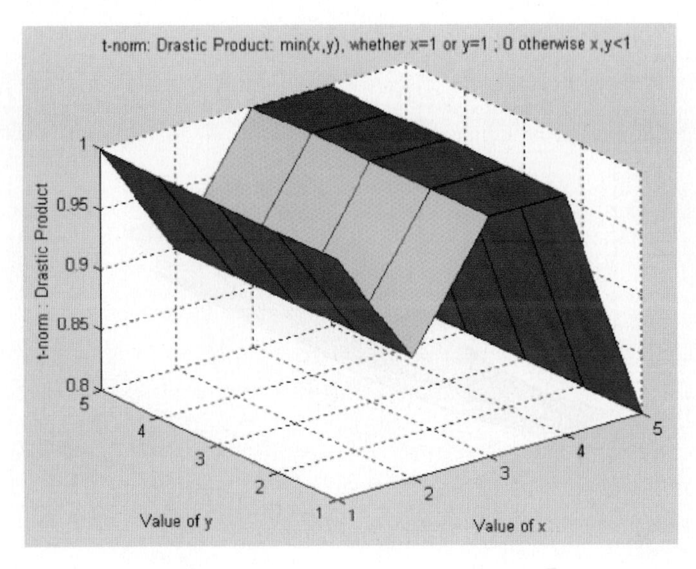


**Fig. 2.16** $t$-norm: drastic product

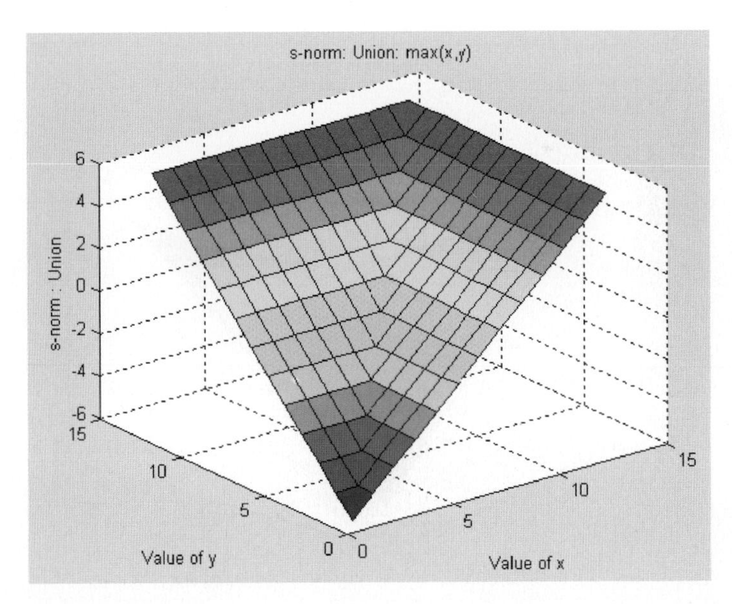


**Fig. 2.17** $s$-norm: union

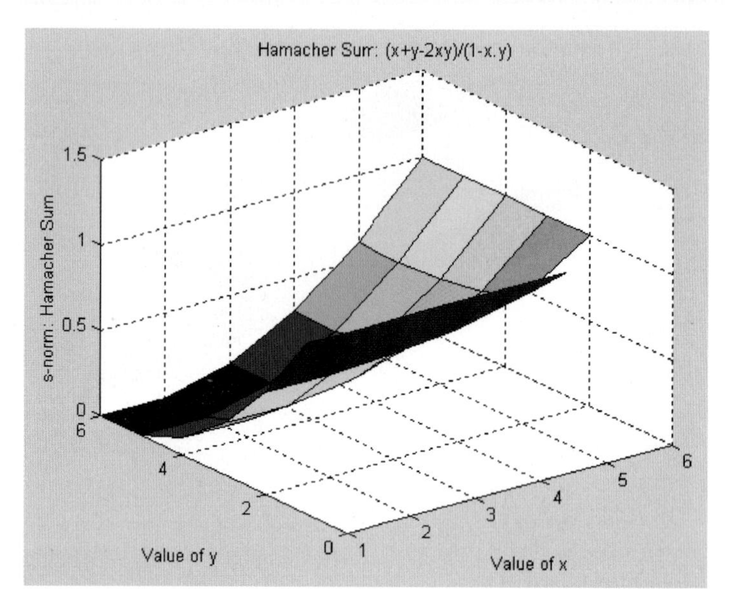


Fig. 2.18 $s$-norm: Hamacher sum

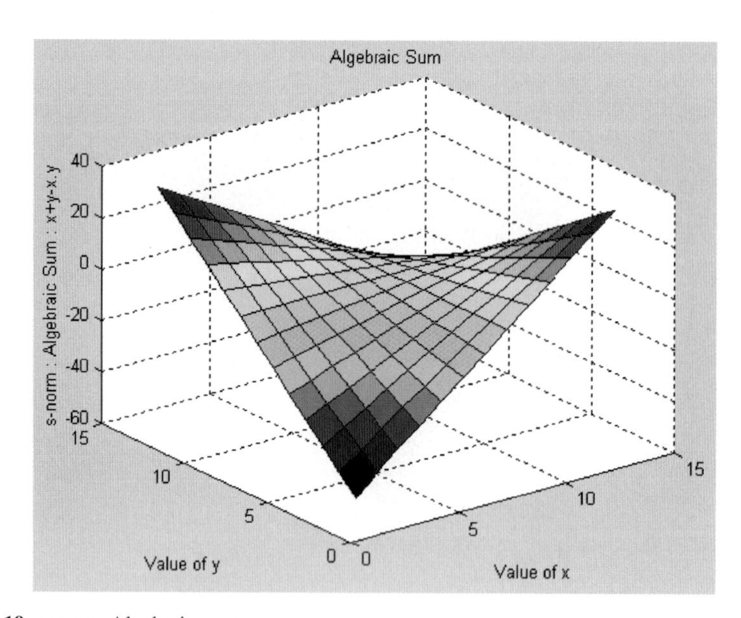


Fig. 2.19 $s$-norm: Algebraic sum

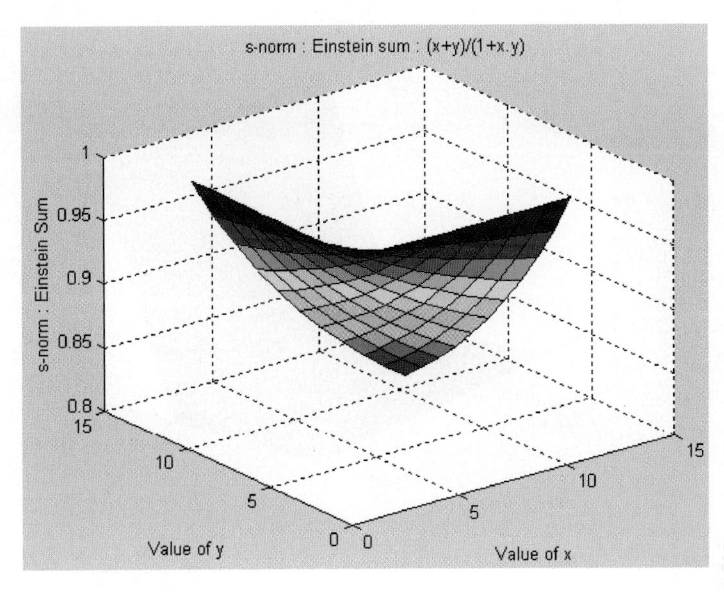


**Fig. 2.20** *s*-norm: Einstein sum

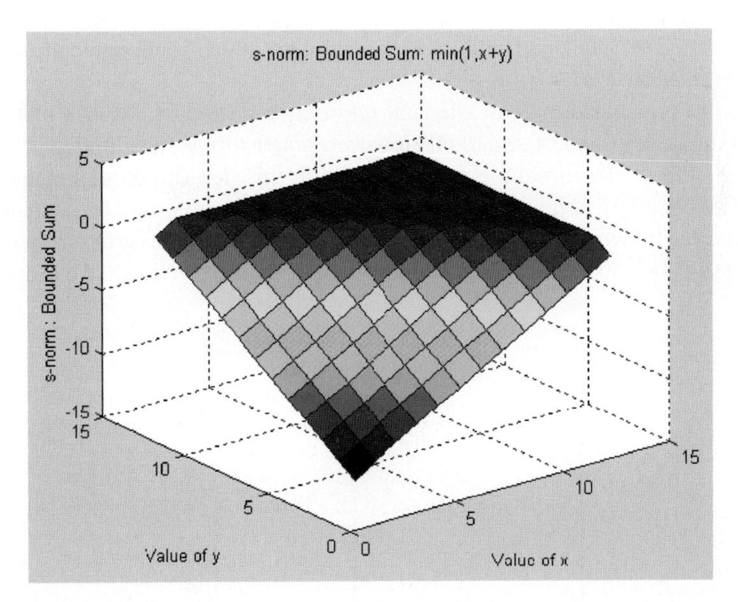


**Fig. 2.21** *s*-norm: bounded sum

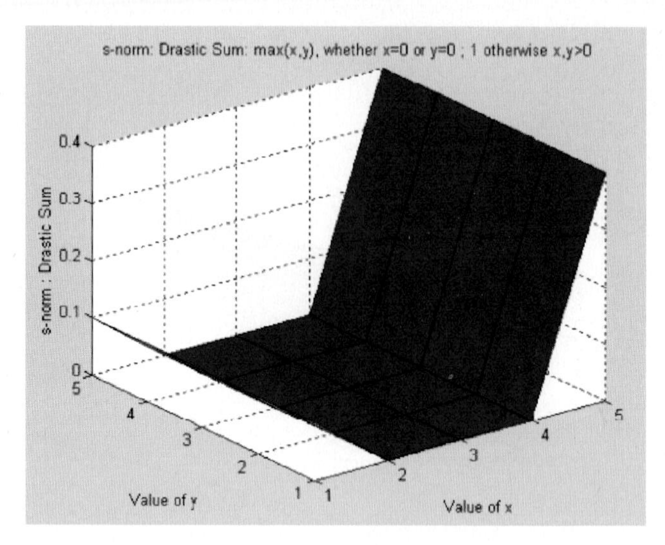


**Fig. 2.22** $s$-norm: drastic sum

## *2.2.4 Classical Set Theory vs. Fuzzy Set Theory*

Fig. 2.23 shows an example graphical representation of the classical set (crisp set) for height. The set shown in Fig. 2.23 has a setpoint of 1.8m above which the heights are considered to be tall, and below which heights are short. Mathematically, this can be represented as $Tall = \{x|x > 1.8\}$.

This is a typical example of classical set or crisp set theory. We can mathematically express this kind of situation. However, practically we can notice the shortcomings of such description. For example, with this description of heights, how should we classify a height of 2.0m and 1.5m? The answers would be tall and short respectively. But what if we ask for a height of 1.799m? With this crisp set definition and classification structure of Fig 2.23, this height should be classified as short! But

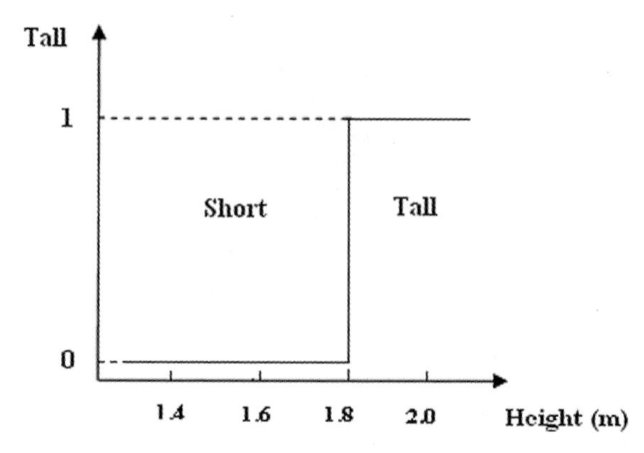


**Fig. 2.23** Crisp diagram for height

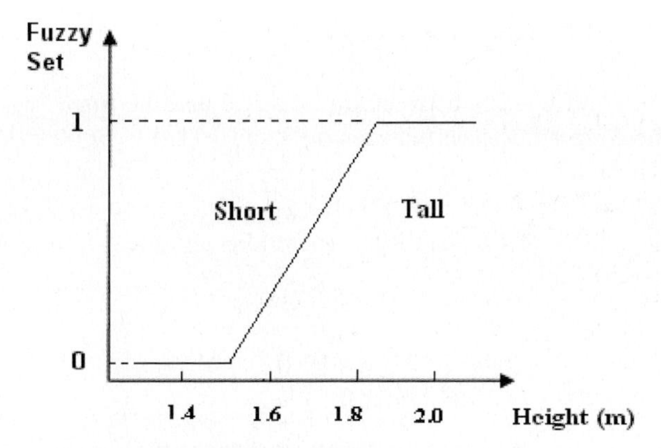


**Fig. 2.24** Fuzzy representation for height

does this satisfy us practically? The obvious answer is no. So, we need a sort of realistic representation which can be mathematically expressed as well as provides an acceptable practical solution. Here comes the fuzzy set theory. The same height set is represented using a fuzzy classification in Fig. 2.24.

With the fuzzy representation, we see a non-binary type rollover from the short to the tall classes. We associate the fuzzy membership function $\mu$. Mathematically, we describe, $Tall = \{x, \mu(x) | x \in X\}$, $X$ being all values of $x$. Hence, we expect a more realistic answer to our question of associating classes for different heights. For the heights of 2.0m and 1.5m, we see from Fig. 2.24 that we get a value of 100% tall and 100% short classes. However, for a height of 1.799m, we get a value of about 90% tall. This is more realistic answer than just saying short. More realistically, instead of just saying tall and short, we can classify the different heights as, 2m: definitely tall, 1.5m: not tall, 1.799m: just tall. That is, with the fuzzy representation, we categorize the heights by assigning a realistic score (membership value) for each points for the entire range of the heights. This score is usually assigned in the normalized scale of [0–1] (which is described as percentage in the foregoing lines for the sake of understanding). We can then utilize this membership description to determine the category of any arbitrary values. An example[5] is shown below.

$$Tall = \frac{\text{Membership Value, } \mu(x)}{\text{Height data point, } x} = \left\{ \frac{0}{1.4}, \frac{0.5}{1.6}, \frac{0.8}{1.7}, \frac{0.9}{1.8}, \frac{0.95}{1.9}, \frac{1.0}{2.0} \right\}.$$

An important difference between the crisp set and fuzzy set representation is that the crisp sets represent the definite probabilities whose total sum is always one (never more than one), while the fuzzy sets represent the possibilities whose total sum is generally more than one.

[5] Here, $\frac{0}{1.4}$ does not indicate a division whose result is 0. It is a symbolic representation which means, for height 1.4m the membership value for Tall class is 0.

#### 2.2.4.1 Standard Logic and Fuzzy Logic

We have seen the differences between the classical and the fuzzy set theory. Synonymously, there are differences between the standard logic (Boolean kind) and the fuzzy logic. However, in a real sense, they are also equivalent (which has to be because otherwise fuzzy logic does not stand a chance to be a valid method). We show four basic kinds of Boolean logic operations and their corresponding fuzzy logic operations in Tables 2.1–2.4.

**Table 2.1** Boolean OR, Fuzzy MAX

| A | B | A OR B | Fuzzy MAX(A,B) |
|---|---|---|---|
| 0 | 0 | 0 | 0 |
| 0 | 1 | 1 | 1 |
| 1 | 0 | 1 | 1 |
| 1 | 1 | 1 | 1 |

**Table 2.2** Boolean AND, Fuzzy MIN

| A | B | A AND B | Fuzzy MIN(A, B) |
|---|---|---|---|
| 0 | 0 | 0 | 0 |
| 0 | 1 | 0 | 0 |
| 1 | 0 | 0 | 0 |
| 1 | 1 | 1 | 1 |

**Table 2.3** Boolean NOR, Fuzzy MAX-COMPLEMENT

| A | B | A NOR B | Fuzzy 1–MAX(A,B) |
|---|---|---|---|
| 0 | 0 | 1 | 1 |
| 0 | 1 | 0 | 0 |
| 1 | 0 | 0 | 0 |
| 1 | 1 | 0 | 0 |

**Table 2.4** Boolean NAND, Fuzzy MIN-COMPLEMENT

| A | B | A NAND B | Fuzzy 1–MIN(A,B) |
|---|---|---|---|
| 0 | 0 | 1 | 1 |
| 0 | 1 | 1 | 1 |
| 1 | 0 | 1 | 1 |
| 1 | 1 | 0 | 0 |

## 2.2.5 Example

We define a universe of discourse as

$$\varepsilon = \{0, 2, 4, 6, 8, 10\}.$$

We define the following fuzzy membership functions for the linguistic variables *large*, *medium* and *small*.

$$large = \mu_l = \left\{\frac{0}{0}, \frac{0}{2}, \frac{0.1}{4}, \frac{0.3}{6}, \frac{0.6}{8}, \frac{1}{10}\right\},$$

$$medium = \mu_m = \left\{\frac{0.3}{0}, \frac{0.6}{2}, \frac{1}{4}, \frac{1}{6}, \frac{0.6}{8}, \frac{0.3}{10}\right\},$$

$$small = \mu_s = \left\{\frac{1}{0}, \frac{0.6}{2}, \frac{0.3}{4}, \frac{0.1}{6}, \frac{0}{8}, \frac{0}{10}\right\}.$$

It is to be noted that large, medium or small numbers from the universe of discourse are given fuzzy membership values. For example, 8, 10 are large numbers hence they receive membership values of 0.6 and 1 respectively for the *large* class. The same numbers are not so medium for the *medium* class, hence they receive membership values 0.6 and 0.3 respectively, while 4 and 6 are quite medium hence get value of 1 in the *medium* class. The numbers 8 and 10 are not small, hence they receive 0 values in the *small* class in which small numbers like 0, 2 get values 1 and 0.6 respectively. This way, practical intuition is represented as knowledge in the fuzzy representation. We will perform certain fuzzy set manipulations considering the above fuzzy membership functions.

#### 2.2.5.1 Complement

$$\overline{large} = not\ large = \mu_l^C = 1 - \mu_l = \left\{\frac{1}{0}, \frac{1}{2}, \frac{0.9}{4}, \frac{0.7}{6}, \frac{0.4}{8}, \frac{0}{10}\right\}.$$

$$very\ large = large^2 = \mu_l^2 = \left\{\frac{0}{0}, \frac{0}{2}, \frac{0.01}{4}, \frac{0.09}{6}, \frac{0.36}{8}, \frac{1}{10}\right\}.$$

$$not\ very\ large = \overline{very\ large} = 1 - \mu_l^2 = \left\{\frac{1}{0}, \frac{1}{2}, \frac{0.99}{4}, \frac{0.91}{6}, \frac{0.64}{8}, \frac{0}{10}\right\}.$$

Similarly, $rather\ large = \mu_l \pm constant$.

#### 2.2.5.2 Union

Union represents the max operation in the fuzzy set theory.

$$
\begin{aligned}
large\ OR\ medium &= large \cup medium = \mu_l \cup \mu_m \\
&= \left\{ \begin{matrix} \max\left(\frac{0}{0}, \frac{0.3}{0}\right), \max\left(\frac{0}{2}, \frac{0.6}{2}\right), \max\left(\frac{0.1}{4}, \frac{1}{4}\right), \ldots \\ \max\left(\frac{0.3}{6}, \frac{1}{6}\right), \max\left(\frac{0.6}{8}, \frac{0.6}{8}\right), \max\left(\frac{1}{10}, \frac{0.3}{10}\right) \end{matrix} \right\} \\
&= \left\{ \frac{0.3}{0}, \frac{0.6}{2}, \frac{1}{4}, \frac{1}{6}, \frac{0.6}{8}, \frac{1}{10} \right\}.
\end{aligned}
$$

#### 2.2.5.3 Intersection

Intersection represents the min operation in the fuzzy set theory.

$$
\begin{aligned}
medium\ AND\ small &= medium \cap small = \mu_m \cap \mu_s \\
&= \left\{ \begin{matrix} \min\left(\frac{0.3}{0}, \frac{1}{0}\right), \min\left(\frac{0.6}{2}, \frac{0.6}{2}\right), \min\left(\frac{1}{4}, \frac{0.3}{4}\right), \ldots \\ \min\left(\frac{1}{6}, \frac{0.1}{6}\right), \min\left(\frac{0.6}{8}, \frac{0}{8}\right), \min\left(\frac{0.3}{10}, \frac{0}{10}\right) \end{matrix} \right\} \\
&= \left\{ \frac{0.3}{0}, \frac{0.6}{2}, \frac{0.3}{4}, \frac{0.1}{6}, \frac{0}{8}, \frac{0}{10} \right\}.
\end{aligned}
$$

#### 2.2.5.4 Other Operations

Using the above-mentioned basic operations, we can perform any other set manipulations following the linguistic description. For example,

$$
\begin{aligned}
large\ but\ not\ very\ large &= large \cap not\ very\ large = \mu_l \cap (1 - \mu_l^2) \\
&= \left\{ \begin{matrix} \min\left(\frac{0}{0}, \frac{1}{0}\right), \min\left(\frac{0}{2}, \frac{1}{2}\right), \min\left(\frac{0.1}{4}, \frac{0.99}{4}\right), \ldots \\ \min\left(\frac{0.3}{6}, \frac{0.91}{6}\right), \min\left(\frac{0.6}{8}, \frac{0.64}{8}\right), \min\left(\frac{1}{10}, \frac{0}{10}\right) \end{matrix} \right\} \\
&= \left\{ \frac{0}{0}, \frac{0}{2}, \frac{0.1}{4}, \frac{0.3}{6}, \frac{0.6}{8}, \frac{0}{10} \right\}.
\end{aligned}
$$

## 2.3 Fuzzy System Design

The typical elements required to develop a fuzzy system are:

1. **Normalization module:** performs a scale transformation which maps the physical values into a normalized (scaled) universe of discourse.
2. **Fuzzification module:** for a crisp input, it calculates its degee of membership using the linguistic variable (represented by the fuzzy membership function) of the rule-antecedent, if the inference is individual-rule-firing based, or it converts a crisp input into a fuzzy set in the case of composition-based inference.
3. **Inferencing module:** it represents in a structured way the control policy of an experienced process operator by combing the fuzzy rules to govern the system operation.

4. **Defuzzification module:** it converts the modified fuzzy set of system output into a single crisp value.
5. **Denormalization module:** it performs a scale transformation on the crisp output value which maps the crisp value onto its physical domain.

We will discuss about the major modules in details.

### *2.3.1 Fuzzification*

As discussed before, a fuzzy set is totally characterized by the membership functions (MF). Fuzzification is the process of converting crisp numerical values into the degrees of membership related to the corresponding fuzzy sets.

A MF will accept as its argument a crisp value and return the degree to which that value belongs to the fuzzy set the MF represents. The degree of membership returned by any MF is always in the range [0, 1]. A value will have a membership degree of 0 if it is wholly outside the fuzzy set, and 1 if completely within the fuzzy set, or any value in between. Since most variables in a fuzzy system have multiple MFs attached to them, fuzzification will result in the conversion of a single crisp value into several degrees of membership.

Variables and MFs often have names attached to them, also referred to as the linguistic variables. This is one of the major advantages of fuzzy rule systems, as the linguistic variables can be much easier to comprehend than an equivalent crisp rule system.

So, the steps in the fuzzification process are:

- Define a universe of discourse,
- Identify and define the linguistic variables,
- Define the membership functions for each linguistic variables bounded by the universe of discourse,
- Represent the membership functions graphically by choosing suitable membership functions (see sect. 2.2.3.3).

Referring to the example discussed in Sect. 2.2.5, we see that first we defined a universe of discourse. Then, we identified three linguistic variables, *large*, *medium* and *small* within this universe. After this, we constructed the respective membership functions for the three linguistic variables. We depict the membership functions by choosing appropriate function categories in Fig. 2.25.

### *2.3.2 Fuzzy Inference*

If we compare the whole fuzzy system with the cooking of a dish, then fuzzification is the step of preparing the ingredients. Mixing and cooking of the ingredients take place in the inference process. Inference is the recipe. The knowledge representation from the fuzzification step are manipulated in the inference process with the help of

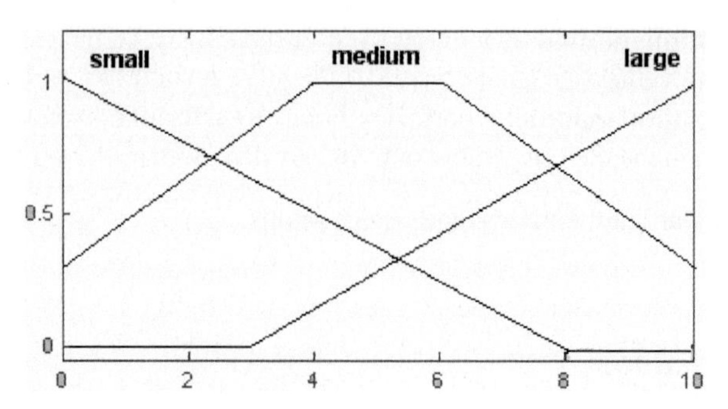


**Fig. 2.25** Fuzzification of the number example

rules. Rules can be thought of as the practical guidelines (much like the cooking recipe) inspired by practical experiences to manipulate the knowledge. In other words, fuzzy logic is a nonlinear technique to map the input space to the output space. And rules are the bridge between the input and the output space.

Let us assume that, $x$ is the input linguistic variable with a membership function $A$, and $y$ is the output linguistic variable with a membership function $B$. Then, the general structure of the rule is,

*IF x is A THEN y is B.*

In the above rule, the portion after the IF and before the THEN ('$x$ is $A$') is called the *premise* or *antecedent* of the rule. And the portion after the THEN ('$y$ is $B$') is called the *conclusion* or *consequent* of the rule. For example, if we are designing a room temperature adjusting fuzzy system, we define an input variable as *room temperature* and an output variable as *fan speed*. We can form rules like,

*IF room temperature is hot THEN fan speed is fast,*

*IF room temperature is cool THEN fan speed is slow.*

Fuzzy inference involves matching fuzzy facts against the antecedents in the rules. When a fuzzy fact matches against a rule, it is termed as '*firing of the rule*' (Zadeh 1973). The degree to which a rule activates depends on the degree to which the facts and antecedents match and the method of fuzzy inference used.

There are two types of fuzzy rules and inference process. The most commonly used one is the Mamdani method (Mamdani 1976, 1977) or commonly known as MAX-MIN method. Another popular fuzzy inference model is the Takagi-Sugeno model (Takagi and Sugeno 1985). The general rule structure shown above is commonly known as Zadeh-Mamdani rule.

#### 2.3.2.1 Mamdani or MAX-MIN Method

The *max-min* algorithm (also known as Mamdani method) operates on each rule (*min* fashion) and combining all the rules (*max* fashion).

In the *min* composition, for each rule, the algorithm matches the membership degrees to the antecedent membership functions, and finds the *minimum*. The minimum function is the equivalent of the AND logical function. This minimum is the degree of activation for that rule. For each output variable, a matrix is constructed where each row corresponds to a rule and each column corresponds to a crisp value in the fuzzy set as described above. The fuzzy set used for each row is the one specified for that variable in the consequent for that rule. The activation value of each rule is used as a cut-off point for the membership functions, that is any degree of membership above the cut-off value is set to that value.

In the rule aggregation step, all individual fuzzy subsets (manipulated using the minimum operation) assigned to each output variable are combined together to form a single fuzzy subset for each output variable. The purpose is to aggregate all individual rule outputs to obtain the overall system output. In the *max* composition, the combined output fuzzy subset is constructed by taking the maximum over all of the fuzzy subsets assigned to the output variable by the inference rule: the maximum of each column is then calculated, which yields a composite or inferred membership function that is then passed to the defuzzification procedure.

To make it more clear, we continue our previous number example. For the universe of discourse $\varepsilon = \{0, 2, 4, 6, 8, 10\}$, we have the three membership functions as

$$large = \mu_l = \left\{\frac{0}{0}, \frac{0}{2}, \frac{0.1}{4}, \frac{0.3}{6}, \frac{0.6}{8}, \frac{1}{10}\right\},$$

$$medium = \mu_m = \left\{\frac{0.3}{0}, \frac{0.6}{2}, \frac{1}{4}, \frac{1}{6}, \frac{0.6}{8}, \frac{0.3}{10}\right\},$$

$$small = \mu_s = \left\{\frac{1}{0}, \frac{0.6}{2}, \frac{0.3}{4}, \frac{0.1}{6}, \frac{0}{8}, \frac{0}{10}\right\}.$$

Now, we define one input variable $x$ and one output variable $y$ utilizing these three membership functions. And we relate the input and the output variables in terms of the following rules.

RULE 1: *IF x is small THEN y is large,*
RULE 2: *IF x is medium THEN y is medium.*

Using the max-min method, we process rule by rule. Firing the first rule, we select the MF $\mu_s$ for the input $x$ (matching the antecedent), and the MF $\mu_l$ for the output $y$ (matching the consequent). We transform the rule 1 into a rule matrix ($R_1$) using the min (AND) operation. The rule matrix is a $6 \times 6$ matrix (corresponding to the 6 elements from universe of discourse on which the input and the output are defined). The rows represent the input ($\mu_s$) and the columns the output ($\mu_l$). The matrix is constructed in the following way.

For row 1, i.e., for $x = 0$, we select the corresponding value from the MF $\mu_s$ which is 1. For row 1, and column 1 to 6, we compare this value 1 (for $x$) against the values from $\mu_l$ (for $y$) and select the minimum. Similarly, for row 2, we compare the value 0.6 (for $x = 2$ from $\mu_s$) against the values from $\mu_l$ (for $y$) and select the minimum. This way, we continue for the six values of the input $x$ and complete the rule matrix. It is to be noted that we concentrate on each value of the input (antecedent) and compare it against the output (consequent), i.e., row-wise computation.

$$R_1 = \begin{bmatrix} \min(1,0) & \min(1,0) & \min(1,0.1) & \min(1,0.3) & \min(1,0.6) & \min(1,1) \\ \min(0.6,0) & \min(0.6,0) & \min(0.6,0.1) & \min(0.6,0.3) & \min(0.6,0.6) & \min(0.6,1) \\ \min(0.3,0) & \min(0.3,0) & \min(0.3,0.1) & \min(0.3,0.3) & \min(0.3,0.6) & \min(0.3,1) \\ \min(0.1,0) & \min(0.1,0) & \min(0.1,0.1) & \min(0.1,0.3) & \min(0.1,0.6) & \min(0.1,1) \\ \min(0,0) & \min(0,0) & \min(0,0.1) & \min(0,0.3) & \min(0,0.6) & \min(0,1) \\ \min(0,0) & \min(0,0) & \min(0,0.1) & \min(0,0.3) & \min(0,0.6) & \min(0,1) \end{bmatrix}$$

$$R_1 = \begin{bmatrix} 0 & 0 & 0.1 & 0.3 & 0.6 & 1 \\ 0 & 0 & 0.1 & 0.3 & 0.6 & 0.6 \\ 0 & 0 & 0.1 & 0.3 & 0.3 & 0.3 \\ 0 & 0 & 0.1 & 0.1 & 0.1 & 0.1 \\ 0 & 0 & 0 & 0 & 0 & 0 \\ 0 & 0 & 0 & 0 & 0 & 0 \end{bmatrix}.$$

For rule 2, we select the MF $\mu_m$ for the input $x$ (matching the antecedent), and the MF $\mu_m$ for the output $y$ (matching the consequent). For row 1, we compare the value 0.3 (for $x = 0$ from $\mu_m$) against the values from $\mu_m$ (for $y$) and select the minimum. Following the method discussed above, we continue for all the values of $x$ to construct the rule matrix $R_2$.

$$R_2 = \begin{bmatrix} \min(0.3,0.3) & \min(0.3,0.6) & \min(0.3,1) & \min(0.3,1) & \min(0.3,0.6) & \min(0.3,0.3) \\ \min(0.6,0.3) & \min(0.6,0.6) & \min(0.6,1) & \min(0.6,1) & \min(0.6,0.6) & \min(0.6,0.3) \\ \min(1,0.3) & \min(1,0.6) & \min(1,1) & \min(1,1) & \min(1,0.6) & \min(1,0.3) \\ \min(1,0.3) & \min(1,0.6) & \min(1,1) & \min(1,1) & \min(1,0.6) & \min(1,0.3) \\ \min(0.6,0.3) & \min(0.6,0.6) & \min(0.6,1) & \min(0.6,1) & \min(0.6,0.6) & \min(0.6,0.3) \\ \min(0.3,0.3) & \min(0.3,0.6) & \min(0.3,1) & \min(0.3,1) & \min(0.3,0.6) & \min(0.3,0.3) \end{bmatrix}$$

$$R_2 = \begin{bmatrix} 0.3 & 0.3 & 0.3 & 0.3 & 0.3 & 0.3 \\ 0.3 & 0.6 & 0.6 & 0.6 & 0.6 & 0.3 \\ 0.3 & 0.6 & 1 & 1 & 0.6 & 0.3 \\ 0.3 & 0.6 & 1 & 1 & 0.6 & 0.3 \\ 0.3 & 0.6 & 0.6 & 0.6 & 0.6 & 0.3 \\ 0.3 & 0.3 & 0.3 & 0.3 & 0.3 & 0.3 \end{bmatrix}$$

Now, for the rule aggregation part, we apply the max rule on the rule matrices $R_1$, $R_2$ and combine them into a single rule matrix $R$ to be used in the defuzzification step. The max rule is applied in the following way.

The resulting final rule matrix $R$ is also a $6 \times 6$ matrix. For the 36 elements, we compare each corresponding elements of the individual rule matrices $R_1$, $R_2$ and select the maximum of them as the candidate (for that position) for the final matrix $R$.

$R = R_1 \cup R_2 =$

$$\begin{bmatrix} \max(0, 0.3) & \max(0, 0.3) & \max(0.1, 0.3) & \max(0.3, 0.3) & \max(0.6, 0.3) & \max(1, 0.3) \\ \max(0, 0.3) & \max(0, 0.6) & \max(0.1, 0.6) & \max(0.3, 0.6) & \max(0.6, 0.6) & \max(0.6, 0.3) \\ \max(0, 0.3) & \max(0, 0.6) & \max(0.1, 1) & \max(0.3, 1) & \max(0.3, 0.6) & \max(0.3, 0.3) \\ \max(0, 0.3) & \max(0, 0.6) & \max(0.1, 1) & \max(0.1, 1) & \max(0.1, 0.6) & \max(0.1, 0.3) \\ \max(0, 0.3) & \max(0, 0.6) & \max(0, 0.6) & \max(0, 0.6) & \max(0, 0.6) & \max(0, 0.3) \\ \max(0, 0.3) & \max(0, 0.3) & \max(0, 0.3) & \max(0, 0.3) & \max(0, 0.3) & \max(0, 0.3) \end{bmatrix}$$

$$R = \begin{bmatrix} 0.3 & 0.3 & 0.3 & 0.3 & 0.6 & 1 \\ 0.3 & 0.6 & 0.6 & 0.6 & 0.6 & 0.6 \\ 0.3 & 0.6 & 1 & 1 & 0.6 & 0.3 \\ 0.3 & 0.6 & 1 & 1 & 0.6 & 0.3 \\ 0.3 & 0.6 & 0.6 & 0.6 & 0.6 & 0.3 \\ 0.3 & 0.3 & 0.3 & 0.3 & 0.3 & 0.3 \end{bmatrix}.$$

#### 2.3.2.2 Takagi-Sugeno Model

The Takagi-Sugeno fuzzy model (Takagi and Sugeno 1985) uses crisp functions as the consequences of the rules. This is the difference between the Takagi-Sugeno and the Mamdani model. Given an input-output data set, Takagi-Sugeno fuzzy model offers a systematic approach to generate fuzzy rules. A Takagi-Sugeno rule set has the general form:

$$RULE_i\text{: } IF\ x \text{ is } A_i\ THEN\ y_i = f_i(x_0),\ i = 1, 2, \ldots, n.$$

The antecedent of each rule is a set of fuzzy propositions connected with the AND operator (for more than one input). The consequent of each rule is a crisp function of the input vector. By means of the fuzzy sets of the antecedent propositions the input domain is softly partitioned into smaller regions where the mapping is locally approximated by the crisp functions $f_i$.

Takagi-Sugeno rule aggregation and their effects considerably differ from the Mamdani method. One variation of the Takagi-Sugeno inference system uses the weighted mean criterion to combine all the local representations in a global approximator, like:

$$y = \frac{\sum_{i=1}^{n} \mu_i . y_i}{\sum_{i=1}^{n} \mu_i}, \tag{2.27}$$

where $\mu_i$ is the degree of fulfillment of the $i$-th rule and $n$ is the number of rules in the rulebase.

### 2.3.3 Defuzzification

Defuzzification is the opposite of the fuzzification step. Defuzzification step transforms back the inferred fuzzy values into the crisp values. Applications dealing with human interface directly, sometimes, do not require the defuzzification step as humans can deal with fuzzified variables like *positive big*, *negative small* etc. However, defuzzification is an important step for fuzzy systems dealing with processes which require as final output the physical variables and values. So, defuzzification step involves chalking out a single (or multiple) output value(s) from the inferred fuzzy set passed on from the inference step.

There are many defuzzification methods. Two common techniques are the Center of Gravity (CoG) or CENTROID method and Mean of Maxima (MoM) or MAXIMUM method.

In the CoG method, the crisp value of the output variable is computed by finding the variable value of the center of gravity of the membership function for the fuzzy value. This is described by the (2.27).

In the MoM method, one of the variable values at which the fuzzy subset has its maximum truth value is chosen as the crisp value for the output variable. There are several variations of the MoM method that differ only in what they do when there is more than one variable value at which this maximum truth value occurs. One of these, the AVERAGE-OF-MAXIMA method, returns the average of the variable values at which the maximum truth value occurs.

Continuing our number example, we start with the inferred fuzzy set, the final rule matrix,

$$R = \begin{bmatrix} 0.3 & 0.3 & 0.3 & 0.3 & 0.6 & 1 \\ 0.3 & 0.6 & 0.6 & 0.6 & 0.6 & 0.6 \\ 0.3 & 0.6 & 1 & 1 & 0.6 & 0.3 \\ 0.3 & 0.6 & 1 & 1 & 0.6 & 0.3 \\ 0.3 & 0.6 & 0.6 & 0.6 & 0.6 & 0.3 \\ 0.3 & 0.3 & 0.3 & 0.3 & 0.3 & 0.3 \end{bmatrix}.$$

We try to obtain a defuzzified output when the input $x = 4$.

- **CoG method**

  In the final rule matrix $R$, the rows correspond to the input and the columns to the output. In the CoG method, for $x = 4$, we select the third row (because 4 is the third element in the universe of discourse). For the third row, we consider columns 1 to 6. So, the output membership function for $x = 4$ is,

$$\mu(x) = \left\{ \frac{0.3}{0}, \frac{0.6}{2}, \frac{1}{4}, \frac{1}{6}, \frac{0.6}{8}, \frac{0.3}{10} \right\}.$$

  We apply (2.27) to get the final crisp output value for input $x = 4$ as,

$$y = \frac{\sum \mu(x).x}{\sum \mu(x)}$$
$$= \frac{(0.3 \times 0) + (0.6 \times 2) + (1 \times 4) + (1 \times 6) + (0.6 \times 8) + (0.3 \times 10)}{0.3 + 0.6 + 1 + 1 + 0.6 + 0.3} = 5.$$

- **MoM method**
For the input $x = 4$, we also select the third row of the final rule matrix (because 4 is the third element in the universe of discourse). So, the output membership function for $x = 4$ is,

$$\mu(x) = \left\{ \frac{0.3}{0}, \frac{0.6}{2}, \frac{1}{4}, \frac{1}{6}, \frac{0.6}{8}, \frac{0.3}{10} \right\}.$$

In the MoM method, for this output membership function, we check the maximum output values. The maximum value (1) occurs for the two central elements, 4 and 6. So, applying the AVERAGE-OF-MAXIMA rule, we get the final output as $(4+6)/2 = 5$. This is same as the result of the CoG method.

However, if the output fuzzy values were asymmetrical, the results of the CoG and the MoM methods would be different. For example, we try to find out the crisp output value for the input $x = 2$. For the input $x = 2$, we select the second row of the final rule matrix (because 2 is the second element in the universe of discourse). So, the output membership function for $x = 2$ is,

$$\mu(x) = \left\{ \frac{0.3}{0}, \frac{0.6}{2}, \frac{0.6}{4}, \frac{0.6}{6}, \frac{0.6}{8}, \frac{0.6}{10} \right\}.$$

Using the CoG method, the output would be,

$$y = \frac{(0.3 \times 0) + (0.6 \times 2) + (0.6 \times 4) + (0.6 \times 6) + (0.6 \times 8) + (0.6 \times 10)}{0.3 + 0.6 + 0.6 + 0.6 + 0.6 + 0.6}$$
$$= 5.45.$$

But using the MoM method, we get the maximum fuzzy value in the output membership function as 0.6 for the elements 2, 4, 6, 8, 10. Using the AVERAGE-OF-MAXIMA rule, we get the final output as $(2+4+6+8+10)/5 = 6$. So, the results are different.

## 2.4 Application Example

Fuzzy logic is a subject which developed around practical applications across various domains. It started with control system applications as a knowledge-based system for direct or supervisory controller. However, over the last few decades, fuzzy logic has emerged as a powerful nonlinear technique to solve different problems in different fields. It has been particularly successful for the tasks which involve human intuition and approach to solve the problem. Hence, fuzzy logic theory has gained its maturity with practical application, hand-in-hand.

Bart Kosko (Kosko 1996) made a significant contribution in the popularization of fuzzy logic. Different industrial and engineering applications of fuzzy logic were presented by Sugeno (Sugeno 1985), Ross (Ross 1995), Mendel (Mendel 1995). Wang (Wang 1994) discussed about the adaptive fuzzy logic for industrial applications. Fuzzy logic has also been successful in the field of pattern recognition, as discussed by Bezdek (Bezdek 1981). One of the upcoming fields is combining fuzzy logic and neural networks popularly known as neuro-fuzzy technique. Applications of neuro-fuzzy technique in soft-computing and machine learning were discussed by Jang and Sun (Jang and Sun 1997).

In this section, we will see an engineering application of fuzzy logic. In the previous sections, we have seen different applications of fuzzy logic. Most of them use fuzzy set theory in a numerical fashion. However, fuzzy systems can be effectively designed and manipulated using the graphical technique as well. In the following example, we will demonstrate the use of graphical technique for designing fuzzy logic-based system.

### 2.4.1 Brake Test Application

In this application, we consider the brake test of car, i.e., calculating the brake effort (output quantity) of the car with the help of a simple fuzzy system. Our system consists of two cars with a certain distance $D$ between them. We are sitting in the second car and driving at a velocity $V$ along a superhighway. We want to calculate the brake effort $K$ in percent in order to maintain a safe drive. More specifically, for any moment, we want to calculate the brake effort for the distance $D = 175\,\text{m}$ and a velocity of $V = 190\,\text{Km/h}$.

This is a classical example of humanoid decision making. We will try to simulate and incorporate our human nature of solving the problem of pressing the brake pedal while driving. This is an example of the sort of real life application examples that fuzzy logic encounters most frequently. We express the human experience in terms of two rules which will be used to design the fuzzy system.

**RULE 1** : *IF distance is average AND velocity is very high*
*THEN brake effort is three - fourth.*
**RULE 2** : *IF distance is low AND velocity is very high*
*THEN brake effort is full*

It is to be noted in this example that we do not incorporate a lot of rules. We only choose the rules needed to reflect the essential human knowledge for the crucial cases. For example, brake effort is extremely important if two cars are going at high speed and getting close. The two rules reflect these situations and the humanoid approach in those cases. However, depending on the complexity of the application, number of rules should be chosen. But whatever be the application, a rule of thumb is to guide the fuzzy system effectively with the most critical situation not every nuts and bolts of the system. We will design the fuzzy system step-by-step. A good tool to use is the Matlab® fuzzy logic toolbox (Mathworks 2002).

### *2.4.2 Fuzzification*

In the fuzzification step, first we identify the required linguistic variables. In this case, we have two input linguistic variables: *distance*, *velocity*, and one output variable, *brake effort*. Choice of linguistic variable also depends on the application. Number of variables should not be too large, but the chosen variables should be the critical ones. We can introduce additional variables if we need more precise simulation of the situation or results. For example, as an assignment on this example (try afterwards), we could extend it by introducing other variables, for example, *road condition*, *weather* as additional input variables.

After the determination of the linguistic fuzzy variables, we need to define the membership functions (MFs) for describing them. These MFs could be given for a problem. However, to gain more insight, for this example we do not assume them. We try to formulate the MFs using intuition (this is the aim of fuzzy logic to simulate human experience). You can try to redo this example later on by defining the MFs differently (say for a different car model).

For this problem, we define the fuzzy MF for the two input variables using five linguistic terms which are:

{***Very-Low, Low, Average, High, Very-High***}.

We can effectively abbreviate these terms to represent them as

{***VL, L, A, H, VH***}.

However, the universe of discourse for the two input variables are different. For *distance*, the universe ranges [0 m, 500 m], and for velocity, [0 Km/h, 200 Km/h]. Regarding the MFs, for the first and the last linguistic terms (*VL* and *VH*) we choose the left- and right-side open trapezoidal approximating functions respectively, and for the rest $(L, A, H)$, the triangular approximating functions. The MFs are shown in Fig. 2.26 and 2.27.

Using these MF descriptions, for the given set of input variable values, i.e., $distance = 175$ m and $velocity = 190$ Km/h, we determine the membership values graphically. This is done as follows.

We show the graphical representation of the MFs. Then, we mark the given input value in the graph (X-axis) and draw vertical line from that point, cutting the MFs. The projections of the intersection points onto the Y-axis determine the corresponding degree of membership for the given set of input values. Figure 2.28 depicts the operation for $distance = 175$ m.

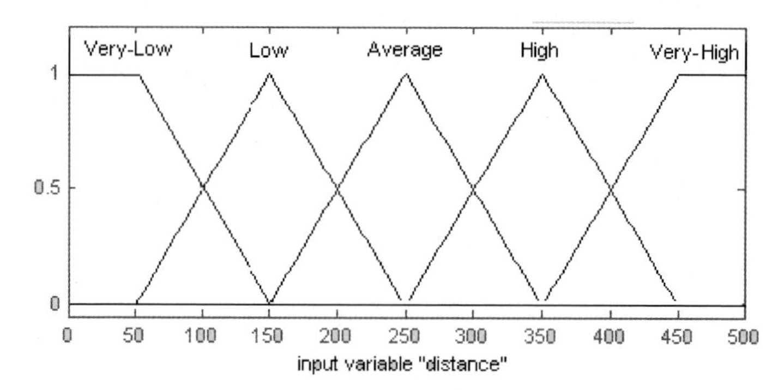


**Fig. 2.26** Membership function for the input variable *distance*

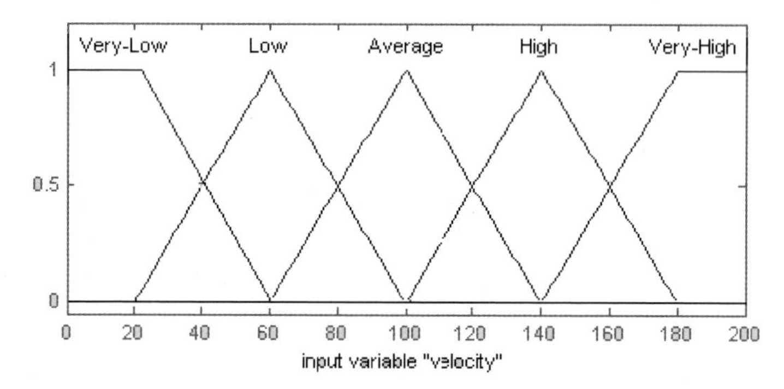


**Fig. 2.27** Membership function for the input variable *velocity*

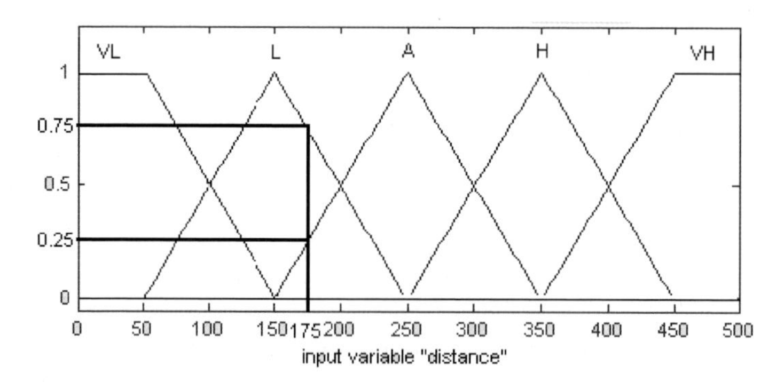


**Fig. 2.28** Membership value for *distance* = 175 m

For *distance* = 175 m value, we mark the point along X-axis in Fig. 2.28 and draw a vertical line which cuts the *Low* ($L$) and *Average* ($A$) categories. From these intersection points we draw the projections on the Y-Axis and get the degree of membership, $L = 0.75$, $A = 0.25$. From this, we get the membership function for the input variable *distance* ($D$) as,

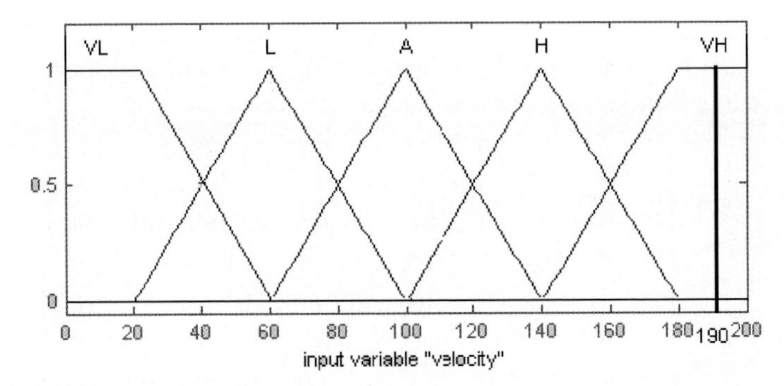


**Fig. 2.29** Membership value for *velocity* = 190 Km/h

$$\mu_D = \left\{ \frac{0}{VL}, \frac{0.75}{L}, \frac{0.25}{A}, \frac{0}{H}, \frac{0}{VH} \right\}.$$

Determination of the membership value for the input variable *velocity* = 190 Km/h is shown in Fig. 2.29.

Like the *distance* variable, for *velocity* = 190 Km/h, we mark the point along X-axis in Fig. 2.29 and draw a vertical line which cuts the *Very-High* (*VH*) category. Projecting this onto the Y-axis we get the fuzzy membership value as, $VH = 1$. We define the membership function for the input variable *velocity* ($V$) as,

$$\mu_V = \left\{ \frac{0}{VL}, \frac{0}{L}, \frac{0}{A}, \frac{0}{H}, \frac{1}{VH} \right\}.$$

For the two input variables and the given values $D = 175$ m, $V = 190$ Km/h, we can represent the MFs $\mu_D$ and $\mu_V$ determined above, in tabular format as shown in Table 2.5.

After defining the input MFs, next we need to define the output MF for the single output variable *brake effort*. We define the fuzzy MF for the output variable using five linguistic terms which are:

{***Zero, One-Fourth, Half, Three-Fourth, Full*** }.

We can effectively abbreviate these terms to represent them as

{ ***Z, OF, H, TF, F*** }.

**Table 2.5** Membership values of the two input variables

| MFs<br>Inputs | VL | L | A | H | VH |
|---|---|---|---|---|---|
| *D* | 0 | 0.75 | 0.25 | 0 | 0 |
| *V* | 0 | 0 | 0 | 0 | 1 |

We divide the range of the percentage brake effort [0%, 100%] in equal steps of 10% to define the five output membership functions utilizing the five above-mentioned linguistic terms for the output variable. These are

$$\mu_Z = \left\{ \frac{1}{0}, \frac{1}{10}, \frac{0.5}{20}, \frac{0}{30}, \frac{0}{40}, \frac{0}{50}, \frac{0}{60}, \frac{0}{70}, \frac{0}{80}, \frac{0}{90}, \frac{0}{100} \right\},$$

$$\mu_{OF} = \left\{ \frac{0}{0}, \frac{0}{10}, \frac{0.5}{20}, \frac{1}{30}, \frac{0.5}{40}, \frac{0}{50}, \frac{0}{60}, \frac{0}{70}, \frac{0}{80}, \frac{0}{90}, \frac{0}{100} \right\},$$

$$\mu_H = \left\{ \frac{0}{0}, \frac{0}{10}, \frac{0}{20}, \frac{0}{30}, \frac{0.5}{40}, \frac{1}{50}, \frac{0.5}{60}, \frac{0}{70}, \frac{0}{80}, \frac{0}{90}, \frac{0}{100} \right\},$$

$$\mu_{TF} = \left\{ \frac{0}{0}, \frac{0}{10}, \frac{0}{20}, \frac{0}{30}, \frac{0}{40}, \frac{0}{50}, \frac{0.5}{60}, \frac{1}{70}, \frac{0.5}{80}, \frac{0}{90}, \frac{0}{100} \right\},$$

$$\mu_F = \left\{ \frac{0}{0}, \frac{0}{10}, \frac{0}{20}, \frac{0}{30}, \frac{0}{40}, \frac{0}{50}, \frac{0}{60}, \frac{0}{70}, \frac{0.5}{80}, \frac{1}{90}, \frac{1}{100} \right\}.$$

The MFs for the output variable *brake effort* is shown in Fig. 2.30.

### 2.4.3 Fuzzy Inference

After the fuzzification step, the structure of our fuzzy system is ready. Now, we activate and manipulate it using the humanoid knowledge represented as the fuzzy rules. The rules are:

RULE 1:
*IF distance is A AND velocity is VH THEN brake effort is TF,*
RULE 2:
*IF distance is L AND velocity is VH THEN brake effort is F.*

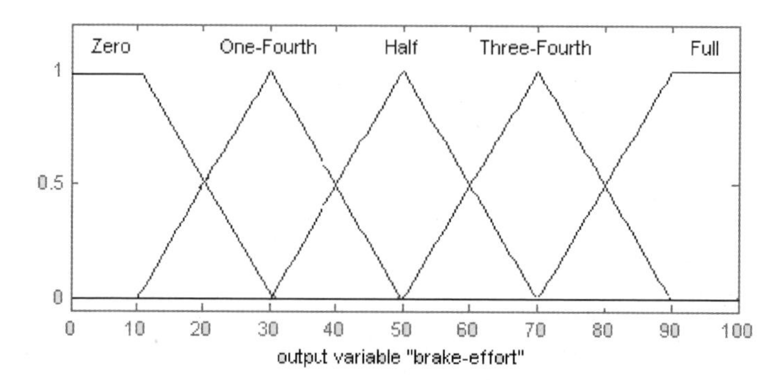


**Fig. 2.30** Membership function for the output variable *brake effort*

Corresponding to these two fuzzy rules and the degree of membership of the two input variables (listed in Table 2.5), we substitute the activation values (using *min* logic following Mamdani or *max-min* argumentation) to get the input and rule activation values as shown in Table 2.6.

**Table 2.6** Fuzzy rule activation

| Inputs<br>Rules | D | V | Activation<br>(Minimum) |
|---|---|---|---|
| 1 | 0.25 | 1 | 0.25 |
| 2 | 0.75 | 1 | 0.75 |

The explanation of this operation is as follows. As in rule 1, *distance* is *A* (*Average*), so from Table 2.5 we get the value 0.25. Similarly, as in rule 2, *distance* is *L* (*Low*), so from Table 2.5 we get the corresponding value of 0.75. For both the rules *velocity* is *VH* (*Very-High*), so its value is 1 (from Table 2.5). Combining these, we get Table 2.6. Now, for each rule we consider the rows and take the minimum value as activation. So, for the rule 1 (1st row) the minimum value (between 0.25 & 1) is 0.25, and for the rule 2 (2nd Row) the minimum value (between 0.75 & 1) is 0.75.

From Table 2.6, we have the composite rule activation for the two input variables for the two rules. Next, we superimpose the rules onto the output variable. Now, from the rules we get that for rule 1 the *brake effort* output is *TF* (*Three-Fourth*), and for rule 2 the output is *F* (*Full*). So, for rule 1, we take the activation value 0.25 (from Table 2.6) and apply it against the MF of the output variable *brake effort* for the *TF* category, i.e., $\mu_{TF}$. For rule 2, we take the activation value 0.75 (from Table 2.6) and apply it against the MF of the output variable *brake effort* for the *F* category, i.e., $\mu_F$. And applying the cut-offs we get the final output rule matrix shown in Table 2.7.

**Table 2.7** Max-Min inference involving the input and the output variables

| Output MF<br>Rules | 0 | 10 | 20 | 30 | 40 | 50 | 60 | 70 | 80 | 90 | 100 |
|---|---|---|---|---|---|---|---|---|---|---|---|
| 1 | 0 | 0 | 0 | 0 | 0 | 0 | 0.25 | 0.25 | 0.25 | 0 | 0 |
| 2 | 0 | 0 | 0 | 0 | 0 | 0 | 0 | 0 | 0.5 | 0.75 | 0.75 |
| Max(1,2) | 0 | 0 | 0 | 0 | 0 | 0 | 0.25 | 0.25 | 0.5 | 0.75 | 0.75 |

The explanation of this aggregation operation shown in Table 2.7 is as follows. In Table 2.7, rule 1 and 2 are represented by row 1 and 2 respectively. For row 1 i.e., rule 1, we consider the activation value 0.25 (from Table 2.6) and compare it against the values of the MF of the output variable, i.e., $\mu_{TF} = \left\{\frac{0}{0}, \frac{0}{10}, \frac{0}{20}, \frac{0}{30}, \frac{0}{40}, \frac{0}{50}, \frac{0.5}{60}, \frac{1}{70}, \frac{0.5}{80}, \frac{0}{90}, \frac{0}{100}\right\}$. We take the minimum values between 0.25 and the MF values for the 11 categories.

Similarly for rule 2, i.e., row 2 in Table 2.7, we consider the activation value 0.75 (from Table 2.6) and compare it against the output MF for rule 2, i.e., $\mu_F =$

$\left\{\frac{0}{0}, \frac{0}{10}, \frac{0}{20}, \frac{0}{30}, \frac{0}{40}, \frac{0}{50}, \frac{0}{60}, \frac{0}{70}, \frac{0.5}{80}, \frac{1}{90}, \frac{1}{100}\right\}$, and take the minimum values between 0.75 and the MF values.

And finally, last row of the Table 2.7 shows the aggregation of the two rules by taking the maximum of the 2 rules (row 1 and 2). From the last row of Table 2.7, we get the output membership function, $\mu_{OUT}$ for defuzzification purpose.

$$\mu_{OUT} = \left\{\frac{0}{0}, \frac{0}{10}, \frac{0}{20}, \frac{0}{30}, \frac{0}{40}, \frac{0}{50}, \frac{0.25}{60}, \frac{0.25}{70}, \frac{0.5}{80}, \frac{0.75}{90}, \frac{0.75}{100}\right\}.$$

This is the numerical way of determining the final output membership function. However, this can also be determined using graphical techniques. To do so, we place the MFs of the two input variables (not the general one but the one with specific values, shown in Fig. 2.28 and 2.29) and the output variable (Fig 2.30). For each rule, from the two input variables, we select the one with the minimum value (minimum graphical area). From the minimum input variable graph, we draw a line extending its upper surface till it cuts the output MF graph. The portion that this extension line cuts in the output MF graph is the output activation for that rule. After computing this way output activation for all the rules, we combine all the output figures to get the final output figure which represents the output membership function. The graphical operation for this example is shown in Fig. 2.31 using the Matlab® fuzzy logic toolbox (Mathworks 2002).

### 2.4.4 Defuzzification

After we determine the output membership function (numerical method) or the output composite figure (graphical method), we want to get back the physical output value from the fuzzified values. This is the defuzzification process. As discussed in Sect. 2.3.3, we can either choose the Center of Gravity (CoG)/CENTROID method or the Mean of Maxima (MoM)/ MAXIMUM method.

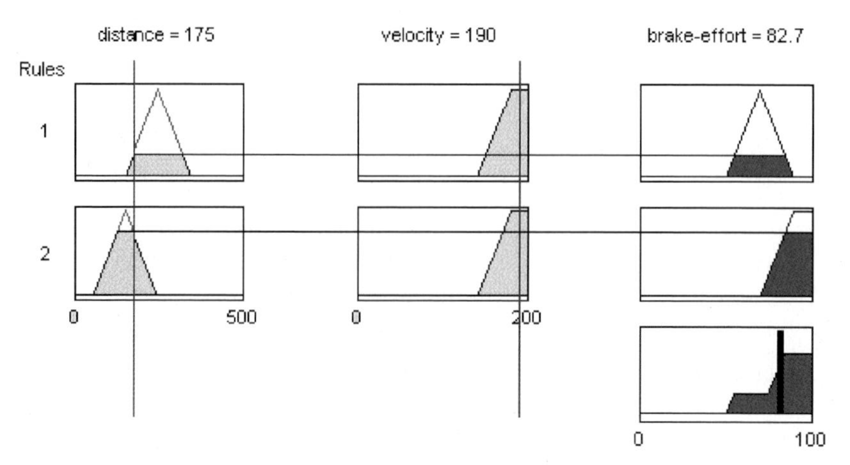


**Fig. 2.31** Graphical determination of the output membership function

Between these two methods, CoG method is used more often. Justification of the name of this method is evident from the graphical operation. Using all the fuzzification and the inferencing process, we arrive at a composite figure which incorporates, assembles and represents all the knowledge, operations etc into a final form. From this figure, we have to determine a single output value (for this example single, might be multiple for other cases). That is, we have to retrieve from this figure a single effective value. So, the first thing that comes to our mind is the center of gravity or the centroid of this figure. The center of gravity or centroid (which is an important factor for the equilibrium of any figure) can be the effective single-value representation of the figure. This is the implication of the Center of Gravity (CoG) or CENTROID method.

Using the CoG method, we calculate the defuzzified output value. We apply (2.27) to get the final defuzzified output as,

$$brake\ effort = \frac{(0\times 0)+(0\times 10)+(0\times 20)+(0\times 30)+(0\times 40)+(0\times 50)+(0.25\times 60)+\ldots (0.25\times 70)+(0.5\times 80)+(0.75\times 90)+(0.75\times 100)}{0+0+0+0+0+0+0.25+0.25+0.5+0.75+0.75} = 86.$$

So, the final output, % brake effort is 86%.

Using the graphical method, we have the final output composite figure as shown in Fig. 2.32.

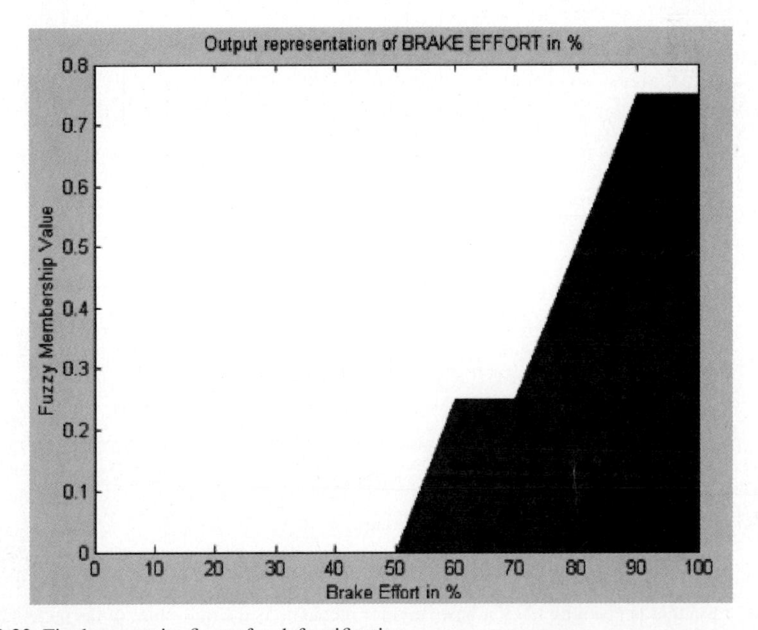


**Fig. 2.32** Final composite figure for defuzzification

To compute the center of gravity (COG)[6], we divide the figure into 4 areas comprising of triangles and rectangles as shown in Fig. 2.33.

From Fig. 2.33, we compute the areas and COG (X-axis value) of the each subdivisions.

Area 1:- $1/2 \times (60 - 50) \times 0.25 = 1.25$ sq. unit,
COG(1) $= 50 + \{2/3 \times (60 - 50)\} = 56.67$;
Area 2:- $0.25 \times (100 - 60) = 10$ sq. unit,
COG(2) $= 80$;
Area 3:- $1/2 \times (90 - 70) \times (0.75 - 0.25) = 5$ sq. unit,
COG(3) $= 70 + \{2/3 \times (90 - 70)\} = 83.33$;
Area 4:- $(100 - 90) \times (0.75 - 0.25) = 5$ sq. unit,
COG(4) $= 95$.

The final COG of a sub-divided figure is determined by

$$COG_{final} = \frac{\sum_{i=1}^{n} Area_i \times COG_i}{\sum_{i=1}^{n} Area_i}, \tag{2.28}$$

where $n$ is the number of sub-divisions.

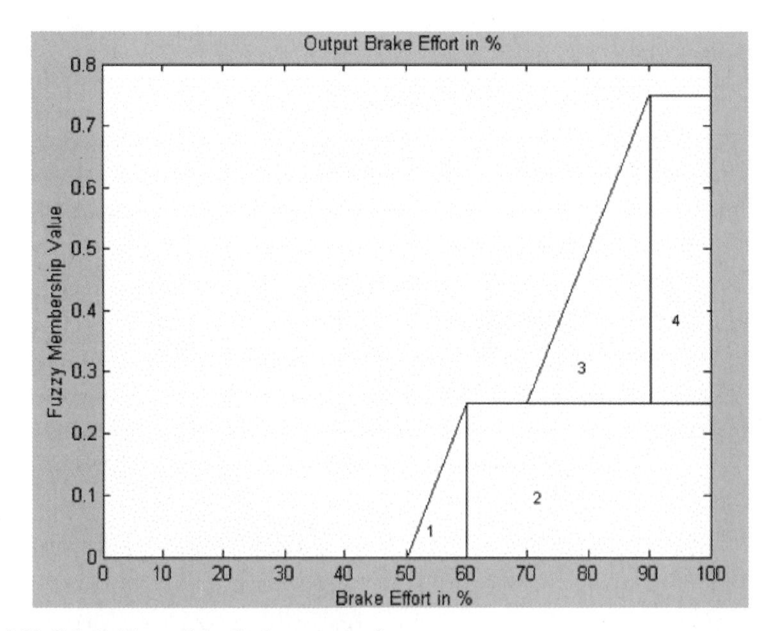


**Fig. 2.33** Sub-divisions of the final composite figure

[6] COG for triangle is given by ($2/3 \times$ base) and for rectangle is the mid-point of the rectangle.

Applying (2.28) for our example in Fig. 2.33 ($n = 4$), we get the final COG as

$$COG = \frac{(1.25 \times 56.67) + (10 \times 80) + (5 \times 83.33) + (5 \times 95)}{(1.25 + 10 + 5 + 5)} = 82.9.$$

So, the final defuzzified output brake effort is 82.9%. This is also indicated in Fig. 2.31 computed using the Matlab® fuzzy logic toolbox (Mathworks 2002).

## 2.4.5 Conclusion

Fuzzy rules, inference techniques etc are utilized to map the nonlinear relationship between the input and the output. If we have single input and single output, we get two-dimensional relationship, while for our example, we have two inputs and one output, hence we get a three-dimensional relationship which is shown in Fig. 2.34. The total number of fuzzy variables (input and output) determines the dimension of the input-output relationship.

Fuzzy logic is a nonlinear technique to model and simulate human approaches in problem solving. In this example, we have seen a fuzzy solution to a real life problem of controlling the brake effort in cars. We incorporate the human experiences in critical situations in fuzzy systems in terms of fuzzy rules. These knowledge representations are manipulated using fuzzy inference techniques (like Mamdani or Takagi-Sugeno). The inferred fuzzy rules are combined into the aggregated output membership function which is defuzzified to get the final crisp physical output value(s).

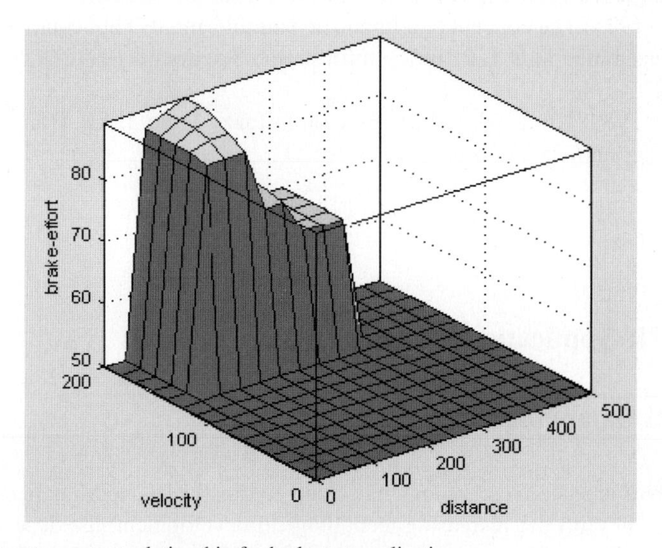


**Fig. 2.34** Inputs – output relationship for brake test application

## References

Bezdek JC (1981) Pattern recognition with fuzzy objective function algorithms. Plenum Press, New York

Driankov D, Hellendoorn H, Reinfrank M (1993) An introduction to fuzzy control. Springer-Verlag, New York

Jang JSR, Sun CT (1997) Neuro-fuzzy and soft computing: a computational approach to learning and machine intelligence. Prentice-Hall Inc., New York

Kosko B (1996) Fuzzy engineering. Prentice-Hall Inc., New York

Mamdani EH (1976) Advances in the linguistic synthesis of fuzzy controllers. International Journal of Man-Machine Studies 8:669–678

Mamdani EH (1977) Applications of fuzzy logic to approximate reasoning using linguistic synthesis. IEEE Transactions on Computers: 26:1182–1191

Mathworks Inc. (2002) MATLAB® documentation–fuzzy logic toolbox, version 6.5.0.180913a, release 13

Mendel JM (1995) Fuzzy logic systems for engineering: a tutorial. Proceedings of IEEE 83:345–377

Ross TJ (1995) Fuzzy logic with engineering applications. McGraw Hill, New York

Sugeno M (1985) Industrial applications of fuzzy control. Elsevier Science Pub. Co., Amsterdam

Takagi T, Sugeno M (1985) Fuzzy identification of systems and its applications to modeling and control. IEEE Trans. Systems, Man Cyber. SMC-15: 116–132

Wang LX (1994) Adaptive fuzzy systems and control: design and stability analysis. Prentice-Hall Inc., New York

Zadeh LA (1965) Fuzzy sets. Information and Control 8:338–353

Zadeh LA (1973) Outline of a new approach to the analysis of complex systems and decision processes. IEEE Trans. Systems, Man, and Cybernetics 3:28–44

Zadeh LA (1975) The concept of a linguistic variable and its application to approximate reasoning, parts 1, 2, and 3. Information Sciences 8:199–249, 8:301–357, 9:43–80

Zadeh LA (1989) Knowledge representation in fuzzy logic. IEEE Transactions on Knowledge and Data Engineering 1:89–100

## Section II: Application Study

## 2.5 Load Balancing

The distribution system technology has changed drastically, both qualitatively and quantitatively. With the increasing technological development, the dependence on electric power supply has increased considerably. While demand has increased, the need for a steady power supply with minimum power interruptions and fast fault restoration has also increased. To meet these demands, automation of the power distribution system needs to be widely adopted. All switches and circuit-breakers involved in the controlled networks are equipped with facilities for remote operation. The control interface equipments must withstand extreme climatic conditions. Also, control equipments at each location must have a dependable power source. To cope with the complexity of the distribution, the latest computer, communication and distribution technologies are needed to be employed. The distribution automation can be defined as an integrated system concept. It includes control, monitoring and some times, decision to alter any kind of loads. The automatic distribution system provides directions for automatic reclosing of the switches and remote monitoring of the loads contributing towards phase balancing (Ukil and Siti 2006).

Load balancing (also known as phase balancing) is very important and usable operation to reduce distribution feeder losses and improve system security. There are a number of normally closed and normally opened switches in a distribution system. By changing the open/close status of the feeder switches, load currents can be transferred from feeder-to-feeder, i.e. from heavily loaded to less loaded feeders. In this application study, we will see the use of fuzzy logic for load balancing purpose.

## 2.5.1 Feeder Representation

The distribution feeder is usually a three–phase, four wire system with a radial or open loop structure. To improve the system phase voltage and current unbalances, the connection between the specific feeder and the distribution transformer should be suitably arranged. The domestic loads are connected, as in most cases, in a single-phase. For the application study, we assume that each feeder contains 50 domestic loads or connections. So, the total load to the three feeders can be 150 connections as shown in Fig. 2.35. In Fig. 2.35, each load, through the tie-switches, can be connected only to the one of the three phases.

In general, distribution loads show different characteristics according to their corresponding distribution lines and line sections. Therefore, at the load levels, each

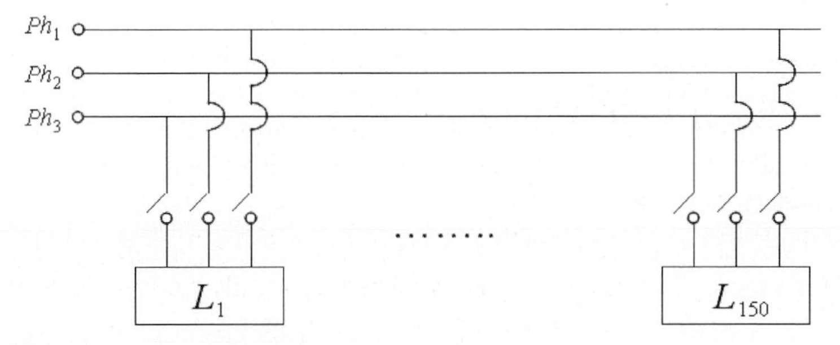


**Fig. 2.35** Example distribution feeder

time period can be regarded as non-identical. In the case of a distribution system, with some overloaded and some lightly loaded branches, there is the need to reconfigure the system such that loads are transferred from the heavily loaded to the less loaded feeders. Here the maximum load current the feeder conductor can take may be taken as the reference (Ukil 2006). Nevertheless, the transfer of load must be such that a certain predefined objective is satisfied. In this case, the objective is for the ensuing network to have minimum real power loss. Consequently, phase balancing may be redefined as the rearrangement of the network such as to minimize the total real power losses arising from the line branches. Mathematically, the total power loss may be expressed as

$$\sum_{i=1}^{n} r_i \frac{P_i^2 + Q_i^2}{|V_i|^2}, \tag{2.29}$$

where $r_i$, $P_i$, $Q_i$, $V_i$ are respectively the resistance, real power, reactive power and voltage of the branch $i$, and $n$ is the total number of branches in the system.

### 2.5.2 Proposed Technique

In this study, we will see a fuzzy logic-based load balancing technique. This can be followed by combinatorial optimization oriented expert system (Ukil and Siti 2006) for implementing the load change decision. This is a typical example of the fuzzy expert system[7] which is most often used in solving various industrial problem. The architecture (Ukil and Siti 2006) of the proposed fuzzy expert system for feeder load balancing is shown in Fig. 2.36.

In Fig. 2.36, the input is the total phase load (for each of the three phases). The average unbalance per phase, calculated according to (2.30), is checked against a threshold of 10 kW. If the average unbalance per phase is below 10 kW, we can assume that the system is more or less balanced and discard any further load balancing. Otherwise, we go for the fuzzy logic-based load balancing. The output from the fuzzy-based load balancing step is the load change values for each phase. A *negative* value indicates that the specific phase has surplus load and should *release* that amount of load, while a *positive* value indicates that the specific phase is less-loaded and should *receive* that amount of load. This load change configuration is the input to the expert system (Ukil and Siti 2006) which tries to optimally shift the specific number of load points. However, sometimes the expert system may not be able to execute the exact amount of load change as directed by the fuzzy step. This is because the actual load points for any phase might not result in an optimum combination which sums up to the exact change value indicated by the fuzzy step. So, we implement the best possible change from the expert system and iteratively

[7] Fuzzy expert system involves fuzzy logic-based knowledge processing followed by expert system which utilizes the fuzzy information and controls the whole process.

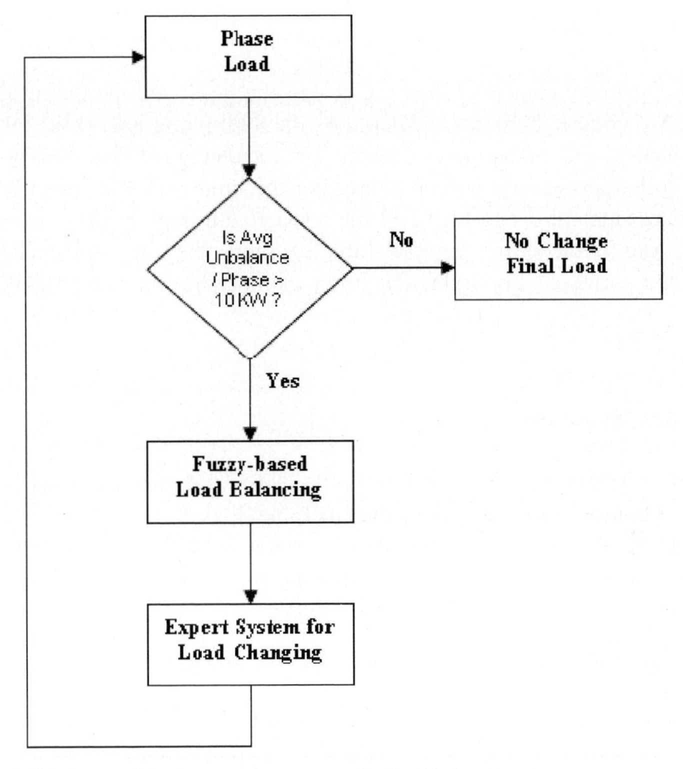


**Fig. 2.36** Architecture of the fuzzy expert system for load balancing

check the system unbalance until we achieve the average unbalance below 10 kW, if achievable. However, in this application study, we concentrate on solving the load unbalance problem using fuzzy logic.

$$Avg\ Unbalance/Ph = \frac{|Load_{Ph1} - Load_{Ph2}| + |Load_{Ph2} - Load_{Ph3}| + |Load_{Ph3} - Load_{Ph1}|}{3}. \tag{2.30}$$

## *2.5.3 Designing Fuzzy Controller*

To design the fuzzy controller, we assume the average per phase capacity of the system to be 150 kW, with 50 load points connected to any specific phase. For designing the fuzzy controller, we further assume the maximum overload capacity of any phase to be 300 kW. Beyond 300 kW the fuzzy controller (this limit is valid only for this example) should not be used for load balancing. Because, in any case, when any phase reaches its 200% overload condition, it should be cut out from the service to prevent power breakdown and severe overloading of the transformer.

#### 2.5.3.1 Fuzzification

To design the fuzzy controller, first we design the input and the output linguistic variables. We choose the input as *Load*, i.e., the total phase load (kW) for each of the three phases, and the output as *Change*, i.e., the change of load (kW, positive or negative) to be performed for each phase. For the input variable, Table 2.8 shows the fuzzy nomenclature, and Fig. 2.37 the respective triangular fuzzy membership functions. And for the output variable, Table 2.9 shows the fuzzy nomenclature, and Fig. 2.38 the corresponding triangular fuzzy membership functions (Ukil and Siti 2006).

#### 2.5.3.2 Fuzzy Inference

In the fuzzy inferencing, we determine the IF-THEN style fuzzy rule set governing the input and output variable as described in Table 2.10.

Corresponding to the fuzzy input, output variables and the associated rule set, the fuzzy surface is shown in Fig. 2.39, depicting the nonlinear relationship between the input and the output variable (Ukil and Siti 2006).

**Table 2.8** Description of the input linguistic variable

| SL. No. | Input (*Load*) Description | Fuzzy Nomenclature | KW Range |
|---|---|---|---|
| 1 | Very Less Loaded | VLL | 0 to 50 |
| 2 | Less Loaded | LL | 35 to 85 |
| 3 | Medium Less Loaded | MLL | 65 to 115 |
| 4 | Perfectly Loaded | PL | 100 to 150 |
| 5 | Slightly Overloaded | SOL | 125 to 175 |
| 6 | Medium Overloaded | MOL | 165 to 215 |
| 7 | Overloaded | OL | 200 to 250 |
| 8 | Heavily Overloaded | HOL | 235 to 300 |

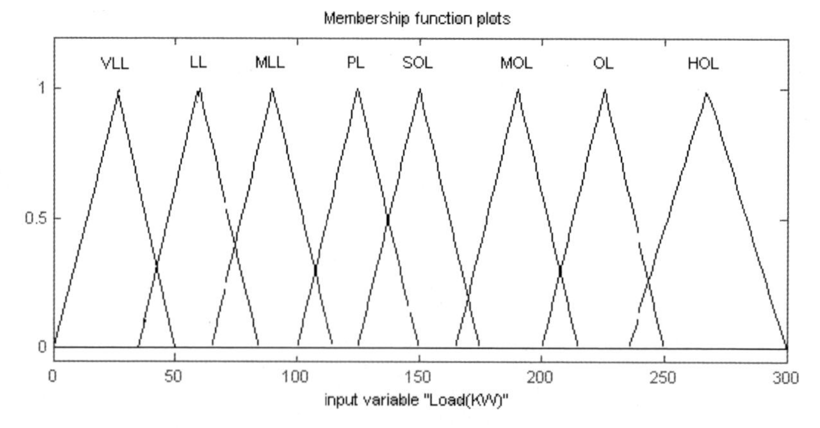


**Fig. 2.37** Fuzzy membership functions describing the input variable

**Table 2.9** Description of the output linguistic variable

| SL. No. | Output (*Change*) Description | Fuzzy Nomenclature | KW Range |
|---|---|---|---|
| 1 | High Subtraction | HS | −150 to −85 |
| 2 | Subtraction | S | −100 to −50 |
| 3 | Medium Subtraction | MS | −65 to −15 |
| 4 | Slight Subtraction | SS | −50 to 25 |
| 5 | Perfect Addition | PA | 0 to 50 |
| 6 | Medium Addition | MA | 35 to 85 |
| 7 | Large Addition | LA | 65 to 115 |
| 8 | Very Large Addition | VLA | 100 to 150 |

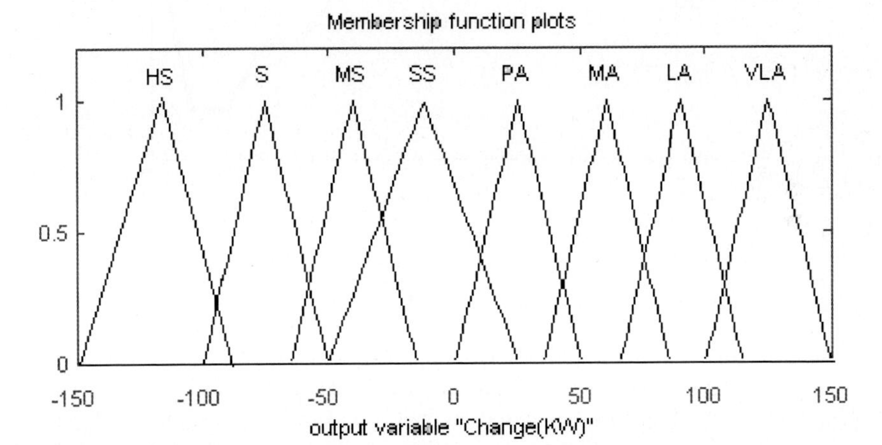


**Fig. 2.38** Fuzzy membership functions describing the output variable

## 2.5.4 Results

In this section, we show the application results using the fuzzy logic-based load balancing technique. Matlab® fuzzy toolbox (Mathworks 2002) was used for the simulation. We have utilized the Mamdani (Mamdani 1977) fuzzy inferencing technique. We take an example input load configuration of [245 120 82] kW for the three

**Table 2.10** Fuzzy rule set

| Rule No. | Rule Description |
|---|---|
| 1 | If Load is VLL then Change is VLA |
| 2 | If Load is LL then Change is LA |
| 3 | If Load is MLL then Change is MA |
| 4 | If Load is PL then Change is PA |
| 5 | If Load is SOL then Change is SS |
| 6 | If Load is MOL then Change is MS |
| 7 | If Load is OL then Change is S |
| 8 | If Load is HOL then Change is HS |

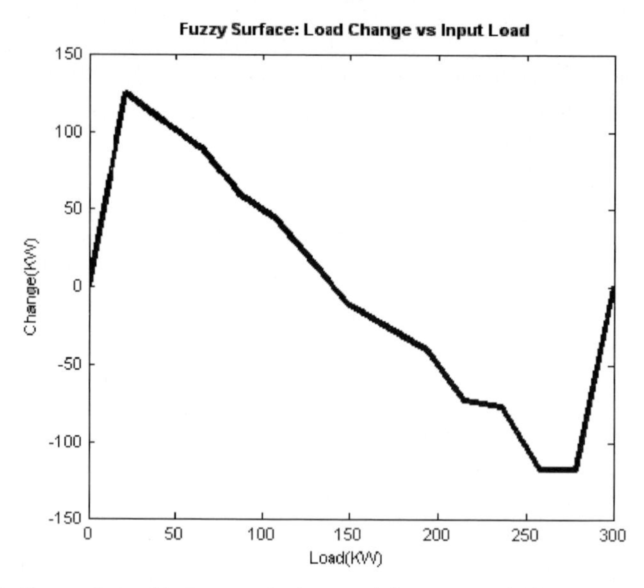


Fig. 2.39 Nonlinear relationship between the input and the output variable

phases. We try to balance it using the fuzzy controller described above. The graphical determination of the output load change for the three phases corresponding to this input load and involving the eight fuzzy rules are shown in Fig. 2.40 to 2.42. The defuzzification operation is based on the CoG technique.

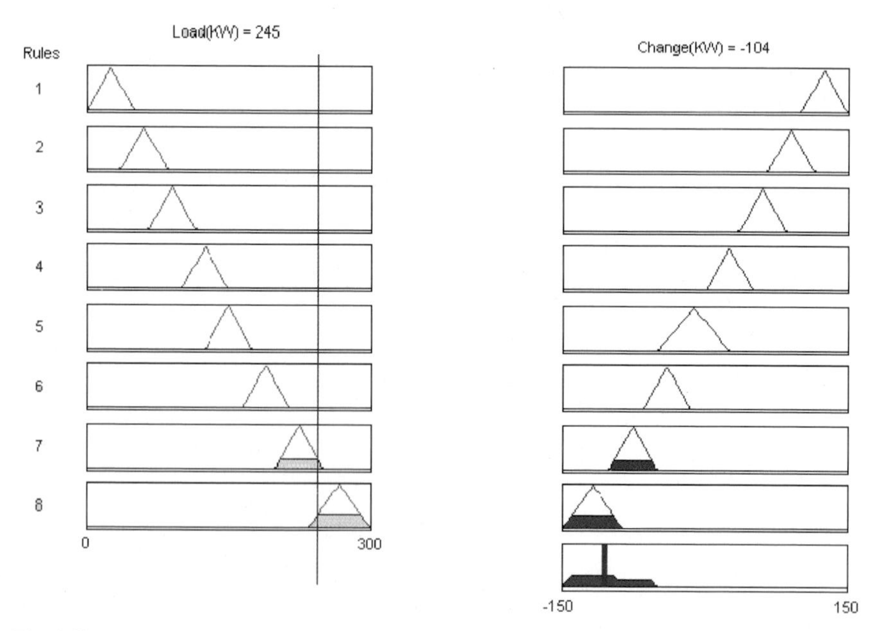


Fig. 2.40 Determination of the output load change for phase 1 of the input load

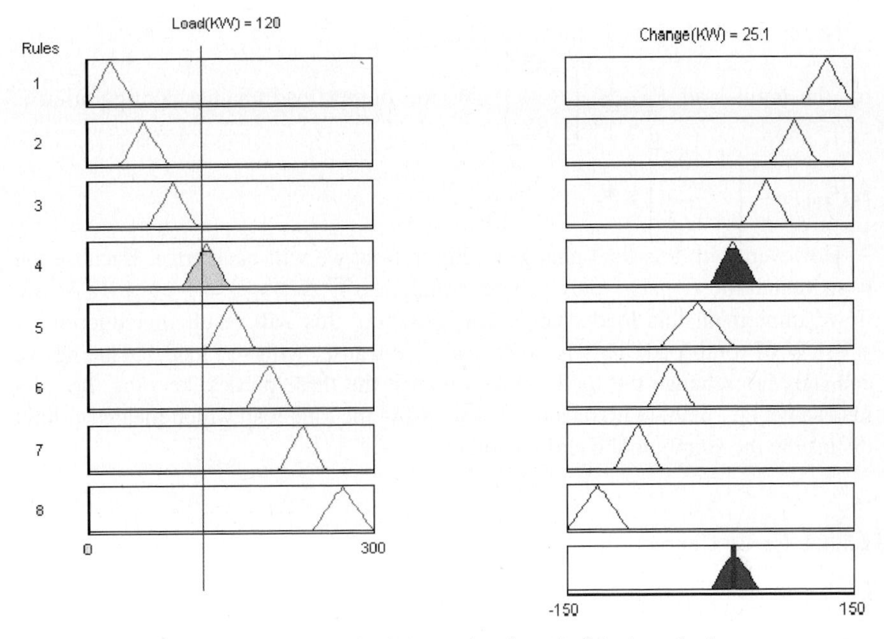


**Fig. 2.41** Determination of the output load change for phase 2 of the input load

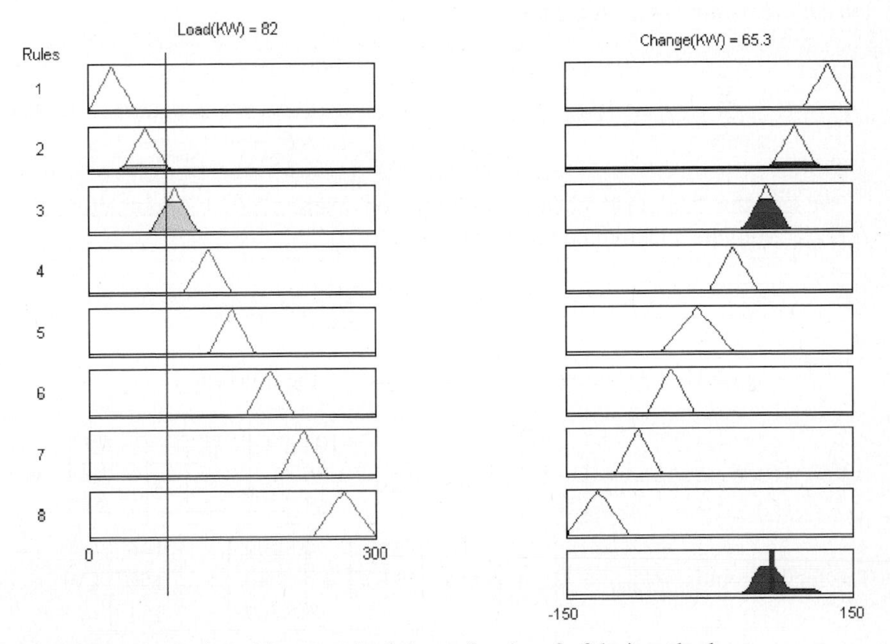


**Fig. 2.42** Determination of the output load change for phase 3 of the input load

So, after rounding the output load change,

for the input load $P_{in} = \begin{bmatrix} 245 \\ 120 \\ 82 \end{bmatrix}$ kW, the output load change configuration is

$\Delta P_{fuzzy} = \begin{bmatrix} -104 \\ 25 \\ 65 \end{bmatrix}$ kW.

However, with this load change configuration, we will have error. Because the positive and the negative totals are not equal, i.e., $\sum \Delta P_{fuzzy} = -14 \neq 0\,\text{kW}$. So, if we implement this load change configuration, this will result in reduction of $-14\,\text{kW}$ of total load. This is not possible, because, with the load balancing we can only interchange the load points amongst the three phases, keeping the total load same, i.e., without increasing or decreasing the total load which means without disturbing the supply to the end-customers.

#### 2.5.4.1 Error Correction

So, we have to perform an error correction. The average error (*AE*) is given as

$$AE = round\left(\frac{\sum \Delta P_{fuzzy}}{3}\right). \quad (2.31)$$

We use the average error to construct the error matrix $\Delta P_{error}$, by distributing the *AE* evenly among the three phases.

$$\Delta P_{error} = \begin{bmatrix} AE \\ AE \\ \sum \Delta P_{fuzzy} - 2^{*} AE \end{bmatrix}. \quad (2.32)$$

We get the final load change configuration matrix $\Delta P$, by subtracting the $\Delta P_{error}$ from the uncorrected fuzzy output $\Delta P_{fuzzy}$.

$$\Delta P = \Delta P_{fuzzy} - \Delta P_{error}, \sum \Delta P = 0. \quad (2.33)$$

Applying (2.31–2.33) in our example case, we get the following.

$$AE = -5\,\text{kW}, \Delta P_{error} = \begin{bmatrix} -5 \\ -5 \\ -4 \end{bmatrix} \text{kW}, \Delta P = \begin{bmatrix} -104 \\ 25 \\ 65 \end{bmatrix} - \begin{bmatrix} -5 \\ -5 \\ -4 \end{bmatrix} = \begin{bmatrix} -99 \\ 30 \\ 69 \end{bmatrix} \text{kW}.$$

The final output is, $P_{final} = P_{in} + \Delta P = \begin{bmatrix} 245 \\ 120 \\ 82 \end{bmatrix} + \begin{bmatrix} -99 \\ 30 \\ 69 \end{bmatrix} = \begin{bmatrix} 146 \\ 150 \\ 151 \end{bmatrix}$ kW.

Applying (2.30) on $P_{in}$ and $P_{final}$, we get respectively,

Table 2.11 Application results showing different phase loading conditions

| Test Case | Initial Load (kW) | IAUB/Ph (kW) | Initial Fuzzy Change (kW) | Error (kW) | Final Fuzzy Change (kW) | Final Load (kW) | FAUB/Ph (kW) |
|---|---|---|---|---|---|---|---|
| | 157 | | −12 | 6 | −18 | 139 | |
| 1 | 134 | 24.67 | 5 | 6 | −1 | 133 | 4 |
| | 120 | | 25 | 6 | 19 | 139 | |
| | 140 | | −1 | −6 | 5 | 145 | |
| 2 | 145 | 10.67 | −12 | −6 | 0 | 145 | 4 |
| | 156 | | −5 | −7 | −5 | 151 | |
| | 205 | | −52 | −28 | −24 | 181 | |
| 3 | 170 | 30 | −21 | −28 | 7 | 177 | 2.67 |
| | 162 | | −12 | −29 | 17 | 179 | |
| | 170 | | −21 | 34 | −55 | 115 | |
| 4 | 95 | 58 | 60 | 34 | 26 | 121 | 6 |
| | 83 | | 64 | 35 | 29 | 112 | |
| | 117 | | 25 | 70 | −45 | 72 | |
| 5 | 74 | 50.67 | 76 | 70 | 6 | 80 | 6 |
| | 42 | | 108 | 69 | 39 | 81 | |

$$Initial\ Absolute\ Average\ Unbalance\ (IAUB)\ /\ Phase = 108.67\,\text{kW},$$
$$Final\ Absolute\ Average\ Unbalance\ (FAUB)\ /\ Phase = 3.33\,\text{kW}.$$

The reduction of unbalance indicates improvement of the phase balancing.

#### 2.5.4.2 Different Results

Table 2.11 (Ukil and Siti 2006) shows more application results for different feeder load configurations. The results presented in Table 2.11 are chosen to represent different phase loading conditions. For each application, the final fuzzy output change configuration is passed onto the expert system for implementing the load change operation.

## References

Ukil A, Siti M, Jordaan JA (2006) Feeder load balancing using neural networks. Lecture Notes in Computer Science 3972:1311–1316

Ukil A, Siti W (2006) Feeder load balancing using fuzzy logic and combinatorial optimization-based expert system. International Journal of Electrical Power and Energy Systems (Under review)

Mamdani EH (1977) Applications of fuzzy logic to approximate reasoning using linguistic synthesis. IEEE Transactions on Computers: 26:1182–1191

Mathworks Inc. (2002) MATLAB® documentation–fuzzy logic toolbox, version 6.5.0.180913a, Release 13

# Section III: Objective Projects

## 2.6 Energy Efficient Operation

With the growing need of power, energy efficient operation is also gaining importance. Energy efficient electricity consumption can reduce the overload conditions often causing disruption in power supply. An energy efficient load controller can optimally switch on and off different power outlets of any premises or institution. Fuzzy logic can be applied to implement such a system.

### *2.6.1 Project Overview*

As a pilot project, any site can be selected, e.g., a domestic house, school, hospital etc. A detailed load profile should be prepared following a detailed load survey. The load survey should indicate the detailed load consumption in terms of power outlets and usage times. All the power supply modes should be carefully reviewed, e.g., normal AC (alternating current) supply, backup supply (generator, inverter etc).

In the next step, a fuzzy logic controller should be designed and implemented to optimally control the loads in an energy efficient manner. The fuzzy controller should optimally cater specific outlets through the specific supply modes. For example, in case of a hospital, the controller should always power in the emergency units. The controller would also look into unused points and switch off the outlets connected therein to save power. For example, in case of school, when all students are off to the classes, the hostel air conditioning system or similar non-emergency power outlets should be cut off.

In the final step, the designed fuzzy controller can be implemented in hardware using the field programmable gate array (FPGA) chip (Sanz 1994). FPGA chip is an IC (integrated circuit) architecture of fixed arrays of complex gates and a fixed mesh of interconnect wires. The interconnections may be user-programmed by RAM (random access memory) or flash memory, or by fusible links.

### *2.6.2 Designing Fuzzy Controller*

As mentioned above, the fuzzy controller should operate in two modes: outlets and operating times. Depending on the specific project site, the operating times could be modeled like *early morning, day, afternoon, evening, night, mid-night*. And the loads can be categorized as *low, medium, high, very high, extra high*. Another additional input could be the emergency factor stating if it is an emergency unit or not. The outputs could be the supply modes: *supply 1, supply 2,..., supply n*; and the operating schedule/status like ON for 2 Hours, or OFF for 1.5 Hours etc.

Depending on the various factors like the project site, load profile, orientation of the emergency and non-emergency units etc, the inputs and the outputs can be related by rules. For example, the same air conditioning load can be classified during the *day* time as *low* for the hostel while *high* for the laboratory. The fan loads of the day outdoor section of a hospital should be *high* from 08:00 Hr to 16:00 Hr during day while should be *low* from 22:00 Hr to 05:00 Hr in the night. Depending on the fuzzy classification of the loads and the emergency category, the outputs should be related. For example, in case of a school, if the load (e.g., classroom fans) is *low* and *non-emergency* it should be *off* for *n* hours. Here, *n* indicates the off period. For the example of the classroom fans, it could be like for 9 Hours, from 20:00 Hr to 05:00 Hr. Similarly, for a hospital, if the load (like operation theater) is *medium* during *mid-night* but *emergency* (but operation still going) and *supply 1* (primary supply) is *off*, then it should be *on* from *supply n* (next available supply in the line). Figure 2.43 shows the schematic diagram of the proposed project.

## 2.6.3 Final Output

After simulating the fuzzy logic controller for the specific project site and the load profile, a detailed energy efficient load scheduling would come out. This software simulation can be done by various software packages like Matlab® fuzzy logic toolbox (Mathworks 2002) and the like.

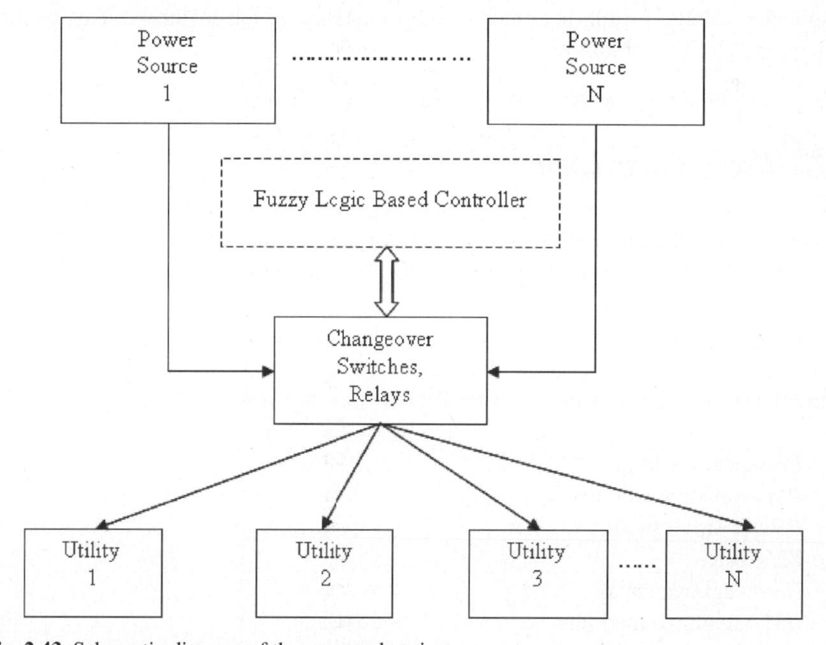


**Fig. 2.43** Schematic diagram of the proposed project

Once the simulation attains an acceptable accuracy level, it could be implemented as real-time controller using the FPGA chips. The FPGA chip could incorporate the final intelligent load scheduling structure as an internal look up table (LUT). The FPGA chip-based real-time controller could drive different power outlet relay driver circuits as a time delay switch (time delay for the 24-hour operation, switch to optimally switch between different supply modes).

## References

Mathworks Inc. (2002) MATLAB® documentation–fuzzy logic toolbox, version 6.5.0.180913a, Release 13

Sanz A (1994) Analog implementation of fuzzy controller. In Proc. of 3rd IEEE Int. Conf. on Fuzzy Systems, FUZZ-IEEE '94, pp 279–283

## 2.7 Stability Analysis

Stability analysis (Kundur 1994) is a much needed operation in power systems. Among the various parameters of power system, stability analysis of voltage (Taylor 1993) is quite critical. Voltage is often found to be quite unstable across the power network due to various factors. These include different types of load configuration, switching in and out of transformers, changes in generation schedules, etc. As a result, the voltage profile becomes heavily unstable which in turn causes overall system instability and failure.

### *2.7.1 Use of Fuzzy Logic*

Fuzzy logic can be effectively used to monitor the stability of the voltage. A fuzzy logic-based controller can be designed to take the necessary actions to improve the voltage stability.

The input to the fuzzy system could be the voltage deviation. And the output could be the degree of operation of the controller. The fuzzy set could be expressed using the common linguistic variables for fuzzy controllers:

*PL* – positive large
*PM* – positive medium
*PS* – positive small
*ZE* – zero
*NS* – negative small
*NM* – negative medium
*NL* – negative large.

The output, i.e. the degree of operation of the fuzzy controller is dependent on the mode of end equipment. For example, if we attempt to adjust the voltage stability using the generation schedule we could incorporate the controller to the generator excitation system. If we try to manipulate using the transformers, we could incorporate the controller with the tap changer of the transformer.

Separate rule bases should be formed for the different output options. The rules should be validated using a large number of test cases for different bus voltage profile across the network. Percentage improvement of the voltage instability could be the sufficient statistic to measure any improvement and the operability of the fuzzy logic-based system.

## References

Kundur P (1994) Power system stability and control. McGraw-Hill, New York
Taylor CW (1993) Power system voltage stability. McGraw-Hill, New York

## 2.8 Demand Side Management

Demand side management (DSM) programs influence the customer electricity usage to such an extent that it produces changes in the utility load shape in ways that benefit both the customer and the utility. Changing the load shape through DSM allows a utility to match its generation capacity with the load more closely. This inevitably reduces the capital and the operating costs for the utility. The cost saving in a way can be passed on to the customers to achieve a mutual benefit. A two-fold DSM activity can be targeted using fuzzy logic.

### *2.8.1 Load Profiling*

Load profiling involves modeling of the load into different categories for efficient modeling of the system. A properly modeled load can provide critical insight into the system operation like load forecasting, system overload, stability and so forth. Fuzzy logic can be effectively applied for clustering the loads measured from the system bus.

Loads (power consumption in terms of MW) can be modeled for 24-hour (short-term) or long-term (one month, one season etc). For the short-term load profiling, a typical load pattern for a substation could be used. The inputs could be the hourly load patterns and the time information, like *early morning, morning, day, afternoon, evening, night, mid-night*: $\{EM, M, D, A, E, N, MN\}$. Additional input could be seasonal information like *summer*, *winter, spring* etc. The inputs could be related to the outputs like *peak*, *off-peak* or *low*, *medium*, *high*, *very high* etc. The fuzzy load clustering model could be utilized to forecast, for example, peak load during the day. Using different clustered load profiles fuzzy regression analysis (Nazarko and Zalewski 1999) could be performed. Fig 2.44 shows a typical clustered 24-hour load profile.

Fuzzy regression analysis of the different clustered load profiles could be utilized to predict system overload conditions during the peak hour. Regression analysis of substation 1 to $n$ could be utilized to prepare a normalized load profile. During any period the normalized load profile can be used to analyze any discrepancy in power flow or generation. The deviations could be used to take necessary actions. Any severe power outage prediction could be helpful in preventing power cuts, eventually blackouts. All these are critical DSM activities which could be effectively addressed using a fuzzy analysis of the load profiles.

## 2.8.2 Energy Consumption Modeling

DSM activities are not restricted only for monitoring the current operating conditions. It also involves insight into futuristic flexible operations. One such sector is modeling of the energy consumption profile. This could be related to different factors like economic conditions etc. Futuristic energy consumption modeling can be quite handy in power system scheduling for generation, distribution systems, future developments of new generation capacity and the like. However, this is a quite nonlinear operation. Hence, fuzzy logic could be effectively used as a solution method.

An example project could be to model the income–energy economic relationship. The relevant fuzzy linguistic variables could be the income: {*low*, *medium*, *standard*, *high*, *very high*}, the energy consumption habit: {*low*, *medium*, *standard*, *high*}. The input and output could be the income and the energy consumption respectively, and vice versa.

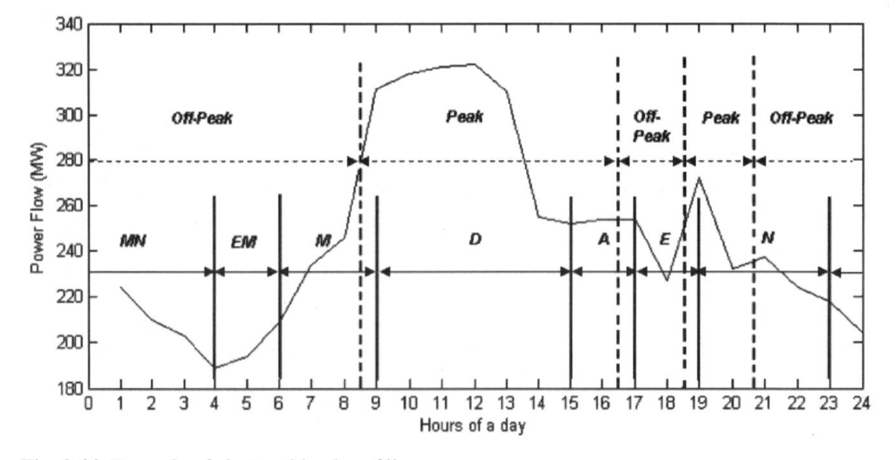


**Fig. 2.44** Example of clustered load profiling

## Reference

Nazarko J, Zalewski W (1999) The fuzzy regression approach to peak load estimation in power distribution systems. IEEE Transactions on Power System 14:809–814

## 2.9 Power Flow Controller

With the growing expansion and complexity of the electrical power network worldwide, power flow study is critically important for various aspects. Power flow analysis provides insight into the bus voltage, voltage magnitude and angle, active and reactive power flow in the network and so on. Besides the standard techniques of the power flow analysis like the Gauss iterative method, Gauss-Seidel iterative method, Newton-Raphson method, fast decoupled load flow method etc, fuzzy logic can also be a useful tool in load flow analysis (Vlachogiannis 2001).

### *2.9.1 System Overview*

The designated fuzzy logic-based power flow analysis system can analyze the bus voltage magnitude and angle, active and reactive power flow. Power flow could be modeled as follows.

$$\frac{\Delta P}{V} = A_{\theta}.\Delta\theta, \tag{2.34}$$

$$\frac{\Delta Q}{\theta} = A_{V}.\Delta V. \tag{2.35}$$

In (2.34–2.35), $P$ and $Q$ indicate the active and reactive power respectively. $V$ and $\theta$ indicate the system bus voltage magnitude and angle respectively. In (2.34), state vector $V$ is fixed while $\theta$ is variable (need to be updated). In (2.35), state vector $\theta$ is fixed while $V$ is variable (need to be updated).

Fuzzy logic-based power flow controller could be used to update the variable parameters in (2.34–2.35). Accordingly the input and the output variables could be formed. For modeling $P$-$\theta$ cycle, the input variable could be the active power $(P)$ and the output could be the bus voltage angle $\theta$. For modeling $Q - V$ cycle, the input variable could be the reactive power $(Q)$ and the output variable could be the bus voltage magnitude $V$.

The fuzzy set could be designed using the standard fuzzy nomenclature for controller as $\{PL, PM, PS, ZE, NS, NM, NL\}$. These are

*PL* – positive large
*PM* – positive medium

*PS* – positive small
*ZE* – zero
*NS* – negative small
*NM* – negative medium
*NL* – negative large

To incorporate the standard power flow techniques, the Gaussian fuzzy membership function (bell-shaped, see Fig. 2.9) could be used. Fuzzy rules, derived from various bus voltage recordings, should be used to map the relationship between the input and the output variables. CoG or MoM technique could be used to get the defuzzified crisp output values for the state vector $\theta$ or $V$ to update the (2.34–2.35).

## Reference

Vlachogiannis JG (2001) Fuzzy logic application in load flow studies. IEE Proc Generation, Transmission, Distribution 148:34–40

## Section IV: Information Section

## 2.10 Research Information

This section is organized towards different applications of fuzzy logic in various power engineering related problems. Power engineering specific fuzzy logic research information in terms of relevant books, publications (journal, conference proceedings), reports etc, have been categorized into different sub-sections depending on the applications.

### *2.10.1 General Fuzzy Logic*

Driankov D, Hellendoorn H, Reinfrank M (1993) An introduction to fuzzy control. Springer-Verlag, New York
George JK, Yuan B (2002) Fuzzy sets & fuzzy logic: theory and applications. Prentice Hall, India
Kosko B (1996) Fuzzy engineering. Prentice-Hall Inc., New York
Ross TJ (1995) Fuzzy logic with engineering applications. McGraw-Hill, New York
Vojislav K (2001) Learning and soft computing–support vector machines, neural networks and fuzzy logic models. The MIT Press, Cambridge

Wang LX (1997) A course in fuzzy systems and control. Prentice Hall, Upper Saddle River

Yager RR, Filev DP (1994) Essentials of fuzzy modelling and control. John Wiley & Sons, New York

### 2.10.2 Fuzzy Logic and Power Engineering

Bansal RC (2003) Bibliography on the fuzzy set theory applications in power systems (1994–2001). IEEE Transactions on Power Systems 18:1291–1299

Dunn RW, Bell KW, Daniels AR (1997) Fuzzy logic and its application to power systems. IEE Colloquium (Digest), pp. 4/1–4/4

El-Hawary ME (1998) Electric power applications of fuzzy systems. IEEE Press, Piscataway, NJ

Hiyama T, Tomsovic K (1993) Current status of fuzzy system applications in power system. Trans. Inst. Elect. Eng. Jpn. 113 B (1):2–6

Momoh JA, Tomsovic K (1995) Overview and literature survey of fuzzy set theory in power systems. IEEE Transactions on Power Systems 10: 1676–1690

Srinivasan D, Liew AC, Chang CS (1995) Applications of fuzzy systems in power systems. Electric Power Systems Research 35:39–43

### 2.10.3 Electrical Load Forecasting

#### 2.10.3.1 Short-term

Al-Kandaria AM, Solimanb SA, El-Hawary ME (2004) Fuzzy short-term electric load forecasting. Electrical Power and Energy Systems 26:111–122

Hsu Y, Ho K (1992) Fuzzy expert system: an application to short-term load forecasting. IEE Proceedings 139:471–477

Kim KH, Youn HS, Kang YC (2000) Short-term forecasting for special days in anomalous load conditions using neural networks and fuzzy inference method. IEEE Transactions on Power Systems 15:559–65

Liang RH, Cheng CC (2000) Combined regression—fuzzy approach for short-term load forecasting. IEE Proc. Gener. Transm. Distrib. 147:261–266

Pandian SC, Duraiswamy K, Rajan CCA, Kanagaraj N (2006) Fuzzy approach for short term load forecasting. Electric Power Systems Research 76:541–548

Rahman S (1990) Formulation and analysis of a rule-based short-term load forecasting algorithm. IEEE Proceedings 78:805–816

Srinivasan D, Swee TS, Cheng CS, Chan EK (1999) Parallel neural network–fuzzy expert system strategy for short-term load forecasting: system implementation and performance evaluation. IEEE Transactions on Power Systems 14:1100–1106

Wu HC, Lu C (1999) Automatic fuzzy model identification for short term load forecast. IEE Proc. Gener. Transm. Distrib. 146:477–82
Yang HT, Huang CM (1998) A new short-term load forecasting approach using self-organizing fuzzy ARMAX models. IEEE Trans. Power Syst. 13:217–225

#### 2.10.3.2 Medium- and Long-term

Angelov P (2004) An approach for fuzzy rule-base adaptation using online clustering. International Journal of Approximate Reasoning 35:275–289
Nazarko J, Zalewski W (1999) The fuzzy regression approach to peak load estimation in power distribution systems. IEEE Transactions on Power Systems 14:809–814
Padmakumari K, Mohandas KP, Thiruvengadam S (1999) Long term distribution demand forecasting using neuro fuzzy computations. International Journal of Electrical Power and Energy System 21:315–322
Wong YK, Chung TS, Kwan YL (1998) Monthly electric load forecasting using fuzzy neural networks. Int. Journal of Power and Energy Systems 18:137–141

### *2.10.4 Fault Analysis*

Aggarwal RK, Joorabian M, Song YH (1997) Fuzzy neural network approach to accurate transmission line fault location. Int. Journal of Engineering Intelligent Systems for Electrical Engineering and Communications 5:251–258
Erenturk K, Altas IH (2004) Fault identification in a radial power system using fuzzy logic. Instrumentation Science and Technology 32:641–653
Hu Q, He ZJ, Zi YY, Zhang ZS (2005) Intelligent fault diagnosis in power plant using empirical mode decomposition, fuzzy feature extraction and support vector machines. Key Engineering Materials 373–381
Mahanty RN, Dutta Gupta PB (2007) A fuzzy logic based fault classification approach using current samples only. Elec. Power Sys. Research 77:501–507
Moshtagh J, Aggarwal RK (2004) A new approach to fault location in a single core underground cable system using combined fuzzy & wavelet analysis. In Proc. 8$^{th}$ IEE Int. Conf. on Dev in Power System Protection, pp. 228–231
Srinivasan D, Cheu RL, Poh YP, Chwee Ng AK (2000) Automated fault detection in power distribution networks using a hybrid fuzzy–genetic algorithm approach. Engineering Applications of Artificial Intelligence 13:407–418
Vasilic S, Kezunovic M (2005) Fuzzy art neural network algorithm for classifying the power system faults. IEEE Transactions on Power Delivery 20:1306–1314
Yeo SM, Kim CH, Hong KS, Lim YB, Aggarwal RK, Johns AT, Choi MS (2003) A novel algorithm for fault classification in transmission lines using a combined adaptive network and fuzzy inference system. Int. Journal of Electrical Power & Energy Systems 25:747–758

Zhang N, Kezunovic M (2005) Coordinating fuzzy ART neural networks to improve transmission line fault detection and classification. In Proc. 2005 IEEE Power Engineering Society General Meeting 1, pp. 734–740

### 2.10.5 Power Systems Protection

Jamil M, Thomas MS, Parmod MK (1999) Improved scheme based on fuzzy logic for EHV transmission line protection. In IEE Conference Publication 1, pp. 1.360.P6–1.364.P6

Kasztenny B, Rosolowski E, Saha MM, Hillstrom B (1997) Fuzzy sets and logic in power system protection. International Journal of Engineering Intelligent Systems for Electrical Engineering and Communications 5:193–203

Liu CW, Tsay SS, Wang YJ, Su MC (1999) Neuro-fuzzy approach to real-time transient stability prediction based on synchronized phasor measurements. Electric Power Systems Research 49:123–127

Rebizant W, Feser K (2001) Fuzzy logic application to out-of-step protection of generators. In Proc. IEEE PES Summer Meeting, Vancouver, Canada

Rebizant W, Bejmert D, Szafran J (2005) Fuzzy logic based overcurrent protection for MV networks. In Proc. PSCC Conf., Liege, Belgium

Wang H, Keerthipala WWL (1998) Fuzzy-neuro approach to fault classification for transmission line protection. IEEE Trans. Power Delivery 13:1093–1104

### 2.10.6 Distance Protection

Das B, Reddy JV (2005) Fuzzy-logic-based fault classification scheme for digital distance protection. IEEE Transactions on Power Delivery 20:609–616

Dash PK, Pradhan AK, Panda G (2000) Novel fuzzy neural network based distance relaying scheme. IEEE Transactions on Power Delivery 15:902–907

Yang HT, Huang CM (2002) Distribution system service restoration using fuzzy petri net models. Int. J. of Electrical Power and Energy Systems 24:395–403

### 2.10.7 Relay

Abyaneh HA, Karegar HK (2003) A new model of overcurrent relay characteristics based on fuzzy logic. International Journal of Power and Energy Systems 23: 163–167

Ferrero A, Sangiovanni S, Zappitelli E (1995) A fuzzy-set approach to fault type identification in digital relaying. IEEE Trans. on Power Delivery 13:169–175

Karegar HK, Abyaneh HA, Al-Dabbagh M (2003) A flexible approach for overcurrent relay characteristics simulation. Electric Power Systems Research 66: 233–239
Kolla SR (1997) Applying fuzzy logic to power system protective relays. InTech 44:42–45
Qureshi F, Saxena A, Hota HS (2005) Design of knowledge base and sensitivity evaluation of relay sensor for power system protection. Modelling, Measurement and Control A78:25–39
Yang H (1994) Analytical structure of a two-input two-output fuzzy controller and its relation to PI and multilevel relay controllers. Fuzzy Sets and Systems 63: 21–33
Youssef OAS (2004) A novel fuzzy-logic-based phase selection technique for power system relaying. Electric Power Systems Research 68:175–184
Youssef OAS (2004) Combined fuzzy-logic wavelet-based fault classification technique for power system relaying. IEEE Trans. Power Deliv. 19:582–589

### *2.10.8 Power Flow Analysis*

Bhattacharyya K, Crow ML (1996) A fuzzy logic based approach to direct load control. IEEE Transactions on Power Systems 6:708–714
Bijwe PR, Raju GKV (2006) Fuzzy distribution power flow for weekly meshed systems. IEEE Transactions on Power Systems 21:1645–1652
Bijwe PR, Hanmandlu M, Pande VN (2006) Fuzzy power flow solutions with reactive limits and multiple uncertainties. Elec. Power Sys. Res. 76: 145–152
Das D, Ghosh S, Srinivas DK (1999) Fuzzy distribution load flow. Elect. Mach. Power Syst. 27:1215–1226
Das D, Satpathy PK, Dattagupta PB (2004) Fuzzy set theory application to load flow analysis. Journal of Institution of Engineers 85:55–59
Edwin Liu WH, Guan X (1996) Fuzzy constraints enforcement and control action curtailment in an optimal power flow. IEEE Trans. Power Syst. 11:639–645
Guan X, Liu WH, Papalexopoulos AD (1995) Application of a fuzzy set method in an optimal power flow. Electric Power Systems Research 34:11–18
Kenarangui R, Seifi A (1994) Fuzzy power flow analysis. Electric Power Systems Research 29:105–109
Lo KL, Lin YJ, Siew WH (1999) Fuzzy-logic method for adjustment of variable parameter in load-flow calculation. IEE Proc. Generation, Transmission, Distribution 146:276–282
Miranda V, Saraiva JP (1992) Fuzzy modeling of power system optimal load flow. IEEE Transactions on Power Systems 7:843–849
Saric AT, Ciric RM (2003) Integrated fuzzy state estimation and load flow analysis in distribution networks. IEEE Transactions Power Delivery 18:571–578
Vlachogiannis JG (2001) Fuzzy logic application in load flow studies. IEE Proceedings: Generation, Transmission, Distribution 148:34–40

### 2.10.9 Power Systems Equipments & Control

Cam E, Koccarslan I (2005) A fuzzy gain scheduling PI controller application for an interconnected electrical power system. Electric Power Systems Research 73:267–274

El-Ela AAA, Bishr MA, Allam SM, El-Sehiemy RA (2007) An emergency power system control based on the multi-stage fuzzy based procedure. Electric Power Systems Research 77:421–429

El-Ela AAA, Bishr MA, Allam SM, El-Sehiemy RA (2005) Optimal preventive control actions using multi-objective fuzzy linear programming technique. Electric Power Systems Research 74:147–155

Hsu YY, Kuo HC (1992) Fuzzy set based contingency ranking. IEEE Transactions on Power Systems 7:1189–1196

Lin BR (1995) Power converter control based on neural and fuzzy methods. Electric Power Systems Research 35:193–206

Noroozian M, Andersson G, Tomsovic K (1996) Robust, near time-optimal control of power system oscillations with fuzzy logic. IEEE Transactions on Power Delivery 11:393–400

Rahideh A, Gitizadeh M, Rahideh A (2006) Fuzzy logic in real time voltage/ reactive power control in FARS regional electric network. Electric Power Systems Research 76:996–1002

Ramakrishna G, Rao ND (1998) Fuzzy inference system to assist the operator in reactive power control in distribution systems. IEE Proceedings: Generation, Transmission and Distribution 145:133–138

Shayeghi H, Shayanfar HA, Jalili A (2006) Multi-stage fuzzy PID power system automatic generation controller in deregulated environments. Energy Conversion and Management 47:2829–2845

Venkatesh B, Sadasivam G, Khan AM (1999) Fuzzy logic based successive LP method for reactive power optimization. Electric Machines and Power Systems 27:1141–1160

Yixin X, On ML, Zhenyu H, Shousun C, Baolin Z (1999) Fuzzy logic damping controller for FACTS devices in interconnected power systems. In Proc. IEEE Conf. 11:591–594

### 2.10.10 Frequency Control

Chang CS, Fu W (1997) Area load-frequency control using fuzzy gain scheduling of PI controllers. Electric Power Systems Research 42:145–152

Ghoshal SP, Goswami S (2003) Application of GA based optimal integral gains in fuzzy based active power-frequency control of non-reheat and reheat thermal generating systems. Electric Power Systems Research 67:79–88

Feliachi A, Rerkpreedapong D (2005) NERC compliant load frequency control design using fuzzy rules. Electric Power Systems Research 73:101–106

Kocaarslan I, Akalin G, Erfidan T (1999) Application of fuzzy reasoning to load-frequency control in two-area power system. In Proc. of European Control Conference ECC'99, Karlsruhe, Germany, pp. 27–31
Talaq J, Al-Basri F (1999) Adaptive fuzzy gain scheduling for load-frequency control. IEEE Transactions on Power Systems 14:145–150
Tesnjak S, Mikus S, Kuljaca O (1995) Load-frequency fuzzy control in power systems. In Proc. Fifth SONT Symposium, Poree, pp. 136–139

### 2.10.11 Harmonic Analysis

Al-Hamadi HM, Soliman SA (2002) Kalman filter for identification of power system fuzzy harmonic components. Elec. Power Systems Research 62:241–248
Chan WL, So ATP (1994) Power harmonics pattern recognition by solving fuzzy equations. Fuzzy Sets and Systems 67:257–266
Soliman SA, Helal I, Al-Kandari AM (1999) Fuzzy linear regression for measurement of harmonic components in a power system. Electric Power Systems Research 50:99–105
Soliman SA, Alammari RA, El-Hawary ME (2003) Frequency and harmonics evaluation in power networks using fuzzy regression technique. Electric Power Systems Research 66:171–177

### 2.10.12 Power Systems Operation

Daneshpooy A, Gole AM, Chapman DG, Davies JB (1997) Fuzzy logic control for HVDC transmission. IEEE Transactions on Power Delivery 12:1690–1697
Das D (2006) Reconfiguration of distribution system using fuzzy multi-objective approach. Electrical Power and Energy Systems 28:331–338
Ghoshal SP (2004) Optimizations of PID gains by particle swarm optimizations in fuzzy based automatic generation control. Electric Power Systems Research 72:203–212
Guimarães ACF, Lapa CMF (2003) Fuzzy inference system for evaluating and improving nuclear power plant operating performance. Annals of Nuclear Energy 31:311–322
Hlebcar B (2001) Forecasting closing overvoltages in high-voltage networks using a fuzzy model. Fuzzy Sets and Systems 118:1–8
Kazemi A, Sohrforouzani MV (2006) Power system damping using fuzzy controlled facts devices. Electrical Power & Energy Systems 28:349–357
Mahmoud M, Dutton K, Denman M (2005) Design and simulation of a nonlinear fuzzy controller for a hydropower plant. Elec. Power Sys. Research 73:87–99
Peiris HJC, Annakkage UD, Pahalawaththa NC (1999) Generation of fuzzy rules to develop fuzzy logic modulation controllers for damping of power system oscillations. IEEE Transactions on Power Systems 14:1440–1445

Ramaswamy R, Nayar KP (2004) Online estimation of bus voltages based on fuzzy logic. Electrical Power and Energy System 26:681–684

Sàrfi RJ, Salama MMA, Chikhani AY (1996) Applications of fuzzy sets theory in power systems planning and operation: A critical review to assist in implementation. Electric Power Systems Research 39:89–101

Soliman SA, Rahman MHA, El-Hawary ME (1997) Application of fuzzy linear regression algorithm to power system voltage measurements. Electric Power Systems Research 42:195–200

Su CT, Lin CT (1996) A new fuzzy control approach to voltage profile enhancement for power systems. IEEE Trans. on Power Systems 11:1654–1659

### *2.10.13 Power Systems Security*

Abdul-Rahman KH, Shahidehpour SM (1995) Static security in power system operation with fuzzy real load conditions. IEEE Trans. Power Systems 10:77–87

Alvarez JMG, Mercado PE (2007) A new approach for power system online DSA using distributed processing and fuzzy logic. Electric Power Systems Research 77:106–118

Boyen X, Wehenkel L (1999) Automatic induction of fuzzy decision trees and its application to power system security assessment. Fuzzy Sets Sys 102:3–19

Liu CW, Chang CS, Su MC (1998) Neuro-fuzzy networks for voltage security monitoring based on synchronized phasor measurements. IEEE Transactions on Power Systems 13:326–332

Su CT, Lin CT (2001) Fuzzy based voltage/reactive power scheduling for voltage security improvement and loss reduction. IEEE Transactions on Power Delivery 16:319–323

Tso SK, Zhu TX, Zeng QY, Lo KL (1995) Fuzzy set approach to dynamic voltage security assessment. IEE Proc. Gener. Trans. Distrib. 142:190–194

Tso SK, Zhu TX, Zeng QY, Lo KL (1997) Evaluation of load shedding to prevent dynamic voltage instability based on extended fuzzy reasoning. IEE Proceedings: Generation, Transmission and Distribution 144:81–86

### *2.10.14 Power Systems Reliability*

Abdelaziz AR (1999) A fuzzy-based power system reliability evaluation. Eletric Machines and Power Systems 27:271–278

Billinton R, Firuzabad MF, Bertling L (2001) Bibliography on the application of probability methods in power system reliability evaluation. IEEE Transactions on Power Systems 16:595–602

Chanda RS, Bhattacharjee PK (1998) A reliability approach to transmission expansion planning using fuzzy fault-tree model. Electric Power Systems Research 45:101–108

Narasimhan S, Asgarpoor S (2000) Fuzzy-based approach for generation system reliability evaluation. Electric Power Systems Research 53:133–138

Saraiva JT et al. (1996) Generation/transmission power system reliability evaluation by Monte-Carlo simulation assuming a fuzzy load description. IEEE Transactions on Power Systems 11:690–695

Verma AK, Ravi Kumar HM, Keshavan BK (2000) A fuzzy logic approach to security-based bulk power system reliability evaluation. Electric Machines and Power Systems 28:45–54

Verma AK, Srividya A, Deka BC (2005) Composite system reliability assessment using fuzzy linear programming. Elec. Power Systems Research 73: 143–149

### 2.10.15 Power Systems Stabilizer

Al-Osaimi SA, Abdennour A, Al-Sulaiman AA (2005) Hardware implementation of a fuzzy logic stabilizer on a laboratory scale power system. Electric Power Systems Research 74:9–15

Chaudhary AK, Kothari ML, Sharma A (2001) A fuzzy logic power system stabilizer. J. Inst. Eng. Elect. Eng. 82:36–42

Dash PK, Mishra S, Liew AC (1995) Fuzzy-logic-based VAR stabilizer for power system control. IEE Proc. Gener. Transm. Distrib. 42:618–624

El-Metwally KA, Hancock GC, Malik OP (1996) Implementation of a fuzzy logic PSS using a micro-controller and experimental test results. IEEE Trans. Energy Convers. 11:91–96

El-Metwally KA, Malik OP (1995) Fuzzy logic power system stabilizer. IEE Proceedings on Generation, Transmission and Distribution 142:277–281

Hariri A, Malik OP (1999) Fuzzy logic power system stabilizer based on genetically optimized adaptive network. Fuzzy Sets and Systems 102:31–40

Hassan M, Malik OP (1993) Laboratory evaluation and test results of rule-based power system stabilizers: a comparative study. Intelligent Systems Engineering, pp. 52–60

Hiyama T (1994) Real time control of micro-machine system using micro-computer based fuzzy logic power system stabilizer. IEEE Trans. Energy Convers. 9: 724–731

Hoang P, Tomsovic K (1996) Design and analysis of an adaptive fuzzy power system stabilizer. IEEE Transactions on Energy Conversion 11:455–461

Hosseinzadeh N, Kalam A (1999) A direct adaptive fuzzy power system stabilizer. IEEE Transactions on Energy Conversion 14:1564–1571

Hsu Y, Cheng C (1990) Design of fuzzy power system stabilizer for multi-machine power system. IEE Proc. C 137, pp. 233–238

Hwang GC, Lin SC (1992) A stability approach to fuzzy control design for nonlinear systems. Fuzzy Sets and Systems 48:279–287

Metwally KAEL, Malik OP (1995) Fuzzy logic power system stabilizer. IEE Proc. Generation, Transmission, Distribution 142:277–281

Narendranath UA, Thukaram D, Parthasarathy K (1999) An expert fuzzy control approach to voltage stability enhancement. International Journal of electrical power and energy systems 21:279–287

Subrahamanyam N, Ramana Rao PV (1998) A new fuzzy logic based power system stabilizer. In Proc. IEEE TENCON'98, 2:531–535

### *2.10.16 Power Quality*

Dash PK, Jena RK, Salama MMA (1999) Power quality monitoring using an integrated Fourier linear combiner and fuzzy expert system. International Journal of Electrical Power and Energy System 21:497–506

Farghal SA, Kandil MS, Elmitwally A (2002) Quantifying electric power quality via fuzzy modelling and analytic hierarchy processing. IEE Proceedings: Generation, Transmission and Distribution 149:44–49

Huang J, Negnevitsky M, Nguyen DT (2002) A neural-fuzzy classifier for recognition of power quality disturbances. IEEE Trans. Power Deliv. 17:609–616

Liao Y, Lee JB (2004) A fuzzy-expert system for classifying power quality disturbances. Int. Journal of Electrical Power & Energy Systems 26:199–205

### *2.10.17 Renewable Energy*

El-Tamaly HH, Mohammed AAE (2006) Impact of interconnection photovoltaic/wind system with utility on their reliability using a fuzzy scheme. Renewable Energy 31:2475–2491

Jurado F, Valverde M (2006) Genetic fuzzy control applied to the inverter of solid oxide fuel cell for power quality improvement. Electric Power Systems Research 76:93–105

Karlis AD, Kottas TL, Boutalis YS (2007) A novel maximum power point tracking method for PV systems using fuzzy cognitive networks (FCN). Electric Power Systems Research 77:315–327

Patcharaprakiti N, Premrudeepreechacharn S, Sriuthaisiriwong Y (2005) Maximum power point tracking using adaptive fuzzy logic control for grid-connected photovoltaic system. Renewable Energy 30:1771–1788

Simoes MG, Bose BK, Spiegel RJ (1997) Design and performance evaluation of a fuzzy-logic-based variable-speed wind generation system. IEEE Transactions on Industry Applications 33:956–965

### *2.10.18 Transformers*

Chen AP, Lin CC (2001) Fuzzy approaches for fault diagnosis of transformers. Fuzzy Sets and Systems 118:139–151

De A, Chartterjee N (2004) A fuzzy ARTMAP fault classifier for impulse testing of power transformer. IEEE Trans. Elect. Insul. 11:1026–1036

Dukarm JJ (1993) Transformer oil diagnosis using fuzzy logic and neural networks. In Proc. Canadian Conf. on Elect. and Computer Engg, pp. 329–332

Hong TY, Chiung CL (1999) Adaptive fuzzy diagnosis system for dissolved gas analysis of power transformers. IEEE Trans. Power Delivery 14:1342–1350

Huang YC, Yang HT, Huang CL (1997) Developing a new transformer fault diagnosis system through evolutionary fuzzy logic. IEEE Transactions on Power Delivery 12:761–767

Kasztenny B, Rosolowski E, Saha MM, Hillstrom B (1997) A self-organizing fuzzy logic based protective relay-an application to power transformer. IEEE Transactions on Power Delivery 12:1119–1127

Mofizul SI, Wu T, Ledwich G (2000) A novel fuzzy logic approach to transformer fault diagnosis. IEEE Trans. Dielectr. Electr. Insul. 7:177–186

Shin MS, Park CW, Kim JH (2003) Fuzzy logic based relaying for large power transformer protection. IEEE Transactions on Power Delivery 18:718–724

Su Q, Mi C, Lai LL, Austin P (2000) A fuzzy dissolved gas analysis method for the diagnosis of multiple incipient faults in a transformer. IEEE Transactions on Power Systems 15:593–598

### *2.10.19 Rotating Machines*

Arnalte S (2000) Fuzzy logic-based voltage control of a synchronous generator. International Journal of Electrical Engineering Education 37:333–343

Chiang CL, Su CT (2005) Tracking control of induction motor using fuzzy phase plane controller with improved genetic algorithm. Electric Power Systems Research 73:239–247

Fonseca J, Afonso JL, Martins JS, Couto C (1999) Fuzzy logic speed control of an induction motor. Microprocessors and Microsystems 22:523–534

Hassan MA, Malik OP, Hope GS (1991) A fuzzy logic based stabilizer for a synchronous machine. IEEE Trans. Energy Convers. 6:407–413

Kim SM, Han WY (2006) Induction motor servo drive using robust PID-like neurofuzzy controller. Control Engineering Practice 14:481–487

Lown M, Swidenbank E, Hogg BW (1997) Adaptive fuzzy logic control of a turbine generator system. IEEE Transactions on Energy Conversion 12:394–399

Nounou HN, Rehman HA (2006) Application of adaptive fuzzy control to ac machines. Applied Soft Computing (In Press)

Spiegel RJ, Turner MW, McCormick VE (2003) Fuzzy-logic-based controllers for efficiency optimization of inverter-fed induction motor drives. Fuzzy Sets and Systems 137:387–401

Ye Z, Sadeghian A, Wu B (2006) Mechanical fault diagnostics for induction motor with variable speed drives using adaptive neuro-fuzzy inference system. Electric Power Systems Research 76:742–752

### 2.10.20 *Energy Economy, Market & Management*

Dondo MG, El-Hawary ME (1996) Application of fuzzy logic to electricity pricing in a deregulated environment. In Proc. Canadian Conf. on Elec. and Comp. Engg. 1, pp. 388–391

LaMeres BJ, Nehrir MH, Gerez V (1999) Controlling the average residential electric water heater power demand using fuzzy logic. Electric Power Systems Research 52:267–271

Liang Z, Yang K, Sun Y, Yuan J, Zhang H, Zhang Z (2006) Decision support for choice optimal power generation projects: fuzzy comprehensive evaluation model based on the electricity market. Energy Policy 34:3359–3364

Nehrir MH, LaMeres BJ (2000) A multiple-block fuzzy logic-based electric water heater demand-side management strategy for leveling distribution feeder demand profile. Electric Power Systems Research 56:225–230

Singh S, Saini JS (2006) Fuzzy FPGA based captive power management. In Proc. 2006 IEEE Power India Conf., New Delhi

### 2.10.21 *Unit Commitment*

Padhy NP, Ramachandran V, Paranjothi SR (1999) Fuzzy decision system for unit commitment risk analysis. Int. J. of Power and Energy Systems 19:180–185

Saneifard S, Prasad NR, Smolleck HA (1997) A fuzzy logic approach to unit commitment. IEEE Transactions on Power Systems 12:988–995

Sen S, Kothari DP (1998) Optimal thermal generating unit commitment: a review. International Journal of Electrical Power and Energy System 20:443–451

Victoire TAA, Jeyakumar AE (2006) A tabu search based hybrid optimization approach for a fuzzy modelled unit commitment problem. Electric Power Systems Research 76:413–425

### 2.10.22 *Scheduling*

Huang SJ (1998) A genetic-evolved fuzzy system for maintenance scheduling of generating units. Int. J. of Electrical Power and Energy System 20:191–195

Li Y, Luh PB, Guan X (1994) Fuzzy optimization-based scheduling of identical machines with possible breakdown. In Proc. IEEE Int. Conf. Robotics and Automation, San Diego, CA, pp. 3447–3452

Srinivasan D, Chang CS, Liew AG (1994) Multiobjective generation scheduling using fuzzy optimal search technique. IEE Proceedings: Generation, Transmission and Distribution 141:233–242

Venkatesh B, Sadasivam G, Khan MA (2000) New optimal reactive power scheduling method for loss minimization and voltage stability margin maximization us-

ing successive multi-objective fuzzy LP technique. IEEE Transactions on Power Systems 15:844–851
Wong KP, Suzannah YWW (1996) Combined genetic algorithm/simulated annealing/fuzzy set approach to short-term generation scheduling with take-or-pay fuel contract. IEEE Transactions on Power Systems 11:128–136
Yan H, Luh PB (1997) A fuzzy optimization-based method for integrated power system scheduling and inter-utility power transaction with uncertainties. IEEE Transactions on Power Systems 12:756–763

### *2.10.23 Power Electronics*

Bose BK (1994) Expert system, fuzzy logic, and neural network applications in power electronics and motion control. Proceedings of the IEEE 82:1303–1323
Bose BK (2000) Fuzzy logic and neural networks in power electronics and drives. IEEE Industry Applications Magazine 6:57–63
Cecati C, Dell'Aquila A, Liserre M, Ometto A (2003) A fuzzy-logic-based controller for active rectifier. IEEE Trans. Industry Applications 39:105–112
Lin BR (1997) Analysis of neural and fuzzy-power electronic control. IEE Proceedings: Science, Measurement and Technology 144:25–33
Lin BR, Hoft RG (1994) Neural networks and fuzzy logic in power electronics. Control Engineering Practice 2:113–121

# Chapter 3
# Neural Network

## Section I: Theory

## 3.1 Introduction

We, humans and other animal beings perform various complex nonlinear tasks with the aid of information processing via biological neural networks. This is done by the huge amount of complex interconnections of the neuronal cells (neurons) which interact amongst each other by exchanging brief electrical pulses or *action potential*. The biological neural network is the motivation of its computer science version, popularly known as artificial neural network (ANN). Many scientists and engineers, however, often drop the initial artificial tag to name it just neural network (NN).

Inspired by the biological nervous system, ANN operates on the principle of largely interconnected simple elements operating as a network function. In doing so, no previous knowledge is assumed, but data, records, measurements, observations are considered. ANN research stands on the fact of *learning* from data to mimic the biological capability of linear and nonlinear problem solving.

Basically, we can design and train the neural networks for solving particular problems which are difficult to solve by the human beings or the conventional computational algorithms. The computational meaning of the training comes down to the adjustments of certain *weights* which are the key elements of the ANN. This is one of the key differences of the neural network approach to problem solving than conventional computational algorithms which work step-by-step. This adjustment of the weights takes place when the neural network is presented with the input data records and the corresponding target values. Depending on the learning method (supervised or unsupervised), the neural network tries to correlate the correspondence between the input and target data by adjusting its weights. Following many iterations of incremental training with many different input–output pairs, the neural network weights are optimally adjusted by comparing its output and the designated target. This depends on various factors, like, the learning algorithm, network architecture etc.

### 3.1.1 History and Background

The first basic neuronal model was proposed in the 1940s by two neuroscientists, Warren McCulloch and Walter Pitts (McCulloch and Pitts 1943). McCulloch and Pitts model of neuron contained the concept of spikes (action potential) carrying information. According to their model, the spikes carry information across the interconnected network of brain: each spike representing a binary 1 (Boolean TRUE), and lack of information or spike representing a binary 0 (Boolean FALSE). The McCulloch-Pitts neuron was much like the operation of a transistor. However, the basic strength and significance of their model was establishment of the neurons as computational elements. The idea behind this model was highly responsible for the design of the modern digital computers by John von Neumann and later on, the ANN. Ideas like the threshold, interconnection etc can be considered to be directly reciprocated from the McCulloch-Pitts model to the modern ANN.

### 3.1.2 Applications

Neural networks are applied widely for solving different problems which in general are difficult to solve by humans or conventional computational algorithms. A comprehensive list of neural network applications including applications like adaptive channel equalizer, word recognizer, process monitor, sonar classifier, risk analysis system and the like was provided by the famous 1988 DARPA neural network study (DARPA 1988). A list is provided below which gives an idea of the versatile uses of the neural networks for different applications. The list is just an idea of the vast application fields.

- Image processing
  - Image recognition
  - Font/letter identification
  - Face, facial expression recognition
  - Hand-written character recognition
- Speech processing
  - Speaker dependent/independent speech recognition
  - Speaker identification/verification
  - Speech coding
- Robotics
  - Motion planning
  - Robotic vision
  - Robotic speech recognition
- Time-series prediction
  - Financial/currency prediction

- Weather forecast
- Load forecast

- Signal processing
  - Signal classification
  - Data compression
  - Signal and object separation
  - Feature extraction
  - Noise classification and reduction
  - Signal conditioning
- Process automation
  - Data filtering
  - Data validation
  - Modeling
  - Diagnosis of production faults
- Automotive
  - Automatic guidance system
  - Valuation of accidents
- Aerospace
  - Flight control
  - Auto-pilot design
  - Flight path simulation
  - Aircraft component fault detection
- Financial
  - Market analysis
  - Market forecasting
  - Investment analysis
  - Trading optimization
  - Product development optimization
- Security
  - Surveillance
  - Intelligent security camera image analysis
  - Fingerprint recognition
- Defense
  - Radar image processing
  - Target tracking
  - Missile guidance
  - Weapon classification.

### 3.1.3 Pros and Cons

Any new field comes with certain advantages and disadvantages. This is also valid for the neural networks. ANN research is motivated towards complex nonlinear problem solving by learning from data. Some associated advantages and disadvantages are:

- Pros
  - Solution for nonlinear and unknown systems
  - Faster modeling than conventional mathematical modeling
  - Robust against faults
  - More insensitive to process interference than conventional systems
  - Black box model, hence usage flexibility and application variability
  - Easy and effective implementation (software and hardware) possible.
- Cons
  - Computation cost increases exponentially with the increase of neurons
  - Data generation is costly and takes long time
  - Data definition sometimes can be difficult
  - Neural networks can produce poor performance when extrapolating since they may diverge quicker than linear functions.

## 3.2 Artificial Neural Networks (ANN)

### 3.2.1 Basic Structure of the Artificial Neural Networks

Biological neuronal network comprises of numerous (in the order of trillion) interconnections among the neurons. The power of neuron comes from its collective behavior in a network where all neurons are interconnected. Hypothetically, it is assumed that in a human brain approximately 10 billion neurons are acting collectively via approximately $10^{14}$ interconnections. Similarly, ANN also develops from the interconnections of several unit neurons or *nodes*. The arrangement of the neurons is quite arbitrary. It depends on several factors, like, the nature of application, inspiration from real biological structure seen under the microscope and so forth. However, most commonly, as a rule, in the artificial networks, the following layers of neurons are used.

- **Input layer**: Number of neurons in this layer corresponds to the number of inputs to the neuronal network. This layer consists of passive nodes, i.e., which do not take part in the actual signal modification, but only transmits the signal to the following layer.

- **Hidden layer**: This layer has arbitrary number of layers with arbitrary number of neurons. The nodes in this layer take part in the signal modification, hence, they are active.
- **Output layer**: The number of neurons in the output layer corresponds to the number of the output values of the neural network. The nodes in this layer are active ones.

The basic structure of the ANN is shown in Fig. 3.1. Note that this is just a basic structure. There are many types of interconnections which we will discuss later.

### *3.2.2 Structure of a Neuron*

A brief description of the typical structure of the biological neuron is given below to compare with the ANN structure. A typical biological neuron consists of three main parts.

**Cell body**. The cell is the most important part of the neuron. It contains the nucleus and is responsible for the information processing.

**Axon**. The axon carries the nerve signals away from the neuron (cell body). Each neuron contains only one axon, usually heavily branched, in order to reach as many other neurons as possible. The axon connects to other neurons through so called synapses.

**Dendrites**. The dendrites are the nerve endings. They carry incoming signals from other neurons to the cell body.

The operation of the biological neuron can be summarized as the signals from other neurons are summed together and compared against a threshold to determine if the neuron shall excite ("fire"). This is the motivation for the structure of the ANN.

For ANN, the structure of a neuron mainly consists of the *sum* and *squash* unit. The inputs pass through the specific weights and then the weighted inputs

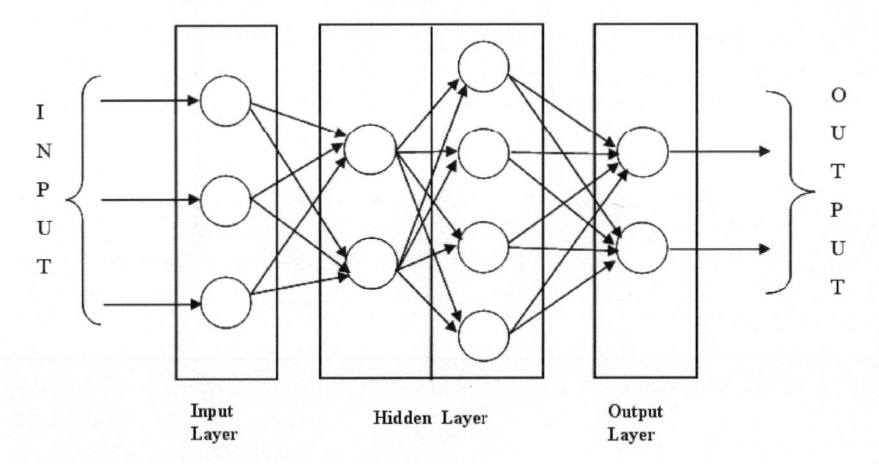


**Fig. 3.1** Basic structure of the artificial neural network

are summed. A *weight* is the strength of the connection between two neurons. The weighted sum is then passed through a transfer function (also often called "squashing" functions, since they compress an infinite input range into a finite output) to produce the final output. The transfer function is chosen to map the input(s) to the output(s).

The structure of an artificial neuron vis-à-vis biological counterpart is shown in Fig. 3.2. In Fig. 3.2, a fully interconnected structure is shown, i.e., where all the input values are passed onto the hidden layer. The operation of the network is as follows. The input values, passed to the hidden layer from the input layer, are multiplied by the weights, the predetermined numbers (processing units of the ANN). The weighted inputs after passing through the transfer function (shown as sigmoid in Fig. 3.2) produce the output (could be more than one). In a network learning operation, this output is compared against the objective target value and the deviation is fed back to modify the weights. This process is iterated until the output correctly approximates the target, i.e., the error is gradually decreased to the minimum. In the end, the neural network gets fully trained, i.e., the weights achieve optimum values (error becomes minimum) to produce the output within a specified accuracy level corresponding to the input data.

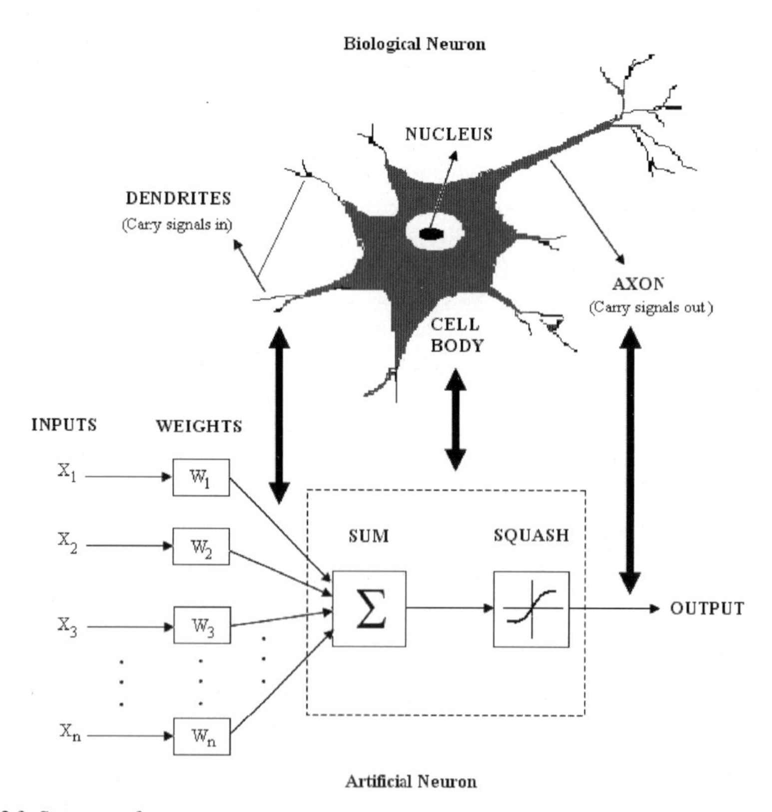


**Fig. 3.2** Structure of a neuron

In Fig. 3.2, the following are assumed.

$j$: Neuron $j$,
$i$: Index of the inputs,
$n$: Number of the inputs,
$X_i$ : Input $i$,
$W_i$ : Weight of the input $X_i$,
$S_j$ : Sum of the weighted inputs for neuron $j$,
$T_j(S)$: Transfer function,
$O_j$: Output of neuron $j$,

$$S_j = \sum_{i=1}^{n} X_i.W_i, \tag{3.1}$$

$$O_j = T_j(S_j). \tag{3.2}$$

Neural networks can be made more flexible by making the input values adjustable (left or right) to the center value. This can be implemented by including a special additional node to the input layer which always has a unity value. When this is multiplied by the weights of the hidden layer, it provides a *bias* (like DC offset), hence, it is called the *bias node*.

### 3.2.3 Transfer Function

Transfer function in the ANN maps the input(s) to the output(s). Hence, it is an important element of the network for a successful network design. But why do we need such a transfer function? The answer is: it is the key element to invoke the nonlinear relationship between the input and the output. Without the transfer function the whole operation is linear and could be solved using linear algebra or similar methods. To simplify the whole neuronal operation, first we produce the weighted sum of the input values which acts like a single lumped input value for the whole input data. And then we apply the transfer function on this lumped input value.

There are different kinds of transfer functions, continuous and discrete. We will discuss the most commonly used ones. We consider $T(S)$ as the transfer function for the weighted sum $S$ (lumped input) of the inputs.

- **Discrete Functions**

1. **Linear Transfer Function**

$$T(S) = \begin{cases} 1, & S > 0 \\ 0, & S \leq 0 \end{cases}.$$

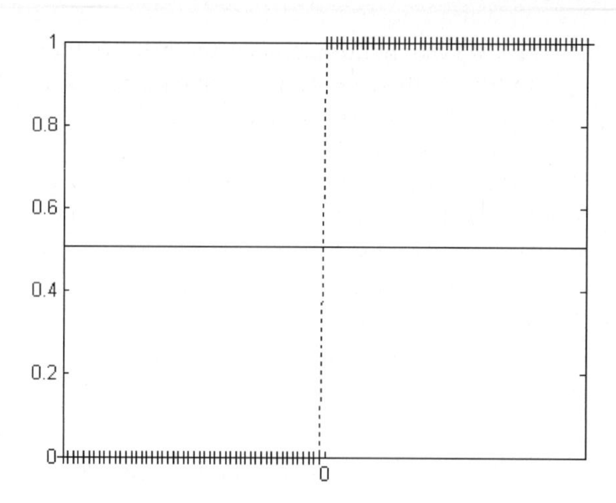

Fig. 3.3 Linear transfer function

2. **Hard-Limit Transfer Function**

$$T(S) = \begin{cases} 1, & S > b \\ S, & a < S < b \\ -1, & S < a \end{cases}.$$

- **Continuous Functions**

1. **Sigmoid Transfer Function**

$$T(S) = \frac{1}{1 + e^{-cS}}, \quad \text{where} \quad 0 < c \leq 1.$$

Sigmoid is an important real function. It is differentiable and quite important in the learning algorithms like, the backpropagation algorithm and the like. Its value lies between 0 and 1, however it is asymptotic to 1.

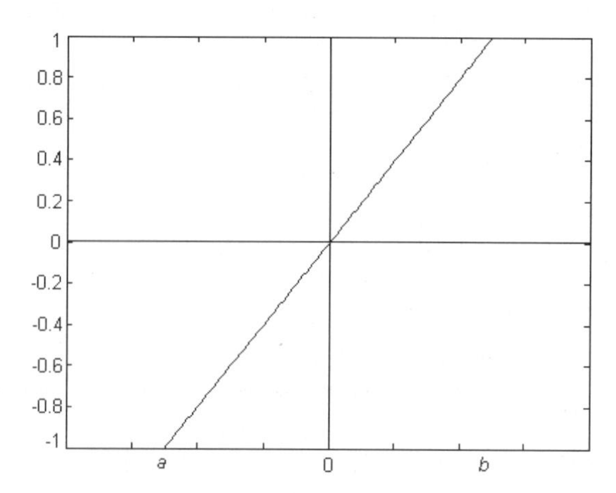

Fig. 3.4 Hard-Limit transfer function

The constant $c$ determines the shape of the sigmoid function. Its reciprocal $1/c$ is called the temperature parameter for the stochastic neural networks. For simplification, we can consider $c = 1$. Higher values of $c$ bring the shape of the sigmoid closer to that of the step function. And in the limit $c \to \infty$, the sigmoid converges to a step function at the origin. Figure 3.5 shows the sigmoid function for the different values of $c$.

We consider the sigmoid function $s(x)$ with $c = 1$. The derivative of the sigmoid function with respect to $x$ is,

$$s'(x) = \frac{d}{dx}s(x) = \frac{e^{-x}}{\left(1 + e^{-x}\right)^2} = s(x)\,(1 - s(x))\,. \tag{3.3}$$

Equation (3.3) is important for quick calculation. For example, for $x = 0$, $s(x) = 0.5$, which gives $s'(x) = 0.5(1 - 0.5) = 0.25$.

2. **Tan-Sigmoid Transfer Function**

$$T(S) = \tanh(S) = \frac{e^S - e^{-S}}{e^S + e^{-S}}.$$

Tan-sigmoid or hyperbolic tangent transfer function is the modification to the sigmoid function, resulting in a symmetrical transfer function. Symmetrical transfer functions have advantages in learning which will be discussed later. Tan-sigmoid transfer function produces an output in the range of $-1$ to 1 as shown in Fig. 3.6.

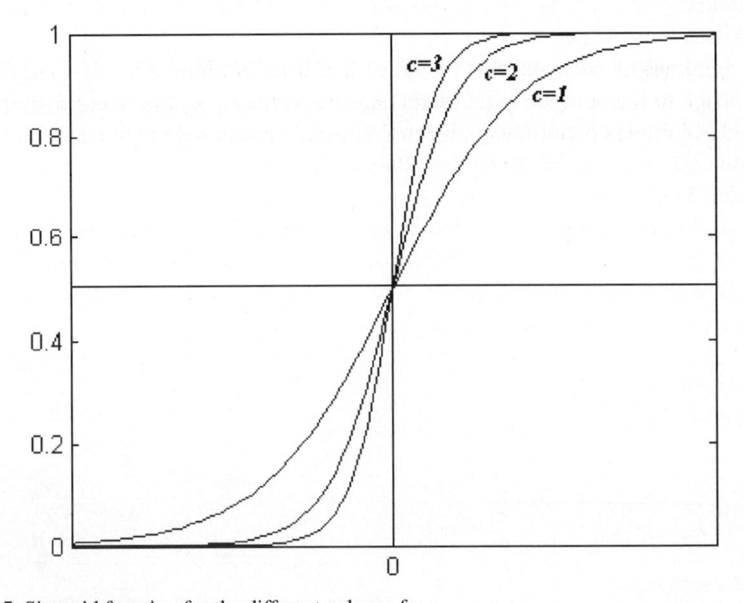


**Fig. 3.5** Sigmoid function for the different values of $c$

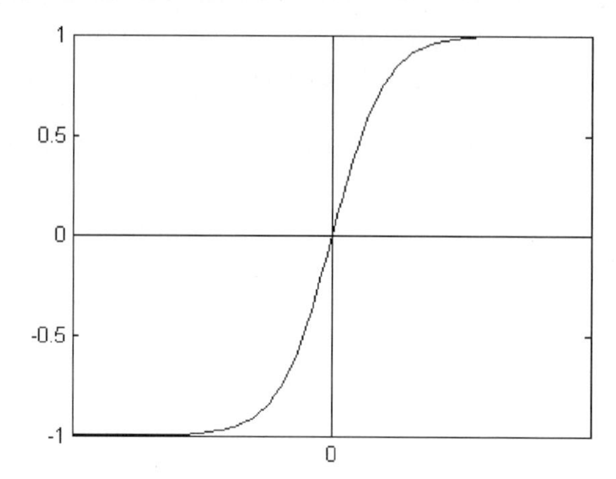


**Fig. 3.6** Tan-Sigmoid transfer function

## *3.2.4 Architecture of the ANN*

The term architecture refers to the connectivity of a network. Depending on the kind of interconnection of the neurons and the training algorithm to fit the weights, different neuronal network types can be defined. These can be divided into three major groups as described below.

1. **Feedforward networks**
   There is no feedback within the network. The coupling takes place from one layer to the next. The information flows, in general, in the forward direction in the feedforward networks. Figure 3.7 shows the structure of the feedforward network.
2. **Feedback networks**
   In this kind of network, the output of a neuron is either directly or indirectly fed back to its input via other linked neurons. This kind of network is frequently used in complex pattern recognition tasks, e.g., speech recognition etc. Figure 3.8 shows the structure of the feedback network.
3. **Lateral networks**
   In this kind of network, there exist couplings of neurons within one layer. There is no essentially explicit feedback path amongst the different layers. This can be thought of as a compromise between the forward and feedback network. Figure 3.9 shows the structure of the lateral network.

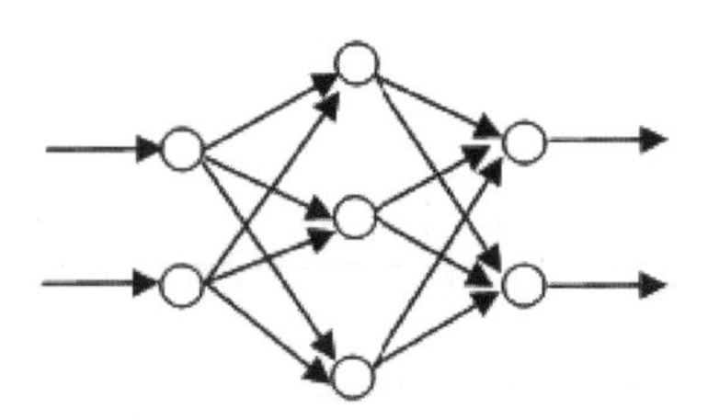

**Fig. 3.7** Feedforward network

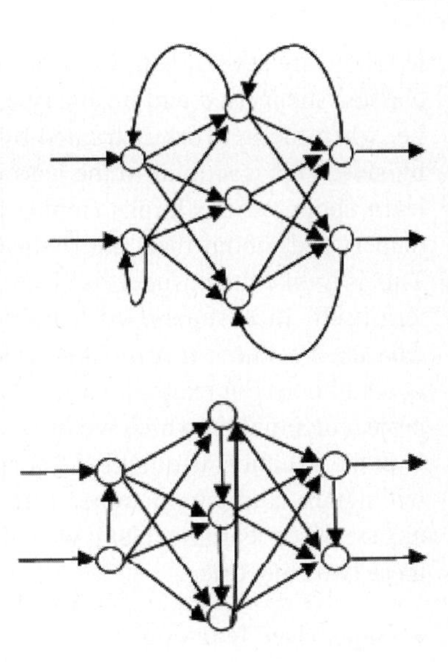

**Fig. 3.8** Feedback network

**Fig. 3.9** Lateral network

### 3.2.5 *Steps to Construct a Neural Network*

Gradual steps to construct the neural networks for performing certain tasks are as follows.

- Identify the associated problem,
- Record data for a long period,
- Determine data which is relevant to the task,
- Preprocess data using mathematical operations to reduce the data redundancy and necessity to calculate,
- Identify the proper network structure,
- Construct the appropriate neural network for the data,
- Train and test the network,
- Save the final weights,
- Implement the network (software or hardware) as per requirement.

## 3.3 Learning Algorithm

Artificial neural networks work through the optimized weight values. The method by which the optimized weight values are attained is called *learning*. In the process of learning we present the neural network with pairs of input and output data and try to teach the network how to produce the output when the corresponding input is presented. When learning is complete, the trained neural network, with the updated optimal weights, should be able to produce the output within desired accuracy corresponding to an input pattern.

There are several learning algorithms. They can be broadly categorized into two classes: supervised and unsupervised. Supervised learning means guided learning, i.e. when the network is trained by showing the input and the desired result side-by-side. This is similar to the learning experience in our childhood. As a child, we learn about various things (input) when we see them and simultaneously are told (supervised) about their names and the respective functionalities (desired result). This is unlike the unsupervised case where learning takes place from the input pattern itself. In unsupervised learning the system learns about the pattern from the data itself without *a priori* knowledge. This is similar to our learning experience in adulthood. For example, often in our working environment we are thrown into a project or situation which we know very little about. However, we try to familiarize with the situation as quickly as possible using our previous experiences, education, willingness and similar other factors. This adaptive mechanism is referred to as unsupervised learning. Next, we will discuss about some important algorithms from these two categories.

- **Supervised learning**
  - Delta rule
  - Gradient descent
  - Backpropagation
- **Unsupervised learning**
  - Hebb's rule

### *3.3.1 The Delta Rule*

The learning algorithm modifies the weight values towards the optimum value by minimizing the difference between the produced output and the desired result. However, two questions remain unanswered: which weight values to change and how much? These are the aims of the learning algorithm.

The delta rule changes the weights by taking into account the difference in error, i.e., the difference between the output and the target. Delta (capital $\Delta$ and small $\delta$) is a Greek letter which is used in mathematics to indicate the changes. Hence comes the name delta rule. Popular variants of the delta rule include the Widrow-Hoff rule (Widrow and Hoff 1960), the gradient descent rule, the perceptron rule etc.

Consider the neuronal description in Sect. 3.2.2. We introduce another variable, $R_j$ the desired result for neuron $j$. The delta rule alters the weight vectors $W_{ij}$ (input $i$ and neuron $j$) in order to minimize the error, that is, the difference between the desired result (target) and the produced output, $(R_j - O_j)$. The simple logic to alter the weights would be to add or subtract a small number $\alpha$ from each of the weights depending on whether the entity $(R_j - O_j)$ is positive or negative. However, this simple rule does not address the magnitude of the error. To incorporate the magnitude of the error we multiply the entity $(R_j - O_j)$ by $\alpha$. This will give a proportionate correction factor. This answers the how much question, but the question of which weights need to be altered still remains unanswered.

If we change all of the weight values using the proportionate correction factor, we might introduce unwanted error which can cause divergence of the optimization of the weight values instead of convergence. To avoid this, we will alter only those weights which have had a significant role in producing the specific output. This can be implemented by incorporating the state determined by the squashing function $T$. Hence, the delta rule stands at,

$$\Delta W_{ij} = \alpha . \left( R_j - O_j \right) . T_{ij} . \tag{3.4}$$

This is followed by the update rule,

$$\Delta W_{ij_{new}} = W_{ij} + \Delta W_{ij} . \tag{3.5}$$

The update takes place iteratively until the weights attain the desired optimum values. $\alpha$ is called the *learning rate* which determines how fast the learning takes place.

### *3.3.2 Gradient Descent*

The delta rule tells us how to alter which weights. However, it does not indicate which is the best possible way to do it. To answer this different algorithms evolved, one of the popular ones is the gradient descent algorithm.

Let us take a real life analogy. Suppose we want to come down (descend) from a high hill (higher error) to a low valley (lower error). To do this, we have to gradually decrease the height (reduce the error) by climbing down. We can start from the hilltop in a random direction and take a step downhill. Then we measure our success parameter, i.e., the downward elevation. If the downhill elevation is negative that means we are going down and we continue the process. Otherwise, if it is positive, that means we are going up. Then we return to the initial position and take a different direction and continue the process. This is a trial-and-error approach. Following this approach we will eventually get down to the valley but this will take a long time. In this method, we try to evolve out the best possible way much like the evolution in nature. Hence, this is also called the *evolution algorithm* (Fogel 1998).

Gradient descent algorithm proposes a faster solution. In this algorithm, instead of proceeding in a random direction, at any point of time, we move along the gradient or slopes of the longitudinal direction. By doing so, we take the *steepest* path to the downhill valley. That is, we descend along the gradient. As we follow the steepest path to the downhill valley, this is also known as the steepest descent algorithm.

The main idea of the steepest descent algorithm is to move along each axis by a distance proportional to the slope of that axis. Now turning to our neural network, we implement this algorithm by calculating the slope for each weight, and then changing each weight by an amount proportional to that slope.

### 3.3.3 Energy Equivalence

The delta rule and the gradient descent algorithm can be also explained effectively using an energy analogy from physics. A learning algorithm changes the weights gradually in the weight space to find the optimum spot in that space. We can define an associated energy field which assigns potential energy for each point in that weight space. Here, potential energy is the analogy for the error. Hence, for the delta rule, the energy function uses sum-squared values, i.e., potential energy function $= \frac{1}{2}\sum(R_j - O_j)^2$.

In the energy analogy, we try to determine the point in the weight space which corresponds to the minimum energy level (using derivatives). Using the energy explanation, the goal is to arrive at a point which maps the input to the output with the minimum potential energy. The choosing of this point in the weight space is called gradient descent if done by following the gradients or slopes. So, the learning algorithm in terms of the energy analogy comes down to energy-minimization. Following the sum-squared expression of the potential energy, we can visualize the energy (error) function as a parabolic curve as shown in Fig. 3.10.

However, we can design algorithms other than following the gradient, which help us to arrive at the minimum energy point. For example, the *genetic algorithm* (Goldberg 1989) jumps around different points in the weight space to find the minimum energy point.

### 3.3.4 The Backpropagation Algorithm

The backpropagation algorithm is one of the well-known algorithms for calculating the neural network weights. It was developed by David Rumelhart, Geoff Hinton and R. J. Williams in 1986 (Rumelhart et al. 1986).

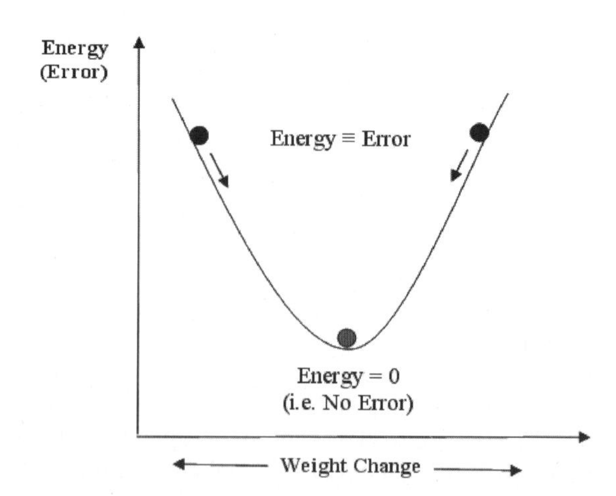


**Fig. 3.10** Energy analogy of the gradient descent

The output values $O_j$ of a given input pattern $X_i$ do not always correspond to their predetermined values $R_j$. The error $E_j$ is given by the difference of $R_j$ and $O_j$ and is to be minimized by the weight changes. In order to achieve this, the error which originates at the output is propagated backwards to the hidden layers. That is, we take the error and back-calculate to arrive at the correct solution. This is the principle of the backpropagation algorithm (the error propagates in the backward direction), which is shown in Fig. 3.11.

Following the energy analogy, the error of a neuron $j$ in the output layer is:

$$E_j = \frac{1}{2}(R_j - O_j)^2. \tag{3.6}$$

The total error $E$ of an output layer is

$$E = \sum_j E_j = \frac{1}{2}\sum_j (R_j - O_j)^2. \tag{3.7}$$

So, we need to minimize the error $E$, with respect to the weight changes ($\Delta W_{ij}$). We follow the delta rule to incorporate the learning rate $\alpha$. Gradient descent algorithm is used to choose the weight change method. We use these two techniques to define the weight change,

$$\Delta W_{kj} = -\alpha.\frac{\partial E}{\partial W_{kj}}; \quad 0 < \alpha \leq 1. \tag{3.8}$$

If the gradient $\partial E/\partial W_{kj}$ is positive then the weight change should be negative and vice versa. Hence, a minus sign is added at the right hand side of (3.8). Considering neuron $j$,

$$\Delta W_{kj} = -\alpha.\frac{\partial E_j}{\partial W_{kj}}. \tag{3.9}$$

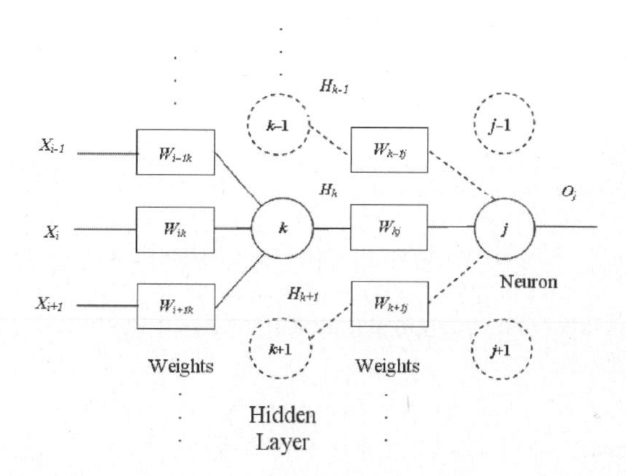


**Fig. 3.11** Example of a backpropagating network

Using a sigmoid transfer function $T_j(S)$, the output $O_j$ is defined as

$$O_j = T_j(S_j) = \frac{1}{1 + e^{-S_j}}. \tag{3.10}$$

For the hidden layer input ($H_k$) to the output layer,

$$S_j = \sum_k W_{kj}.H_k. \tag{3.11}$$

From (3.8) and (3.9),

$$\Delta W_{kj} = -\alpha.\frac{\partial E_j}{\partial W_{kj}} = -\alpha.\frac{\partial E_j}{\partial O_j}.\frac{\partial O_j}{\partial W_{kj}} = -\alpha.\frac{\partial E_j}{\partial O_j}.\frac{\partial O_j}{\partial S_j}.\frac{\partial S_j}{\partial W_{kj}}. \tag{3.12}$$

From (3.6) we get,

$$\frac{\partial E_j}{\partial O_j} = \frac{-2}{2}\left(R_j - O_j\right) = -\left(R_j - O_j\right). \tag{3.13}$$

Using (3.3) and (3.10) we get,

$$\frac{\partial O_j}{\partial S_j} = O_j\left(1 - O_j\right). \tag{3.14}$$

From (3.11) we get,

$$\frac{\partial S_j}{\partial W_{kj}} = \frac{\partial\left(\sum W_{kj}.H_k\right)}{\partial W_{kj}} = H_k. \tag{3.15}$$

Combining (3.12) to (3.15), we finally get,

$$\Delta W_{kj} = \alpha.\left(R_j - O_j\right).O_j\left(1 - O_j\right).H_k. \tag{3.16}$$

So, new weights are,

$$W'_{kj} = W_{kj} + \Delta W_{kj}. \tag{3.17}$$

The error of the output layer is backpropagated to the weights of the hidden and the input layer. $\Delta W_{kj}$ is the change in weights from the output layer to the hidden layer. The backpropagated error $E_k$ of the hidden layer is given by:

$$E_k = \sum_j E_j = \frac{1}{2}\sum_j \left(R_j - O_j\right)^2 \tag{3.18}$$

Corresponding to (3.7) is the weight change $\Delta W_{ik}$. We introduce a new learning rate $\alpha'$.

$$\Delta W_{ik} = -\alpha'.\frac{\partial E_k}{\partial W_{ik}}; \quad 0 < \alpha' \leq 1; \quad \alpha' \neq \alpha. \tag{3.19}$$

Using (3.10) and (3.11),

$$\Delta W_{ik} = -\alpha'.\frac{\partial E_k}{\partial W_{ik}} = -\alpha'.\frac{\partial E_k}{\partial H_k}.\frac{\partial H_k}{\partial W_{ik}}. \tag{3.20}$$

$$\frac{\partial E_k}{\partial H_k} = \frac{\partial \sum\limits_j E_j}{\partial H_k} = \sum_j \frac{\partial E_j}{\partial H_k}. \tag{3.21}$$

$$\frac{\partial E_j}{\partial H_k} = \frac{\partial E_j}{\partial O_j}.\frac{\partial O_j}{\partial S_j}.\frac{\partial S_j}{\partial H_k}. \tag{3.22}$$

From (3.13) and (3.14),

$$\frac{\partial E_j}{\partial O_j}.\frac{\partial O_j}{\partial S_j} = \left(R_j - O_j\right).O_j\left(1 - O_j\right). \tag{3.23}$$

For the hidden layer we have,

$$\frac{\partial S_j}{\partial H_k} = \frac{\partial \left(\sum\limits_k W_{kj}.H_k\right)}{\partial H_k} = W_{kj}, \tag{3.24}$$

$$H_k = \frac{1}{1 + e^{-S_k}}, \tag{3.25}$$

$$S_k = \sum_i W_{ik}.X_i. \tag{3.26}$$

Using (3.3) and (3.25) we get,

$$\frac{\partial H_k}{\partial S_k} = H_k\left(1 - H_k\right). \tag{3.27}$$

From (3.26),

$$\frac{\partial S_k}{\partial W_{ik}} = X_i. \tag{3.28}$$

So,

$$\frac{\partial H_k}{\partial W_{ik}} = \frac{\partial H_k}{\partial S_k}.\frac{\partial S_k}{\partial W_{ik}} = H_k\left(1 - H_k\right).X_i. \tag{3.29}$$

Using (3.20) to (3.29) we finally get the weight changes in the hidden layer as,

$$\Delta W_{ik} = \alpha'.\sum_j \left(R_j - O_j\right).O_j\left(1 - O_j\right).W_{kj}.H_k(1 - H_k).X_i. \qquad (3.30)$$

### 3.3.5 *The Hebb Rule*

Hebb's rule (Hebb 1949) is one of the most important concepts in computational neuroscience. Originally, it explained the biological neural network architectures. Also, it was the first learning rule which explained how a collection of neuronal cells might form a concept. Neuroscientists believe that the Hebbian learning takes place in the nervous system.

Though originated from the biological domain, the Hebb rule, like the concept of neural network itself, is also successfully applied in the ANN and other scientific domains. Following the Hebb's rule of cell assemblies, the Hebbian rule when applied to the nonbiological fields, forms the basis for some sort of storage or memory. Sometimes, this is called *associative*, *autoassociative* or *heteroassociative* memories.

The Hebbian rule is often used for unsupervised learning, i.e., to learn from the data pattern itself. It helps the neural network or neuron assemblies to remember specific patterns much like the memory. From that stored knowledge, similar sort of incomplete or spatial patterns could be recognized. This is even faster than the delta rule or the backpropagation algorithm because there is no repetitive presentation and training of input–output pairs. Among various algorithms to produce Hebb assemblies, the most important is the Hopfield algorithm (Hopfield 1982) which we will discuss later.

Hebb's rule can be easily arithmetically expressed using the outer product of two vectors. The outer product uses a column vector (size $C$) times a row vector (size $R$) to get a matrix of size $C \times R$. We multiply the entire column vector in turn by each element of the row vector to get the outer product, $W = s_{column} \times s_{row}$. The associative matrix formed by the outer product possesses the qualities of the content-addressable memories which can be used to recognize partial or imperfect input patterns from the memorized patterns. We show a simple example of the outer-product below.

$$C = [-1\ -1\ 1\ 1\ 1]^T, \qquad R = [1\ -1\ 1]^T,$$

$$W = C \times R = \begin{bmatrix} -1 & -1 & 1 & 1 & 1 \\ 1 & 1 & -1 & -1 & -1 \\ -1 & -1 & 1 & 1 & 1 \end{bmatrix}.$$

The outer product associative matrix $W$ could be used further for pattern recognition. This is equivalent to the learning process of the delta rule or the backpropagation algorithm.

## 3.4 Different Networks

Most commonly used neural network models are

- Perceptron
- Multilayer Perceptrons (MLP)
- Backpropagation (BP) network
- Radial Basis Function (RBF) network
- Hopfield network
- Adaline
- Kohonen network.

### *3.4.1 Perceptron*

Perceptron is the earliest kinds of artificial neural network, dating back to the McCulloch-Pitts model in the 1940s. Rosenblatt (Rosenblatt 1961) created many versions of the perceptron. Perceptron is the simplest kind of single-layer network whose weights and biases could be trained to produce the correct target vector corresponding to the input vector. Perceptron became popular due to its ability to generalize. A simple single-layer perceptron is shown in Fig. 3.12.

In the perceptron model, each external input is weighted with an appropriate weight, and the sum of the weighted inputs is sent to the linear transfer function, which also has an input of unity as the bias. The linear transfer function is used to return output as a 0 or 1. The perceptron neuron produces the output 1 if the net input into the transfer function is equal to or greater than 0; otherwise it produces 0.

$$f(x) = \begin{cases} 1 & x \geq 0 \\ 0 & \text{otherwise} \end{cases} \tag{3.31}$$

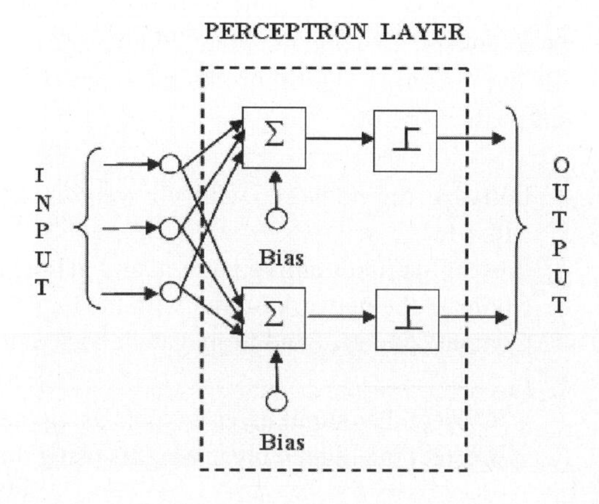


**Fig. 3.12** Perceptron neuron model

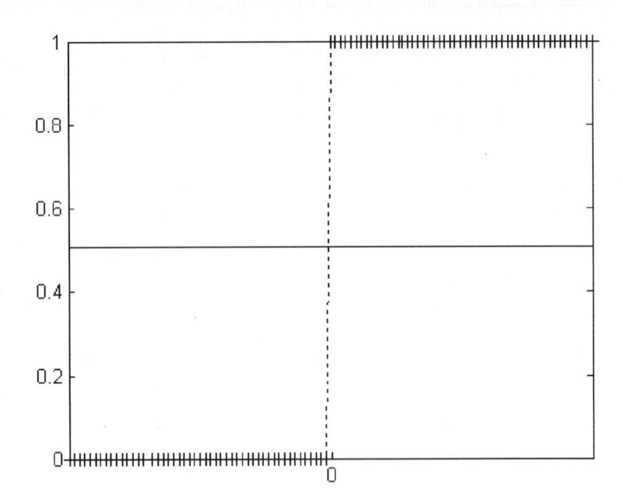


**Fig. 3.13** The linear activation function for the perceptron

The activation function is shown in Fig. 3.13.

Due to the nature of the output (1 or 0), the perceptron can be considered to be a linear classifier. Perceptron can clearly divide the input class into two classification regions formed by the decision boundary line. This line is perpendicular to the weight matrix. Hard-limit neurons without a bias will always have a classification line going through the origin. Addition of the bias nodes solves this problem, by allowing the classification line to be shifted away from the origin.

### *3.4.2 Multilayer Perceptrons (MLP)*

The principle weakness of the perceptron is that it can only solve problems that are linearly separable. However, most of the real-world problems are nonlinear in nature. A better solution is to learn the weights by standard optimization techniques by minimizing the error function. Multilayer perceptrons (MLP) involve more than one layer of neurons, usually fully interconnected. A typical example of MLP is the three-layered network, the input, hidden and output layer as shown in Fig. 3.1. MLP involves learning algorithm like the gradient descent. The basic MLP algorithm is summarized below.

1. Initialize the network, with all weights set to random numbers between $-1$ and $+1$.
2. Present the first training pattern, and get the output.
3. Compare the network output with the target output.
4. Compute the error and propagate it backwards to the hidden and input layer.

   - Correct the output layer weights using the formula as in (3.16).
   - Correct the hidden layer weights using the formula as in (3.30).

5. Calculate the error, by taking the average difference between the target and the output vector.
6. Repeat from step 2 for each pattern in the training set to complete one epoch (iteration).
7. Repeat from step 2 for a designated number of epochs, or until the error drops down to a specified limit.

## *3.4.3 Backpropagation (BP) Network*

The backpropagation algorithm is the most popular training algorithm. The backpropagation network involves multilayer neurons which use the backpropagation algorithm (see Sect. 3.3.4) for learning. Most frequently, multilayer feedforward networks (see Fig. 3.7) are employed for the backpropagation algorithm. However, feedback (see Fig. 3.8) and lateral networks (see Fig. 3.9) are also used depending on the application.

The basics of the backpropagation algorithm have been described in Sect. 3.3.4. There exist many versions of the backpropagation algorithm. All these algorithms mainly contribute to faster training. Some of these are mentioned below.

### 3.4.3.1 Variable Learning Rate

Instead of keeping the learning rate $\alpha$ constant, as done generally in the steepest descent algorithm, we can make it variable. This will improve the performance of the steepest descent algorithm. The variable (adaptive) learning rate will try to keep the learning step size as large as possible while keeping the learning stable. The learning rate is made responsive to the complexity of the local error surface.

### 3.4.3.2 Resilient Backpropagation

The resilient backpropagation training algorithm (Riedmiller and Braun 1993) eliminates the harmful effects of the magnitudes of the partial derivatives of the backpropagation algorithm. Only the sign of the derivative is used to determine the direction of the weight update; the magnitude of the derivative has no effect on the weight update. The size of the weight change is determined by a separate update value.

### 3.4.3.3 Conjugate Gradient Algorithm

The steepest descent algorithm produces rapid function decrease along the negative of the gradient. In the conjugate gradient algorithms (Hagen et al. 1996), a search is performed along the conjugate directions, which produces generally faster convergence than the steepest descent directions. The step size is generally adjusted at each iteration.

#### 3.4.3.4 Quasi-Newton Algorithm

Newton's method is an alternative to the conjugate gradient methods for fast optimization. Newton's method involves computation of the Hessian matrix (second derivatives) of the performance index at the current values of the weights and biases. However, this is very computation intensive for the feedforward networks. The quasi-Newton (or secant) method bypasses the expensive second derivative computation by updating an approximate Hessian matrix at each iteration of the algorithm.

#### 3.4.3.5 Levenberg-Marquardt Algorithm

When the performance function has the form of a sum of squares (in case of feedforward networks) the Hessian matrix can be approximated by the Jacobian matrix (first derivatives of the network error). Computation of the Jacobian matrix is much less intensive than that of the Hessian matrix. The Levenberg-Marquardt algorithm (Levenberg 1944, Marquardt 1963) uses this approximation to the Hessian matrix by involving a scalar update factor $\mu$ in the Newton-like update. When the scalar $\mu$ is zero, this becomes the Newton's method.

#### 3.4.3.6 Reduced Memory Levenberg-Marquardt Algorithm

The main drawback of the Levenberg-Marquardt algorithm (Levenberg 1944; Marquardt 1963) is that it requires the storage of some matrices that can be quite large for certain problems. The approximate Hessian can be computed by summing a series of subterms. Once one subterm has been computed, the corresponding submatrix of the Jacobian can be computed. This logic is utilized in the reduced memory algorithm to reduce the memory consumption.

### 3.4.4 Radial Basis Function (RBF) Network

An alternative model to the multilayer perceptron for the identification and time series prediction is the neural network employing the radial basis functions (RBFs). A RBF is a multidimensional function which depends on the Euclidian distance $r$.

$$r = \|x - c\| ; \quad x, c \in \mathbf{IR}^n, \tag{3.32}$$

where,

$c$ is the center of the RBF
$x$ is the input vector, and
$\|\bullet\|$in (3.32) denotes a vector norm.

One of the simplest approaches to the approximation of a nonlinear function is to represent it by a linear combination of the fixed nonlinear basis functions $\varphi_i(x)$, so that,

$$F(x) = \sum_{i=1}^{h} W_i.\varphi_i(x). \tag{3.33}$$

$F(x)$ is the approximated function with the weights $W_i$.

#### 3.4.4.1 Choices for the Radial Basis Functions

The purpose of an associated adaptive neural network will be to estimate the set of weights $W_i$. However, the number of required basis functions usually grows exponentially with the dimension of the function. Radial basis functions provide a powerful method for the multidimensional approximation or fitting which essentially does not suffer from the problem of the proliferation of adjustable parameters as the dimensionality of the problem increases. Typical choices for the radial basis function $\varphi(x) = \Phi\left(\|x - c\|\right)$ are mentioned below.

1. Piecewise linear approximations (see Fig. 3.14),

$$\varphi(r) = r. \tag{3.34}$$

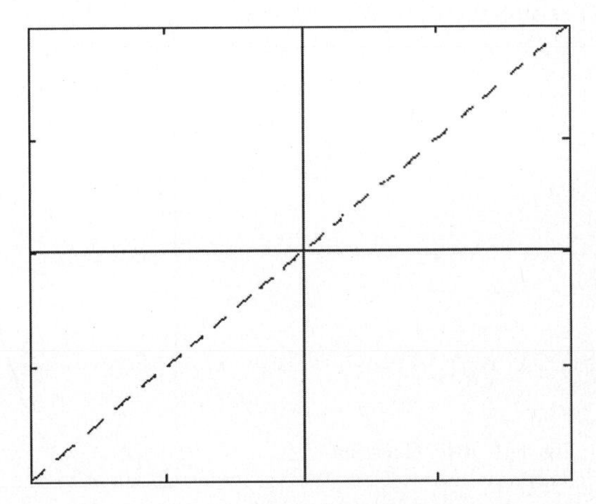

**Fig. 3.14** RBF: Piecewise linear approximations

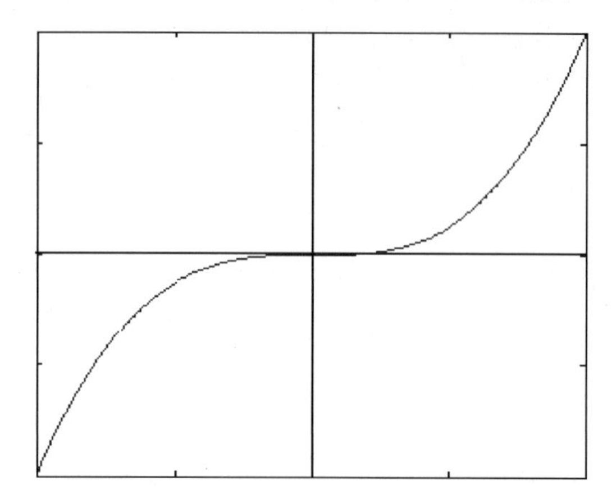

**Fig. 3.15** RBF: Cubic approximations

2. Cubic approximations (see Fig. 3.15),

$$\varphi(r) = r^3. \tag{3.35}$$

3. Gaussian function (see Fig. 3.16),

$$\varphi(r) = \exp\left(-r^2/\sigma^2\right). \tag{3.36}$$

4. Thin plate splines (see Fig. 3.17),

$$\varphi(r) = r^2 \log(r). \tag{3.37}$$

5. Multiquadratic function (see Fig. 3.18),

$$\varphi(r) = \sqrt{r^2 + \sigma^2}. \tag{3.38}$$

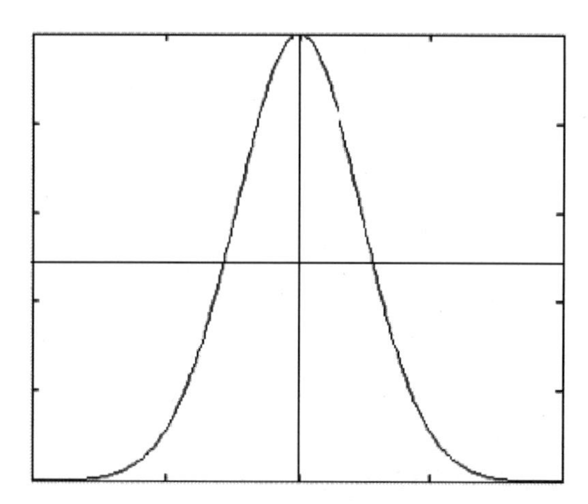

**Fig. 3.16** RBF: Gaussian functions

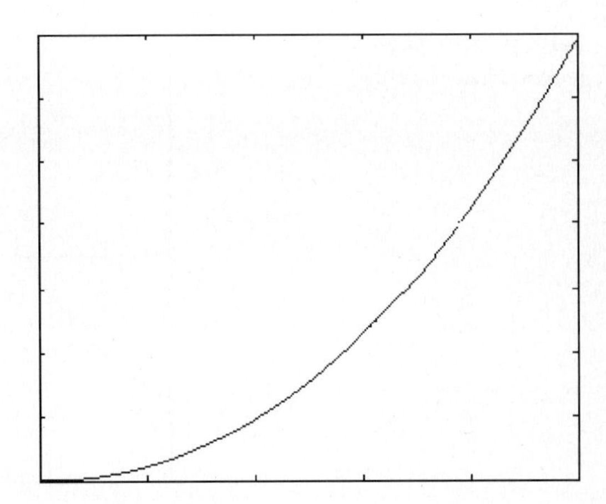

**Fig. 3.17** RBF: Thin plate splines

6. Inverse multiquadratic function (see Fig. 3.19),

$$\varphi(r) = \frac{1}{\sqrt{r^2 + \sigma^2}}, \tag{3.39}$$

where, $\sigma$ is a real coefficient called the width or scaling parameter. From the above described functions the most popular and widely used one is the Gaussian function which has a peak at the center $c$ and decreases monotonically as the distance from the center increases.

#### 3.4.4.2 RBF Network Architecture

The RBF network was proposed by Broomhead and Lowe (Broomhead and Lowe 1988) and extended by many other researchers. Figure 3.20 shows the basic structure of the RBF network. The network contains only one hidden layer, but in contrast to the perceptron, the RBF network does not extend to more than hidden layers.

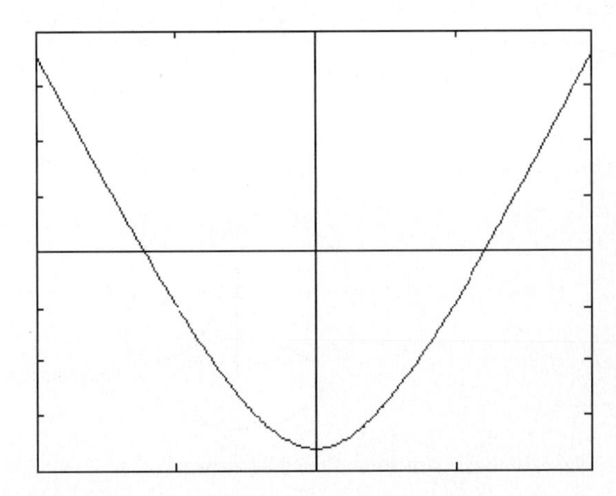

**Fig. 3.18** RBF: multiquadratic function

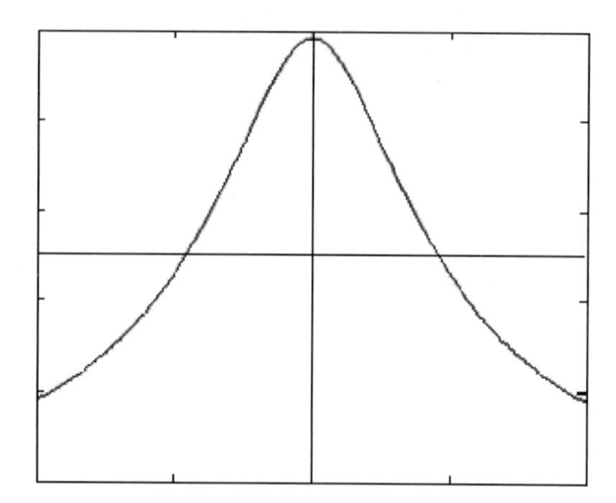

**Fig. 3.19** RBF: Inverse multiquadratic function

Moreover, note that the hidden layer connections are not weighted, i.e., each hidden node (neuron) receives each input value unaltered. The hidden nodes are the processing units which perform a radial basis functionality. Furthermore, the activation functions (RBFs) are, in general, nonmonotonic, compared to the monotonic sigmoid function of the perceptron. However, similar to the perceptron, the output unit performs simple weighted summation of its inputs.

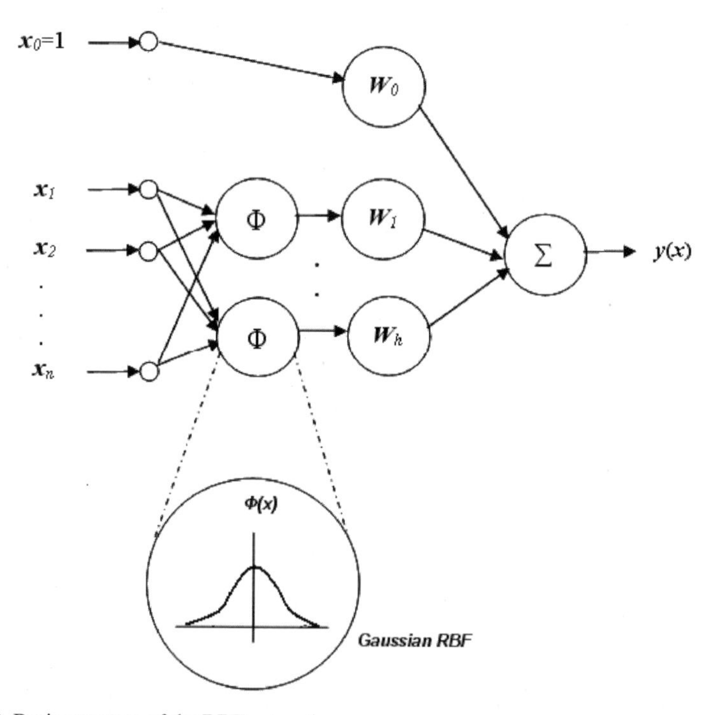


**Fig. 3.20** Basic structure of the RBF network

The RBF network computes a linear combination of the radial basis functions to give the estimate

$$\hat{y}(k) = W_0 + \sum_{i=1}^{h} W_i.\Phi\left(\|x(k) - c_i\|\right), \tag{3.40}$$

where

$x(k) = [x_1, x_2, \ldots, x_n]^T = \left[y(k-1), y(k-2), \ldots, y(k-n)\right]^T$,
$c_i \in \mathbf{IR}^n (i = 1, 2, \ldots, h)$ are the centers,
$W_i \in \mathbf{IR}(i = 1, 2, \ldots, h)$ are the weights,
$n$ is the input number
$h$ is the hidden layer number.

In a basic RBF network, the centers $c_i$ and the distance scaling (width) parameters are usually fixed (i.e., they are not adjustable during the learning process), only the coefficients $W_i$ (synaptic weights) are the adjustable parameters. It has been shown experimentally that if a sufficient number of hidden units are used and the centers $c_i$ are suitably distributed in the input domain, the RBF network is able to approximate a wide class of nonlinear dimensional functions. Moreover, it was found that the choice of the nonlinearity is not crucial for the performance of the network. The performance of the RBF network critically depends on the chosen centers.

#### 3.4.4.3 Parameters of the RBF Network

In general, a RBF network is specified by three sets of parameters,

- the centers $c_i \in \mathbf{IR}^n$,
- the width or distance scaling parameters $\sigma_i$, and
- the synaptic weights $W_i \in \mathbf{IR}(i = 1, 2, \ldots, h)$.

The parameters for the radial basis function can be determined in three successive steps. First, the $h$ vectors $c_i$ are determined. The simplest technique is to choose these vectors randomly from the subset of the training data. However, in such a case the number of hidden units must be relatively large in order to cover the entire output domain. Another improved approach is to apply the "$K$-means clustering algorithm" (Seber 1984). The basic idea of this algorithm is to distribute the centers $c_i$ according to the natural measure of the attractor, i.e., if the density of data points is high so is the density of the centers. The $K$-means clustering algorithm finds a set of cluster centers and a partition of the training data into subsets. Each cluster center is associated with one of the $h$ hidden units in the RBF network. The data is partitioned in such a way that the training points are assigned to the cluster with the nearest center.

Once the cluster centers have been determined, the distance scaling parameters $\sigma_i$ are determined heuristically from the $p$-nearest neighbor.

$$\sigma_i = \frac{1}{p}\sum_{j=1}^{p}\left(\left\|c_i - c_j\right\|^2\right)^{1/2}, \tag{3.41}$$

where $c_j$ is the $p$-nearest neighbor of $\sigma_i$. Having obtained the centers $c_i$ and the distance scaling parameter $\sigma_i$, we would be able to adjust the synaptic weights $W_i$ by solving the linear regression

$$\hat{y}(k) = W_0 + \sum_{i=1}^{h} W_i.\Phi_i\left[x(k)\right], \tag{3.42}$$

where,

$$\varphi_i\left[x(k)\right] = \Phi\left(\left\|x(k) - c_i\right\|\right) \tag{3.43}$$

and $x(k) = [x_1, x_2, \ldots, x_n]^T := \left[y(k-1), y(k-2), \ldots, y(k-n)\right]^T$.

Applying, for example, the standard linear least-squares criterion, we obtain the simple least mean squares (LMS) algorithm (Widrow and Hoff 1966).

$$\frac{dW_i}{dt} = \mu_i e_k \mu_i\left[x(k)\right], \tag{3.44}$$

where $\mu_i > 0$ and $e_k = y(k) - \hat{y}(k)$.

The important advantage of the RBF network is that if offers training times one to three orders of magnitude faster than the standard backpropagation algorithm.

#### 3.4.4.4 Alternative Derivation of the Learning Rule

A generalized RBF network is represented as

$$\hat{y} = W_0 + \sum_{i=1}^{h} W_i.\Phi_i, \tag{3.45}$$

$$\Phi_i = \Phi_i(r_i) = e^{-r_i}, \tag{3.46}$$

$$r_i = r_i\left(x, c_i, W^{(i)}\right) = \frac{1}{2}\left\|W^{(i)}\left(x - c_i\right)\right\|_2^2 = \frac{1}{2}\sum_{j=1}^{n}\left[\sum_{p=1}^{n} W_{jp}^{(i)}\left(x_p - c_{ip}\right)\right]^2, \tag{3.47}$$

where $x$ is an $n$-dimensional vector of the center for the $i$-th unit.

The problem can then be stated as: derive the learning rules for the network parameters: $W_i$, $W_{jp}^{(i)}$ and $c_i = [c_{i1}, c_{i2}, \ldots, c_{in}]^T$ $(i = 1, 2, \ldots, h)$ based on the backpropagation algorithm.

Solution: The network parameters can be derived by applying the gradient descent algorithm to the error function

$$E_k = \frac{1}{2}e_k^2 = \frac{1}{2}\left[y(k) - \hat{y}(k)\right]^2. \tag{3.48}$$

For convenience, the subscript $k$ representing the $k$-th pattern is omitted in the following derivations.

- For the output neuron we obtain

$$\frac{dW_i}{dt} = -\mu_1 \frac{\partial E}{\partial W_i} = -\mu_1 \frac{\partial E}{\partial \hat{y}} \cdot \frac{\partial \hat{y}}{\partial W_i} = \mu_1 . e . \Phi_i(r_i), \tag{3.49}$$

$$\frac{dW_0}{dt} = -\mu_1 . e, \tag{3.50}$$

where $\mu_1 > 0$.

- For the hidden layer we get

$$\frac{dW_{jp}^{(i)}}{dt} = -y_2 \frac{\partial E}{\partial W_{jp}^{(i)}} = -y_2 \frac{\partial E}{\partial \Phi_i} \cdot \frac{\partial \Phi_i}{\partial r_i} \cdot \frac{\partial r_i}{\partial W_{jp}^{(i)}}, \tag{3.51}$$

$$\frac{\partial E}{\partial \Phi_i} = \frac{\partial E}{\partial \hat{y}} \cdot \frac{\partial \hat{y}}{\partial \Phi_i} = -e . W_i, \tag{3.52}$$

$$\frac{\partial \Phi_i}{\partial r_i} = -\Phi_1, \tag{3.53}$$

$$\frac{\partial r_i}{\partial W_{jp}^{(i)}} = u_j^{(i)} \left(x_p - c_{jp}\right). \tag{3.54}$$

Hence we obtain

$$\frac{dW_{jp}^{(i)}}{dt} = -\mu_2 . e . W_i . \Phi_i(r_i) . u_j^{(i)} \left(x_p - c_{jp}\right), \tag{3.55}$$

where

$$u_j^{(i)} = \sum_{t=1}^{n} W_{jt}^{(i)} \left(x_t - c_{it}\right). \tag{3.56}$$

In a similar way, we can derive the learning rule for the updating parameter $c_{ij}$.

$$\frac{dc_{ip}}{dt} = -\mu_3 \frac{\partial E}{\partial c_{ip}} = -\mu_1 \frac{\partial E}{\partial \Phi_i} \cdot \frac{\partial \Phi_i}{\partial r_i} \cdot \frac{\partial r_i}{\partial c_{ip}} = \mu_3 . e . W_i . \Phi_i (r_i) . \left( \sum_{j=1}^{n} W_{jp}^{(i)} u_j^{(i)} \right) . \tag{3.57}$$

### 3.4.5 Hopfield Network

In the 1980s John Hopfield (Hopfield 1982) showed that models of physical systems could be used to solve computational problems which led to the creation of neural networks named after him. Hopfield networks are typically used for classification problems with binary pattern vectors although the applicability is limited due to theoretical limitations of the network structure. Hopfield networks are sometimes called *associative network* since they associate a class pattern to each input pattern.

The Hopfield network can be created using input data (pattern) vectors, of the different classes, called *class patterns*. The class patterns should have $n$ components, typically represented as $\{1, -1\}$ in an $n$-dimensional space. The network has the ability to classify partial or imperfect input patterns into these classes. When a partial or imperfect input pattern is presented, the network can then associate it with another pattern.

There are two types of Hopfield networks, continuous-time and discrete-time. For both types of the Hopfield network, the weight matrix $W$ is defined as,

$$W = \frac{1}{n} \sum_{i=1}^{K} \sum_{j=1}^{K} \alpha_i^T \alpha_j , \tag{3.58}$$

where $K$ is the number of pattern class $\{\alpha_1, \alpha_2, \ldots, \alpha_K\}$ and $n$ is the number of components, the dimension of the class pattern vectors. The patterns are stored as $-1$ and 1 elements which make the Hopfield networks perform better instead of storing them as 0 and 1.

The continuous-time Hopfield network is defined as

$$\frac{dx(t)}{dt} = -x(t) + W.\sigma \{x(t)\} , \tag{3.59}$$

where $x(t)$ is the state-vector of the network, $W$ is parametric weights and $\sigma$ is nonlinear function operating on the state-vector.

The discrete-time Hopfield network is defined as

$$x[t+1] = sign(W.x[t]) . \tag{3.60}$$

At each iteration, a randomly chosen state is updated using (3.60).

The weight matrix of the Hopfield network is calculated using the Hebb rule based on the outer product calculation technique (see Sect. 3.3.5). The Hopfield network is used to store one or more stable target vectors which can be viewed as memories that the network recalls when provided with similar vectors that act as a cue to the network memory. A structure of the Hopfield network is shown in Fig. 3.21.

### 3.4.6 Adaline

The Adaline network is an adaptive linear network. The Adaline network mostly employs linear network like the perceptron (see Sect. 3.4.1) in a multilayer fashion. This network is sometimes called a Madaline for Many Adaline. The Adaline network, much like the perceptron, can only solve linearly separable problems. Nevertheless, the Adaline has been and is today one of the most widely used neural networks found in practical applications. Adaptive filtering is one of its major application areas which will be discussed later.

### 3.4.7 Kohonen Network

Kohonen networks, popularly known as Kohonen Self Organizing Feature Maps, or SOFMs were created by Teuvo Kohonen (Kohonen 1987, 1997). SOFMs learn to recognize groups of similar input vectors in such a way that neurons physically near each other in the neuron layer respond to the similar input vectors. Such networks can learn to detect regularities and correlations in their input and adapt their future responses to that input accordingly.

The objective of a Kohonen network is to map the $n$-dimensional input vectors (patterns) onto a discrete map with 1 or 2 dimensions. Patterns close to one

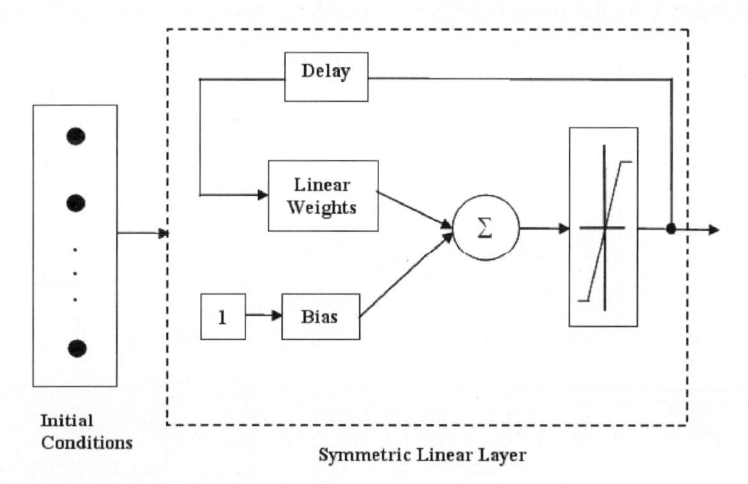


**Fig. 3.21** Structure of the Hopfield network

another in the input space should be close to one another in the map: they should be topologically ordered. A Kohonen network is composed of a grid of output units and $N$ input units. The input pattern is fed to each output unit. The input lines to each output unit are weighted. These weights are initialized to small random numbers. Figure 3.22 shows a schematic diagram of the Kohonen network.

The ability of the SOFMs to represent multidimensional data in much lower dimensional spaces like one or two dimensions, is utilized for the problem of dimensionality reduction. This process, of reducing the dimensionality of vectors, is essentially a data compression technique known as *vector quantization*. In addition, the Kohonen technique creates a network that stores information in such a way that any topological relationships within the training set are maintained.

One of the most interesting aspects of the SOFMs is that they follow unsupervised learning. Training a SOFM however, requires no target vector. A SOFM learns to classify the training data without any external supervision whatsoever.

### 3.4.7.1 Operating Principle of the Kohonen Network

The basic idea behind the Kohonen network is competitive learning. The neurons are presented with the inputs, which calculate their weighted sum and neuron with the closest output magnitude is chosen to receive additional training. Training, though, does not just affect the one neuron but also its neighbors. One way to judge the 'closest output magnitude' is to find the distance between the input and weighted sum of the neuron:

$$d = \sqrt{(x_i - W_{ij})^2}. \tag{3.61}$$

When applied to our two-dimensional network, it reduces down to the standard Euclidean distance formula. So, if we want the output which closely represents the input pattern, it is the neuron with the *smallest* distance. Let us call the neuron

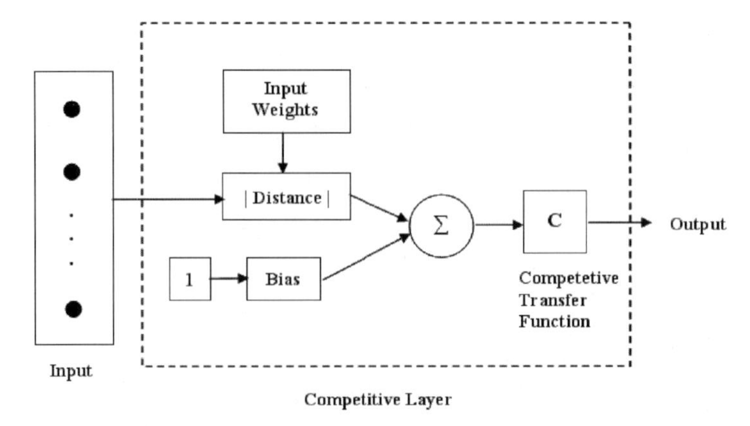


**Fig. 3.22** Schematic diagram of the Kohonen network

with the least distance $x_{d_{(\min)}}$. Now, we change both the neuron and the neurons in its neighborhood which is not constant. The neighborhood can range between the entire network to just eight adjacent neurons.

#### 3.4.7.2 Learning in the Kohonen Network

The learning process in the Kohonen network is roughly as follows:

- initialize the weights for each output unit,
- loop until weight changes are negligible, *for* each input pattern

  - present the input pattern,
  - find the winning output unit,
  - find all units in the neighborhood of the winner,
  - update the weight vectors for all those units,

  reduce the size of neighborhoods if required.

The winning output unit is simply the unit with the weight vector that has the smallest Euclidean distance to the input pattern (see (3.61)). The neighborhood of a unit is defined as all units within some distance of that unit on the map (not in the weight space). The weights of every unit in the neighborhood of the winning unit (including the winning unit itself) are updated using

$$W_{ij_{(new)}} = W_{ij} + \alpha . \left(x_i - W_{ij}\right), \tag{3.62}$$

where $\alpha$ is the learning coefficient. The learning coefficient is not constant but adjustable during the training operation. This is a subject of greater research. However, Kohonen suggested to split the training operation into two phases. Phase 1 will reduce down the learning coefficient from 0.9 to 0.1 (or similar values), and the neighborhood will reduce from half the diameter of the network down to the immediately surrounding cells. Following that, phase 2 will reduce the learning coefficient from perhaps 0.1 to 0.0 but over double or more the number of iterations than in phase 1.

#### 3.4.7.3 Applications of the Kohonen Networks

As mentioned earlier, Kohonon networks, namely, the SOFMs are greatly used in the dimension reduction problems. Kohonen networks have been also successfully applied to speech recognition. Kohonen networks can also be well applied to gaming applications.

### 3.4.8 Special Networks

In this section we will briefly discuss about some special networks which are important from the point of view of their usage.

#### 3.4.8.1 Tapped Delay Line (TDL)

The tapped delay line (TDL) operates on the principle of backward shift (or delay) operator. For example, the backward operator $q$ can be applied on a time-dependent or sampled function as follows,

$$q^{-n}y(t) = y(t-n), \quad n = 1, 2, 3, \ldots. \tag{3.63}$$

This can be utilized to format the input of a neural network using a delay line. The main philosophy is to format the input utilizing the previous samples. The input signal enters from the left, and passes through $n-1$ delay elements. The output of the TDL is an $n$-dimensional vector, formed by the input signal at the current time, the previous input signal, etc.

Tapped delay line is frequently used in conjunction with linear and adaptive filtering tasks. For example, time-series prediction, speech recognition, image processing, adaptive process filters etc where the output performance depends heavily and critically on the input signals at different level. This can be effectively done with the TDL. Figure 3.23 shows an example of the TDL.

#### 3.4.8.2 Filtering and TDL

As described in the previous section, the tapped delay line can be used to format the input to be used in conjunction with linear or adaptive filtering purpose. Figure 3.24 shows the structure of the network for filtering purpose involving the TDL. Filtering with the TDL is analogous to the finite impulse response (FIR) filters in the field of digital signal processing.

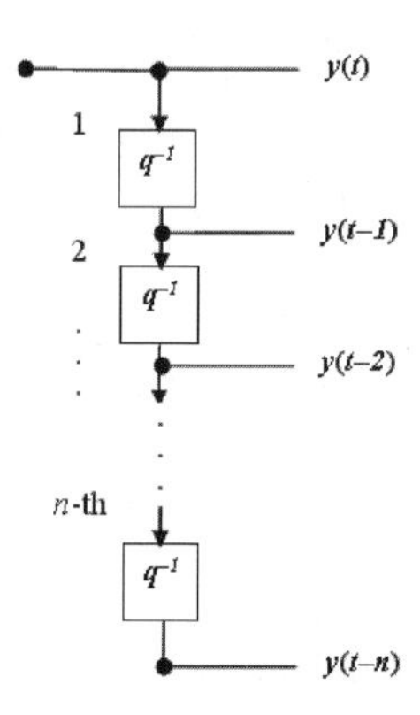


**Fig. 3.23** Tapped delay line with $n$ delay elements

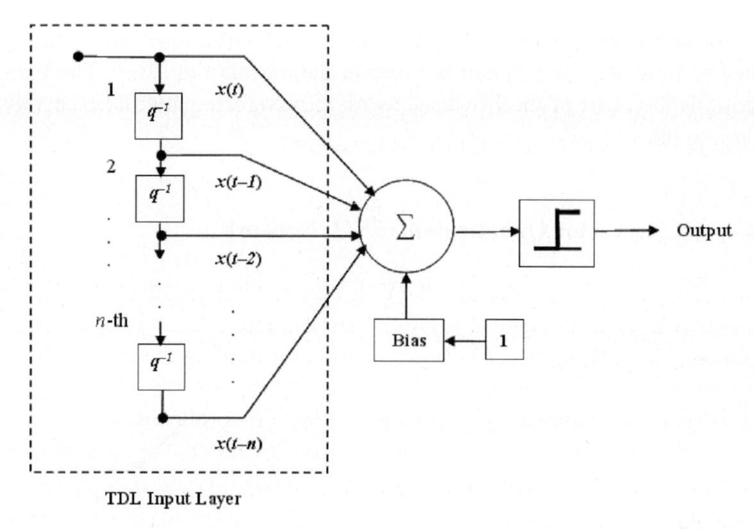


**Fig. 3.24** Filtering architecture along with TDL

### 3.4.8.3 Elman Network

Elman network (Elman, 1990) is a type of recurrent network. It is a two-layer back-propagation network, with the addition of a feedback connection from the output of the hidden layer to its input. The Elman networks learn to recognize and generate temporal patterns, as well as spatial patterns due to this feedback path. A two-layer Elman network is shown in Fig. 3.25.

Although Elman network is also a two-layered network, it differs from the conventional two-layered network in that the first layer has a recurrent connection.

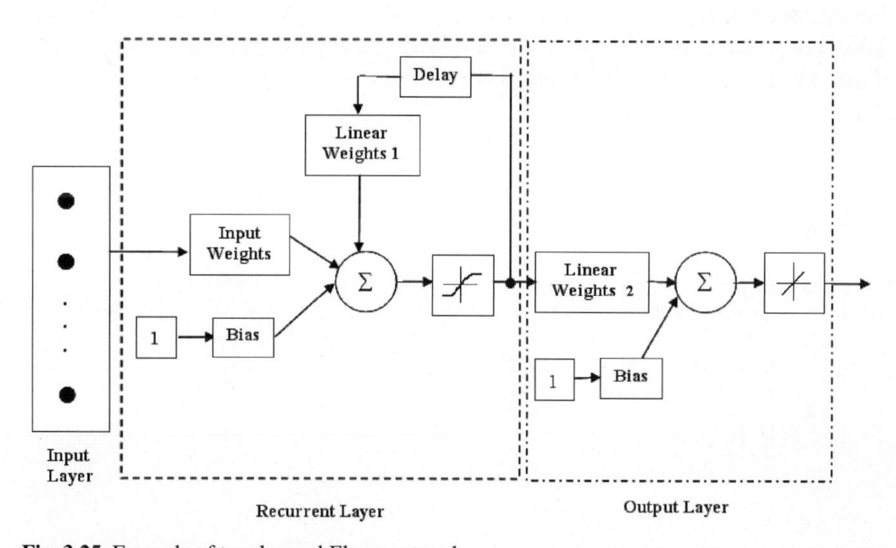


**Fig. 3.25** Example of two-layered Elman network

Application of the delay elements in this recurrent connection stores values from the previous time step, which can be used in the current time step. The recurrent connection is the heart of the Elman network to recognize patterns, especially the time-varying ones.

#### 3.4.8.4 Learning Vector Quantization (LVQ) Network

Many neural network structures, like the Kohonen network and the SOFMs, involve a competitive layer to recognize groups of similar input vectors. Learning vector quantization (LVQ) (Kohonen 1987) is a method for training the competitive layer neurons in a supervised manner.

An LVQ network comprises of a competitive layer first followed by a linear layer as shown in Fig. 3.26. The competitive layer learns to classify the input vectors and the linear layer transforms the competitive layer's classes into the user-defined target classifications.

#### 3.4.8.5 Probabilistic Neural Network

Probabilistic neural network is a kind of multilayer neural network. The first layer (radial basis) maps the given input to the training input set, i.e., the distances between the given input and the input training vectors are measured and a vector is produced which indicates how close the given input pattern is to the training input sets. The next layer produces an output vector of probabilities taking into account the sum of the contributions of the given class of inputs. In the final layer, a transfer function selects the maximum of the probabilities from the output of the second layer. This is done in a competitive manner. The final output produced resembles a binary classification, 1 for the class and 0 for other classes. Principle use of probabilistic neural networks is in the field of classification problems. Figure 3.27 shows the architecture of the probabilistic neural network.

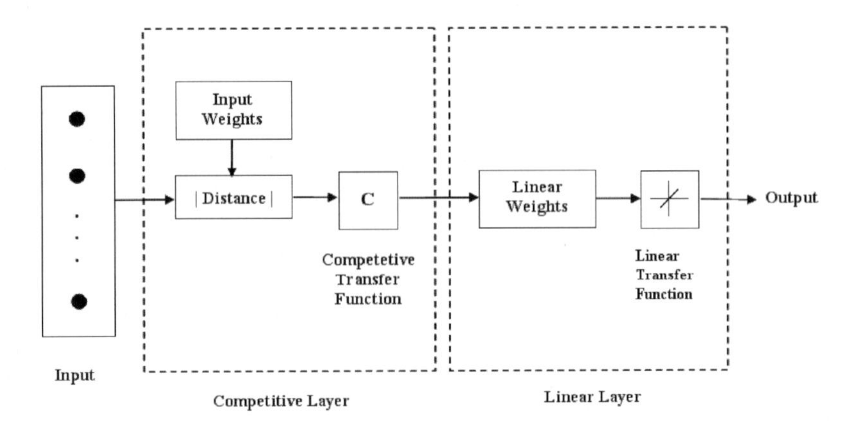


**Fig. 3.26** Structure of an LVQ network

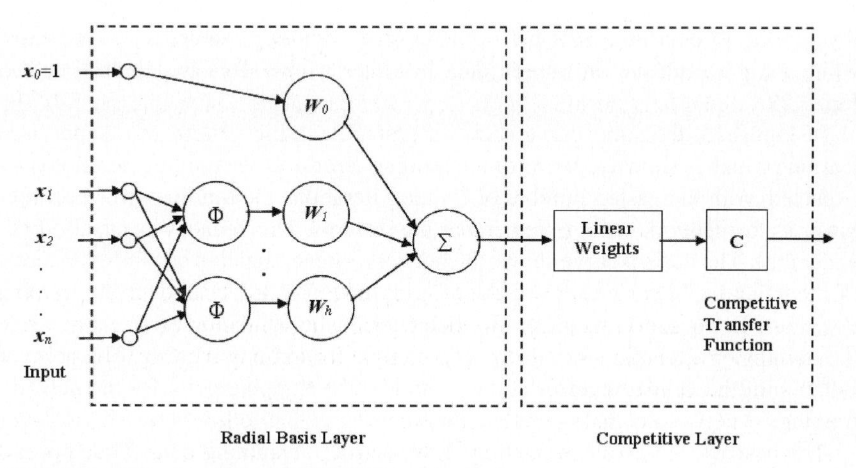


**Fig. 3.27** Architecture of the probabilistic neural network

## *3.4.9 Special Issues in NN Training*

We have seen so far different architectures, learning algorithms, network topologies and so on. In the end it comes to training the network (i.e., adjusting the weights and bias) for effectively mapping the input-output relationship. However, there are some special concerns regarding the optimization of the training procedure.

The critical issue in developing a neural network is generalization: how well will the network make predictions for cases that lie outside the training set? NNs, like other machine learning techniques like kernel regression, smoothing splines, etc can suffer from either underfitting or overfitting. Overfitting is especially dangerous because it can easily lead to some wild predictions that are far beyond the range of the training data. We discuss these along with some other important issues below.

### 3.4.9.1 Overfitting

From statistical point of view, overfitting means fitting a model with too many parameters. If the model has enough and surplus complexity compared to the amount of data available, it could result in absurd and false predictions when presented with unseen test data. This is particularly important for neural network training.

In a supervised training, the learning algorithm is trained using some set of training examples, i.e. exemplary situations for which the desired output is known. The learner is assumed to reach a state where it will also be able to predict the correct output for other examples, thus generalizing to situations not presented during training. However, especially in cases where learning was performed too long or where training examples are rare, the learner may adjust to very specific random features of the training data, that have no causal relation to the target function. In this process of overfitting, the performance on the training examples still increases while the performance on unseen data becomes worse. This is demonstrated in Fig. 3.28.

In Fig. 3.28, the solid curve shows a typical learning pattern in a supervised learning which is shown in terms of the training error over the training iterations. As expected, with increasing number of training iterations the training error becomes lower as the network starts to remember the patterns. This reduces its capability to generalize. The dashed curve shows the network's generalization capability in terms of the validation error of performance for an unseen test dataset over the training iterations. If we notice an increasing behavior of the validation error curve while the training error continues to decrease we might lead into overfitting. The point of inflation of the validation error curve should be the stopping point for the network training.

The best way to avoid overfitting is to use lots of training data. The approximate rule of thumb is to have at least thirty times as many training cases as there are weights in the network to avoid the risk of overfitting. Given a fixed amount of training data, the possible approaches to avoiding overfitting are mentioned below.

- **Model selection**. We should select the proper model as per the data orientation. This particularly concerns about the number of weights, and accordingly the number of hidden layers and units. However, if we keep the number of weights too low we might introduce the error of underfitting in order to avoid overfitting.
- **Early stopping**. As indicated in Fig. 3.28, the network should not be trained for too long. A good practice is to stop at the inflation point of the validation error (on test set). However, if the validation set is not available, it is always good to stop the training early to be on the safe side. Again, we cannot stop too early in fear of underfitting.

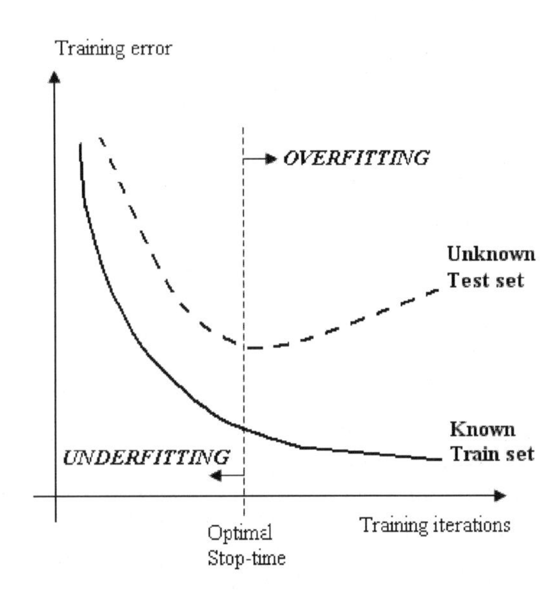


**Fig. 3.28** Overfitting in neural network

- **Decaying weights**. If we do not introduce weights that decays a monotonic growth or random behavior of the weight change, that might lead to the network acting on the noise. This would result in wild predictions out of the training data range when presented with unseen test data.
- **Bayesian learning**. Unlike the conventional training methods which assume that there is exactly one model that generates the data, the Bayesian approach presumes that several models may have produced the dataset with a certain probability.
- **Combination of network**. There is no particular rule about this, however this approach is certainly beneficial when appropriate models are combined together depending on the data orientation and application.

For more detailed discussion on overfitting see Smith (1996). Geman, Bienenstock, and Doursat (Geman et al. 1992) discussed the issue of overfitting from the bias/variance trade-off point of view. Tetko, Livingstone and Luik (Tetko et al. 1995) discussed the issues of overfitting and overtraining in a comparative manner.

#### 3.4.9.2 Underfitting

Compared to the overfitting problem, which is more dangerous, an underfitting happens when the model fit is not sufficiently complex to detect fully the signal in a complicated dataset (test). Underfitting often takes place in order to avoid overfitting. As mentioned above, to avoid overfitting, the number of weights, hidden layer units might be lowered. This in a way is good for keeping the overfitting prossibility low. However, a network with few weights or hidden layer/units might fail to map fully the complexity of the input-output relationship. This might also happen with premature and too early stopping of the training before the network reaches an optimum learning and generalization stage. Underfitting might also be due to improperly chosen network model. A very simple network model for a complex process might be too simple to catch the complexity of the input-output relationship. Hence, it might fail to produce result close to the range of the data set. Therefore, selection of a proper network model is particularly important for tackling the underfitting problem.

#### 3.4.9.3 Number of Hidden Layer Units

As already mentioned, number of hidden layer units vis-à-vis weights, have a great influence on the network performance. In general, it is better to have too many hidden units than too few. With too few units, the network is prone to insufficient capability to capture the nonlinearities in the data. A rule of thumb is to have 5 to 100 hidden units. This might increase with number of inputs and training sets. However, the rule of thumb is not to have more weights than number of training samples available. Choice of optimal number of hidden layer, in principle, is best determined by experimentation. It is generally a good practice to employ multiple hidden layers in a hierarchical fashion instead of using only one hidden layer with bulk amount of weights.

#### 3.4.9.4 Weight Initialization

If the starting values of the weights are close to zero, the output of the sum unit produces rather low value. As a result the squashing unit would not be able to impart enough nonlinearity and the neural network model collapses to approximately linear model, losing its power to model the nonlinear relationship. Usually the weights are chosen as random low numbers. If zero-valued weights are chosen, the training algorithm might get stuck, failing to move. On the other hand, large weights are not good for generalizations.

#### 3.4.9.5 Scaling of the Inputs and Outputs

Scaling of the inputs, in most cases, has influence on the quality of the output produced. This is an important step because the network learns better with a regularized and bounded input range rather than arbitrary values. This is also important for improving the generalization capability of the network. A standard practice is to regularize (also termed as normalize) the inputs to zero mean and unity standard deviation.

Also, if sigmoid or tan-sigmoid transfer functions are used, they are particularly flat if the data range lies beyond the range $[0, 1]$ or $[-1, 1]$ respectively. Hence, it is absolutely important to use the regularization step in these cases. Uniform scaling of the inputs equalizes their relative importance. Also, with normalized inputs, a typical practice is to initialize weights with random uniform values over the range $[-0.8, +0.8]$.

Other effective normalization techniques include dividing the entire data range by the absolute maximum value (ensuring range bound of $[-1, 1]$), mean, median etc.

In many cases, it is also advisable to regularize the output ranges in a certain bounds. Oftentimes, a better generalization performance is observed with regularized output values during the network training. A regularization is must if the output layer employes tan-sigmoid transfer function, instead, if linear or hard-limit transfer functions are used it is optional.

## 3.5 Examples

The following step-by-step application examples would provide a practical understanding of the neural network theories described in the previous sections.

### 3.5.1 Linear Network: Boolean Logic Operation

In the first example we are going to see how linear neural networks can be used to implement the Boolean logic operations[8] like, OR, AND. For implementing the Boolean logic operation we employ a single layer neuron having two inputs. Table 3.1 lists the Boolean OR (output is 1 if either Input 1 or Input 2 or both are 1, otherwise 0) and AND (output is 1 if both the Input 1 and Input 2 are 1, otherwise 0) operations.

Figure 3.29 shows the linear network structure to implement these Boolean operations.

The weighted sum $S$ involves the inputs $I_i$ with the weights $W_i$. The transfer function $T(s)$ used in the network shown in Fig. 3.29, is a modified version of the linear transfer function.

$$S = \sum_{i=0}^{2} I_i . W_i , \tag{3.64}$$

$$T(S) = \begin{cases} 1, & S > 1 \\ 0, & S \leq 1 \end{cases} \tag{3.65}$$

The transfer function is shown in Fig. 3.30.

For both the operations, we start with an initial weight matrix $W = [0\ 0]^T$. And we iteratively train the neural network for all combinations of the input vectors. At each iteration (epoch), first, the weighted sum is computed. Application of the transfer function on the weighted sum produces the output which is then compared

**Table 3.1** Boolean OR, AND operation

| $Input_1$ | $Input_2$ | Output | $Input_1$ | $Input_2$ | Output |
|---|---|---|---|---|---|
| 0 | 0 | 0 | 0 | 0 | 0 |
| 0 | 1 | 1 | 0 | 1 | 0 |
| 1 | 0 | 1 | 1 | 0 | 0 |
| 1 | 1 | 1 | 1 | 1 | 1 |
| OR operation | | | AND operation | | |

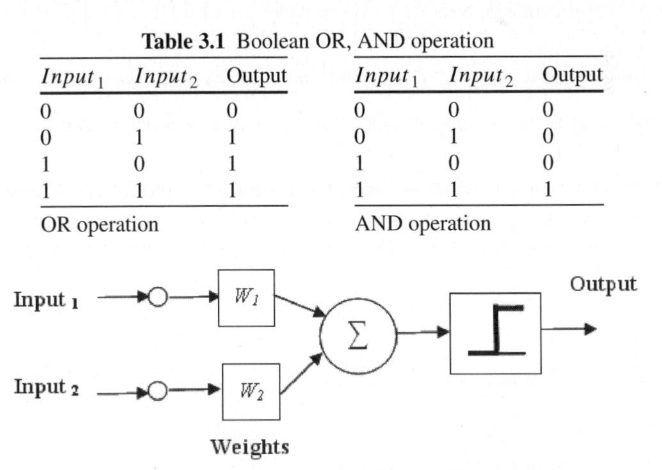


**Fig. 3.29** Single-layer linear network structure for Boolean logic implementation

[8] OR, AND are examples of linear Boolean operations which can be solved by the linear network. For nonlinear Boolean operations like the exclusive-OR (XOR) operation linear network cannot be used.

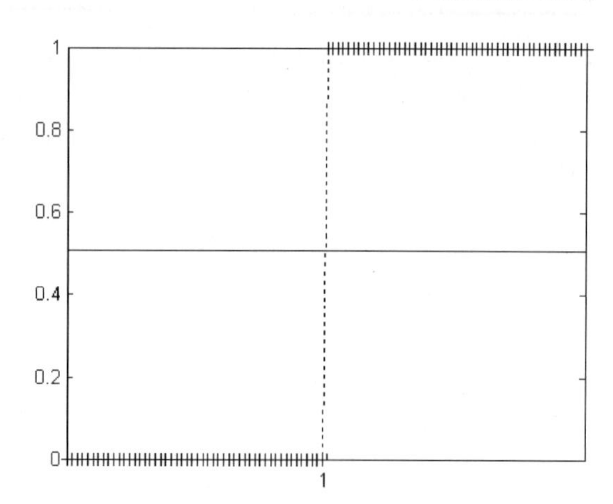


**Fig. 3.30** Linear transfer function for implementation of the Boolean logic

against the desired target value (from Table 3.1). Difference between the target and the produced output is the change in the weight matrix $\Delta W$, which is added to the initial weight matrix to update the weights. This process is continued for all the input combinations, after which the weights achieve the optimum value. The step-by-step operations for each epoch are described in the Tables 3.2 and 3.4 for the OR and AND operations respectively.

So, the final weight matrix becomes, $W = [2\ 2]^T + [0\ 0]^T = [2\ 2]^T$, for OR operation. Table 3.3 shows the simulation of the results for the Boolean OR operation using the final weight matrix. The result in Table 3.3 matches exactly with that shown in Table 3.1.

The final weight matrix becomes, $W = [0\ 0]^T + [1\ 1]^T = [1\ 1]^T$, for AND operation. Table 3.5 shows the simulation of the results for the Boolean AND operation using this final weight matrix, which matches exactly with that in the Table 3.1.

**Table 3.2** Neural network implementation steps for Boolean OR operation

| Epoch | Input | Weight (W)<br>($W_{new} = W + \Delta W$) | Sum | Output | Target | $\Delta W$<br>(Target - Output) |
|---|---|---|---|---|---|---|
| 1 | [0 0] | $[0\ 0]^T$ | 0 | 0 | 0 | $[0\ 0]^T$ |
| 2 | [0 1] | $[0\ 0]^T$ | 0 | 0 | 1 | $[1\ 1]^T$ |
| 3 | [1 0] | $[1\ 1]^T$ | 1 | 0 | 1 | $[1\ 1]^T$ |
| 4 | [1 1] | $[2\ 2]^T$ | 4 | 1 | 1 | $[0\ 0]^T$ |

**Table 3.3** Result simulation of the Boolean OR operation

| Trial | Input | Weight (W) | Sum | Output |
|---|---|---|---|---|
| 1 | [0 0] | $[2\ 2]^T$ | 0 | 0 |
| 2 | [0 1] | $[2\ 2]^T$ | 2 | 1 |
| 3 | [1 0] | $[2\ 2]^T$ | 2 | 1 |
| 4 | [1 1] | $[2\ 2]^T$ | 4 | 1 |

**Table 3.4** Neural network implementation steps for Boolean AND operation

| Epoch | Input | Weight (W) ($W_{new} = W + \Delta W$) | Sum | Output | Target | $\Delta W$ (Target - Output) |
|---|---|---|---|---|---|---|
| 1 | [0 0] | $[0\ 0]^T$ | 0 | 0 | 0 | $[0\ 0]^T$ |
| 2 | [0 1] | $[0\ 0]^T$ | 0 | 0 | 0 | $[0\ 0]^T$ |
| 3 | [1 0] | $[0\ 0]^T$ | 0 | 0 | 0 | $[0\ 0]^T$ |
| 4 | [1 1] | $[0\ 0]^T$ | 0 | 0 | 1 | $[1\ 1]^T$ |

**Table 3.5** Result simulation of the Boolean AND operation

| Trial | Input | Weight (W) | Sum | Output |
|---|---|---|---|---|
| 1 | [0 0] | $[1\ 1]^T$ | 0 | 0 |
| 2 | [0 1] | $[1\ 1]^T$ | 1 | 0 |
| 3 | [1 0] | $[1\ 1]^T$ | 1 | 0 |
| 4 | [1 1] | $[1\ 1]^T$ | 2 | 1 |

## *3.5.2 Pattern Recognition*

In this section, we discuss about a simple image pattern recognition example. We consider two types of image patterns, dot and line. A $2 \times 2$ matrix is used to indicate the color values, 1 for black and 0 for white which define the dots and the lines. The image patterns and the corresponding binary color values are shown in Fig. 3.31.

Four inputs $I_i$ with weights $W_i$ and an output $O$ are considered. The transfer function is the linear transfer function (see Fig. 3.3). The output has the value 0 for the dots and 1 for the lines. An additional input $I_0$ is used for the bias node. The weighted sum is similar in nature as in (3.64), only difference being the number of nodes. The initial values are: bias $I_0 = 1$, and weight matrix $W_i = 0$. The neuron shall be trained for the given input patterns using the modified delta rule. The change of the input weights is defined as

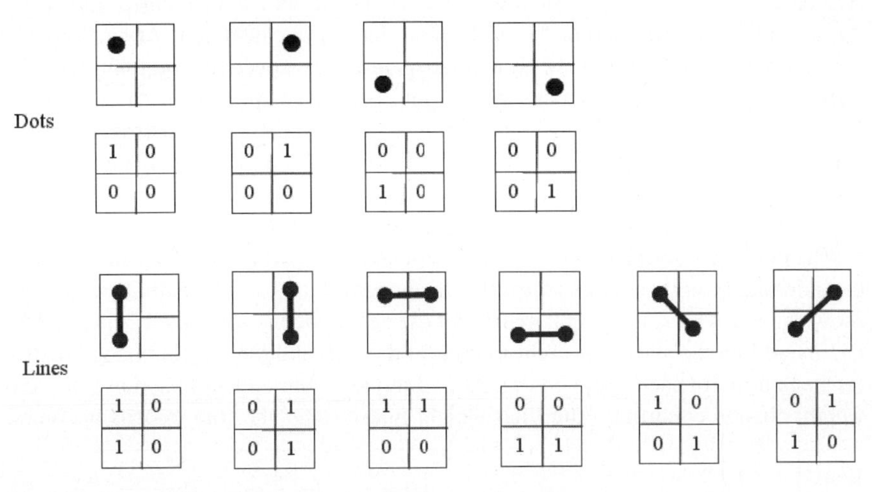


**Fig. 3.31** Dots and lines definitions using the binary color values

$$W_{i_{new}} = W_i + \Delta W_i, \tag{3.66}$$

$$\Delta W_i = -\alpha \sum_{j=1}^{n} (O_j - R_j).I_{ij}, \tag{3.67}$$

where,

$\alpha$ is the learning rate ($\alpha = 1$ in this example),
$i$ is the index for input, $0 \leq i \leq 4$
$j$ is the index for the pattern, $0 \leq j \leq n$ ($n = 10$ in the example)
$O_j$ is the output for the pattern $j$
$R_j$ is the desired output for the pattern $j$
$I_{ij}$ is the input $i$ of the pattern $j$.

In general, for $m$ inputs ($0 \leq i \leq m$) the change in weight matrix for $n$ patterns becomes

$$\begin{bmatrix} \Delta W_0 \\ \Delta W_1 \\ \vdots \\ \Delta W_m \end{bmatrix} = \begin{bmatrix} -\alpha \sum_{j=1}^{n} (O_j - R_j).I_{0j} \\ -\alpha \sum_{j=1}^{n} (O_j - R_j).I_{1j} \\ \vdots \\ -\alpha \sum_{j=1}^{n} (O_j - R_j).I_{mj} \end{bmatrix}. \tag{3.68}$$

The training is continued until the weight changes become 0. After each epoch the weights are updated using (3.66). At each epoch, for each input patterns (indicated by $j$, $0 \leq j \leq n$, $n = 10$) the weighted sum is calculated and using the weighted sum the output for each input pattern is calculated. At each epoch, the outputs for each pattern are compared against the respective target outputs. The differences are used to calculate the weight changes as per (3.67). At any epoch, for each input pattern, the weight change is calculated by multiplying the input pattern with the negative of the difference between the output and the target. The total weight change at any epoch is calculated by adding (notice the summation in (3.67)) all the individual weight changes for each input patterns.

The iterative training of the neuron is depicted in details in Fig. 3.32. In Fig. 3.32, the individual weight change procedure, the weight changes and the final weights are shown epoch-wise. The weight matrix for the next epoch is calculated as per (3.66), i.e., by adding the starting weight matrix (at the top) and the weight change matrix (at the bottom) of each epoch in Fig. 3.32. It requires four epochs to achieve the zero weight change condition. The final weight matrix after the four epochs becomes,

$$W = \begin{bmatrix} -2 & \\ 1 & 1 \\ 1 & 1 \end{bmatrix}.$$

| Epoch | | | 1 | | 2 | | 3 | | 4 | |
|---|---|---|---|---|---|---|---|---|---|---|
| Weights $W_{new}=W+\Delta W$ | | | 0; 0 0 / 0 0 | | -4; -1 -1 / -1 -1 | | 2; 2 2 / 2 2 | | -2; 1 1 / 1 1 | |
| Input Patterns $j$ | | Targets $(R)$ | Sum | Output $(O)$ | Sum | Output $(O)$ | Sum | Output $(O)$ | Sum | Output $(O)$ |
| 1 | 1; 1 0 / 0 0 | *0* | 0 | *1* | -5 | *0* | 4 | *1* | -1 | *0* |
| 2 | 1; 0 1 / 0 0 | *0* | 0 | *1* | -5 | *0* | 4 | *1* | -1 | *0* |
| 3 | 1; 0 0 / 1 0 | *0* | 0 | *1* | -5 | *0* | 4 | *1* | -1 | *0* |
| 4 | 1; 0 0 / 0 1 | *0* | 0 | *1* | -5 | *0* | 4 | *1* | -1 | *0* |
| 5 | 1; 1 0 / 1 0 | *1* | 0 | *1* | -6 | *0* | 6 | *1* | 0 | *1* |
| 6 | 1; 0 1 / 0 1 | *1* | 0 | *1* | -6 | *0* | 6 | *1* | 0 | *1* |
| 7 | 1; 1 1 / 0 0 | *1* | 0 | *1* | -6 | *0* | 6 | *1* | 0 | *1* |
| 8 | 1; 0 0 / 1 1 | *1* | 0 | *1* | -6 | *0* | 6 | *1* | 0 | *1* |
| 9 | 1; 1 0 / 0 1 | *1* | 0 | *1* | -6 | *0* | 6 | *1* | 0 | *1* |
| 10 | 1; 0 1 / 1 0 | *1* | 0 | *1* | -6 | *0* | 6 | *1* | 0 | *1* |
| Weight Changes $\Delta W$ | | | -4; -1 -1 / -1 -1 | | 6; 3 3 / 3 3 | | -4; -1 -1 / -1 -1 | | 0; 0 0 / 0 0 | |

**Fig. 3.32** Iterative training of the neuron for dot and line pattern recognition

In the final weight matrix, −2 indicates the bias weight and the four 1s indicate the weights for the four input nodes. In the fourth epoch, using this final weight matrix, outputs for each input patterns matches exactly with the targets, which indicates the completed training.

The weight change procedure for the individual input patterns ($j$) and the final weight change at epoch 1 are shown as example below.

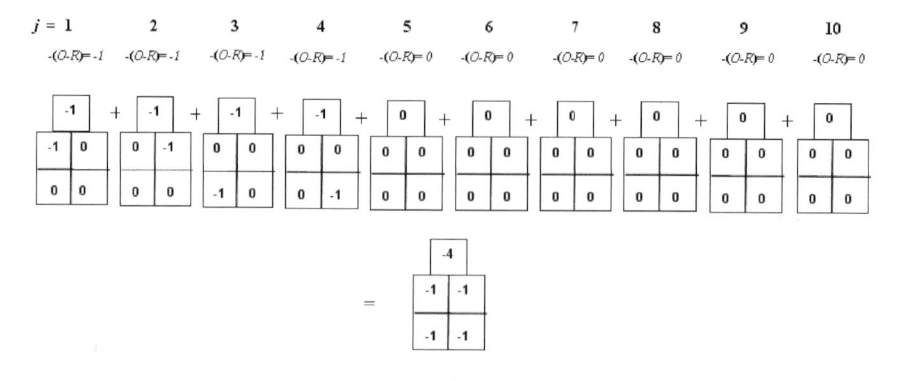


## 3.5.3 Incomplete Pattern Recognition

Pattern recognition has traditionally been one of the main application areas of neural networks. However, oftentimes we find incomplete or partial patterns which are, in general, difficult to recognize using traditional pattern recognition techniques. Nevertheless, a robust incomplete pattern recognition technique would be closer to real life scenario in many aspects. In this section, we will see an application study on incomplete pattern recognition involving generalized regression neural networks (GRNN) which is a special form of radial basis function (RBF) (see Sect. 3.4.4) network. The network is trained robustly with complete patterns. Then it is demonstrated that the network can effectively detect incomplete patterns. Therefore, it will be closer to be applicable in real applications.

### 3.5.3.1 Application Example

The input patterns in this example are the decimal digits 0 to 9 displayed using the seven segment display (Malvino and Leach 1994) kinds of display unit shown in Fig. 3.33. This is quite similar to the seven segment display using LEDs (light-emitting diodes) (Malvino and Leach 1994), commonly used in electronics as a method of displaying decimal numeric. In Fig. 3.33, the numbers 1 to 7 denotes the seven displaying units of the seven segment display. Properly activating and deactivating these seven displaying units, we can display the decimal digits 0 to 9.

The decimal digits 0 to 9 are represented in the binary form as the output, shown in Table 3.6. The inputs are the corresponding decimal seven segment displays.

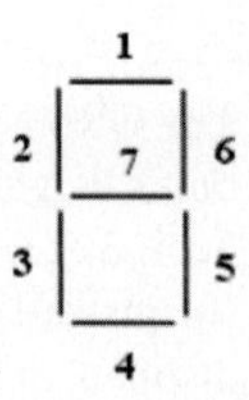


**Fig. 3.33** Decimal digit display using seven segment display

In our example, to represent the seven segment displays, we use 1 to indicate the brightness and $-1$ to indicate the darkness. This is because many networks perform better with 1 and $-1$ vectors than 1 and 0 vectors. Another reason for better performance might be the classified (positive and negative) nature of the input patterns. Hence, the display brightness and darkness input values are chosen as 1 and $-1$ instead of 1 and 0.

#### 3.5.3.2 Generalized Regression Neural Network

Generalized regression neural network (GRNN) is a kind of radial basis function network. As the name suggests, this type of network is often used for function approximation and pattern matching. Details about the RBF network could be found in Sect. 3.4.4.

A GRNN comprises of two layers of regression network. The first layer has radial basis neurons, and the second layer has linear neurons. Only the first layer has biases. The second layer also has as many neurons as the input/target vectors. The network acts like it is taking a weighted average between target vectors whose design input vectors are closest to the new input vector (Mathworks 2002). Architecture of the regression neural network is shown in Fig. 3.34.

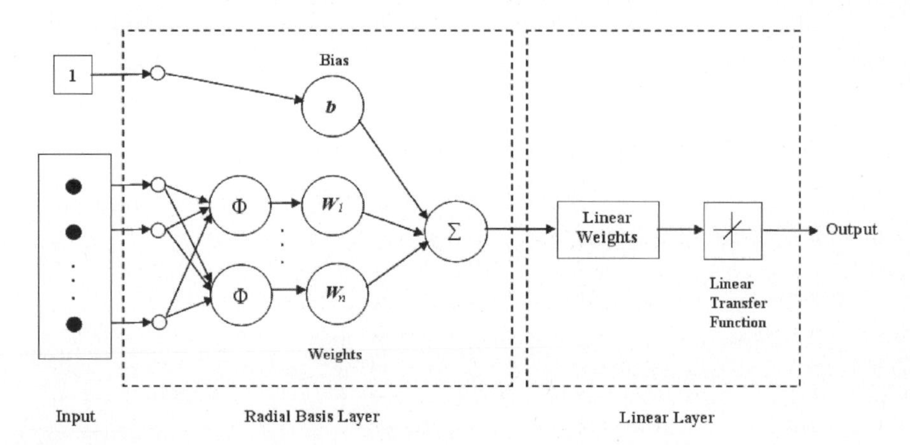


**Fig. 3.34** Architecture of the generalized regression neural network (GRNN)

#### 3.5.3.3 Network Training

The GRNN employed has seven inputs corresponding to the seven segment display unit's brightness and darkness values for the seven decimal digits. As mentioned in Sect. 3.5.3.1, the display brightness and darkness input values are chosen as 1 and $-1$ instead of 1 and 0. Figure 3.35 shows the input formation. In the seven segment displays in Fig. 3.35, the bold lines indicate the active units and the non-bold lines indicate the inactive units.

The input matrix $I$ follows the representations of the seven segment display for the decimal digits 0 to 9 as shown in Fig. 3.35. It is to be noted that the display values are indicated as unit 7 to unit 1, i.e., left to right manner. In $I$ matrix, these are represented columnwise. And the target matrix $T$ comes from the binary values of the decimal digits 0 to 9 as per Table 3.6. The only thing to be noted is that the zeros of Table 3.6 are represented as $-1$ in the target matrix $T$.

$$I = \begin{bmatrix} -1 & -1 & 1 & 1 & 1 & 1 & 1 & -1 & 1 & 1 \\ 1 & 1 & 1 & 1 & 1 & -1 & -1 & 1 & 1 & 1 \\ 1 & 1 & -1 & 1 & 1 & 1 & 1 & 1 & 1 & 1 \\ 1 & -1 & 1 & 1 & -1 & 1 & 1 & -1 & 1 & 1 \\ 1 & -1 & 1 & -1 & -1 & -1 & 1 & -1 & 1 & -1 \\ 1 & -1 & -1 & -1 & 1 & 1 & 1 & -1 & 1 & 1 \\ 1 & -1 & 1 & 1 & -1 & 1 & 1 & 1 & 1 & 1 \end{bmatrix}, \quad T = \begin{bmatrix} -1 & -1 & -1 & -1 \\ -1 & -1 & -1 & 1 \\ -1 & -1 & 1 & -1 \\ -1 & -1 & 1 & 1 \\ -1 & 1 & -1 & -1 \\ -1 & 1 & -1 & 1 \\ -1 & 1 & 1 & -1 \\ -1 & 1 & 1 & 1 \\ 1 & -1 & -1 & -1 \\ 1 & -1 & -1 & 1 \end{bmatrix}.$$

The GRNN is trained using Matlab® Neural Network toolbox (Mathworks 2002) and the input-output matrix. We will go step-by-step demonstrating how to use the Matlab® Neural Network toolbox. The following '*newgrnn*' command (Mathworks 2002) is used to create the trained GRNN object. However, it is to be noted that some of the functions might be different depending on the version of the Matlab®. Hence, the user is requested to check the compatibility of the functions with his/her Matlab® version.

**Table 3.6** Output binary representation of the decimal digits 0 to 9

| Decimal | Binary | | | |
|---|---|---|---|---|
| 0 | 0 | 0 | 0 | 0 |
| 1 | 0 | 0 | 0 | 1 |
| 2 | 0 | 0 | 1 | 0 |
| 3 | 0 | 0 | 1 | 1 |
| 4 | 0 | 1 | 0 | 0 |
| 5 | 0 | 1 | 0 | 1 |
| 6 | 0 | 1 | 1 | 0 |
| 7 | 0 | 1 | 1 | 1 |
| 8 | 1 | 0 | 0 | 0 |
| 9 | 1 | 0 | 0 | 1 |

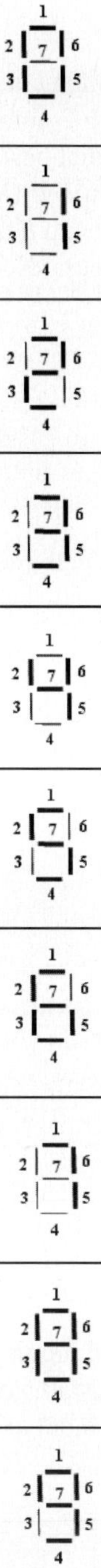

| Decimal | LED No. 7 | 6 | 5 | 4 | 3 | 2 | 1 |
|---|---|---|---|---|---|---|---|
| 0 | –1 | 1 | 1 | 1 | 1 | 1 | 1 |
| 1 | –1 | 1 | 1 | –1 | –1 | –1 | –1 |
| 2 | 1 | 1 | –1 | 1 | 1 | –1 | 1 |
| 3 | 1 | 1 | 1 | 1 | –1 | –1 | 1 |
| 4 | 1 | 1 | 1 | –1 | –1 | 1 | –1 |
| 5 | 1 | –1 | 1 | 1 | –1 | 1 | 1 |
| 6 | 1 | –1 | 1 | 1 | 1 | 1 | 1 |
| 7 | –1 | 1 | 1 | –1 | –1 | –1 | 1 |
| 8 | 1 | 1 | 1 | 1 | 1 | 1 | 1 |
| 9 | 1 | 1 | 1 | 1 | –1 | 1 | 1 |

**Fig. 3.35** Input binary patterns for the digits 0 to 9 using seven segment display

```
>> network = newgrnn(I, T′)
```

It is to be noted that to make the input and the output matrix of equal columns for training, transpose of the target matrix ($T^t$, Matlab® notation T′) is used. It can be noted that after the training, the weights of the radial basis and the linear layer and the bias of the radial basis layer are formulated in *cell* (Mathworks 2002) format. These are given usually by the cell arrays *IW*, *LW* and *b* (Mathworks 2002), which are the design variables for GRNN in Matlab® Neural Network toolbox. These can be viewed as follows.

```
>> cell2mat (network.IW)
>> cell2mat (network.LW)
>> cell2mat (network.b)
```

For this application example, the values of the cell arrays *IW*, *LW* and *b* (Mathworks 2002) are given as

$$IW = I^t, \quad LW = T^t,$$

$$b = \begin{bmatrix} 0.8326 \\ 0.8326 \\ 0.8326 \\ 0.8326 \\ 0.8326 \\ 0.8326 \\ 0.8326 \\ 0.8326 \\ 0.8326 \\ 0.8326 \end{bmatrix},$$

where, the superscript $t$ indicates transpose.

#### 3.5.3.4 Results

The trained network is first tested for the standard inputs ($I$) used for the training. The result is simulated using the '*sim*' command (Mathworks 2006), which has two inputs: the trained GRNN network object and the input to simulate.

```
>> R = sim (network, I)′
```

It is to be noted that the result matrix $R$ is presented in a transposed form to match with the target matrix $T$. The simulation result $R$ for the given input is given as

$$R = \begin{bmatrix} -0.8760 & -0.9918 & -0.9913 & -0.9913 \\ -1.0000 & -0.8755 & -0.8824 & 0.9926 \\ -0.9918 & -0.9995 & 0.9913 & -0.9917 \\ -0.8769 & -0.9846 & 0.8683 & 0.9842 \\ -0.9918 & 0.9835 & -0.9990 & -0.9831 \\ -0.8828 & 0.8755 & -0.8828 & 0.8819 \\ -0.8828 & 0.8751 & 0.7656 & -0.8824 \\ -0.9995 & 0.8746 & 0.8814 & 0.9990 \\ 0.7716 & -0.8888 & -0.8827 & -0.8827 \\ 0.7712 & -0.8824 & -0.8885 & 0.8759 \end{bmatrix}.$$

To make the result consistent to the binary form as in Table 3.6, we use a simple filter to round off the values.

```
>> R = ceil (R)
```

$$R = \begin{bmatrix} 0\,0\,0\,0 \\ 0\,0\,0\,1 \\ 0\,0\,1\,0 \\ 0\,0\,1\,1 \\ 0\,1\,0\,0 \\ 0\,1\,0\,1 \\ 0\,1\,1\,0 \\ 0\,1\,1\,1 \\ 1\,0\,0\,0 \\ 1\,0\,0\,1 \end{bmatrix}.$$

So, we see that the network perfectly follows the input pattern. Next, we test the network with incomplete patterns. Figures 3.36 to 3.39 show four such examples.

Input for the incomplete pattern in Fig. 3.36 is $iI = [-1\ 1\ -1\ 1\ 1\ -1\ 1]^t$, in transposed form to make the columns equal. We use the following command to simulate the incomplete input *iI*.

```
>> R = sim (network, iI)'
```

The result is $R = [-0.9923 \quad -0.9922 \quad 0.8755 \quad -0.9845]$. Filtering (round) it with '*ceil*' command (Mathworks 2006) gives $R = [0 \quad 0 \quad 1 \quad 0]$. From Table 1, this turns out to be the decimal digit 2 which is the most likely one for the incomplete pattern shown in Fig. 3.36.

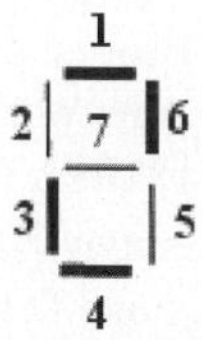

1
2 7 6
3 5
4

**Fig. 3.36** Example 1 of incomplete pattern recognition using GRNN

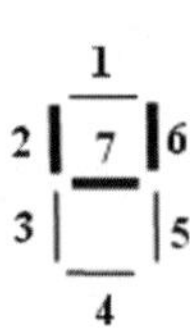


**Fig. 3.37** Example 2 of incomplete pattern recognition using GRNN

**Fig. 3.38** Example 3 of incomplete pattern recognition using GRNN

**Fig. 3.39** Example 4 of incomplete pattern recognition using GRNN

$iI = [-1 \quad 1 \quad 1 \quad 1 \quad -1 \quad -1 \quad 1]^t$ is the input for the incomplete pattern shown in Fig. 3.37. The input is in transposed form to make the columns equal.

Using the simulation procedure mentioned above for the example 1, the initial result is $R = [-0.9396 \quad -0.0868 \quad 0.8222 \quad 0.9356]$. Filtering producces the final result $R = [0 \quad 0 \quad 1 \quad 1]$. From Table 3.6, this turns out to be the decimal digit 3 which seems to be the closest representation for the incomplete pattern shown in Fig. 3.37.

Input for the incomplete pattern in Fig. 3.38 is $iI = [1\ 1\ -1\ -1\ -1\ 1\ -1]^t$, in transposed form to make the columns equal.

Using the simulation procedure mentioned above for the example 1, the initial result is $R = [-0.9918 \quad 0.9830 \quad -0.9985 \quad -0.9831]$. Filtering produces the final result $R = [0 \quad 1 \quad 0 \quad 0]$. From Table 3.6, this turns out to be the decimal digit 4 which is the nearest representation for the incomplete pattern shown in Fig. 3.38.

Input for the incomplete pattern in Fig. 3.39 is $iI = [-1\ -1\ 1\ -1\ -1\ -1\ \ 1]^t$, in transposed form to make the columns equal.

Using the simulation procedure mentioned above for the example 1, the initial result is $R = [-0.9995 \quad 0.8751 \quad 0.8746 \quad 0.9986]$. Filtering produces the final result $R = [0 \quad 1 \quad 1 \quad 1]$. From Table 3.6, this turns out to be the decimal digit 7 which seems to be the most likely target value for the incomplete pattern shown in Fig. 3.39.

## References

Broomhead DS, Lowe D (1988) Multivariable functional interpolation and adaptive networks. Complex Systems 2:321–355

DARPA (1988) DARPA neural network study. Lexington, MA: MIT Lincoln Lab

Elman JL (1990) Finding structure in time. Cognitive Science 14:179–211

Fogel D (1998) Evoluationary computation – the fossil record. IEEE Press, Piscataway, NJ

Geman S, Bienenstock E, Doursat R (1992) Neural networks and the bias/variance dilemma. Neural Computation 4:1–58

Goldberg D (1989) Genetic algorithms in search, optimization, and machine learning. Addison-Wesley, Reading, MA

Hagan MT, Demuth HB, Beale MH (1996) Neural network design. PWS Publishing, Boston

Hebb DO (1949) The organization of behavior. John Wiley & Sons, New York

Hopfield JJ (1982) Neural networks and physical systems with emergent collective computational abilities. In: Proceedings of the National Academy of Sciences 79:2554–2558

Kohonen T (1987) Self-organization and associative memory, 2nd Edition. Springer-Verlag, Berlin

Kohonen T (1997) Self-organizing maps, 2nd Edition. Springer-Verlag, Berlin

Levenberg K (1944) Method for the solution of certain problems in least squares. Quart. Appl. Math. 2:164–168

Malvino AP, Leach DP (1994) Digital Principles and Applications, 5th Edition. McGraw-Hill, Singapore

Marquardt D (1963) An algorithm for least-squares estimation of nonlinear parameters. SIAM J. Appl. Math. 11:431–441

Mathworks Inc. (2002) MATLAB® Documentation–Neural Network Toolbox, Version 6.5.0.180913a, Release 13

McCulloch W, Pitts W (1943) A logical calculus of ideas immanent in nervous activity. Bulletin of Mathematical Biophysics 5:115–133

Riedmiller M, Braun H (1993) A direct adaptive method for faster backpropagation learning: The RPROP algorithm. In: Proceedings of the IEEE International Conference on Neural Networks

Rosenblatt F (1961) Principles of neurodynamics. Spartan Press, Washington DC

Rumelhart DE, Hinton GE, Williams RJ (1986) Learning representations by backpropagating errors. Nature 323:533–536

Seber GAF (1984) Multivariate observations. John Wiley & Sons, New York

Smith M (1996) Neural networks for statistical modeling. International Thomson Computer Press, Boston

Tetko IV, Livingstone DJ, Luik AI (1995) Neural network studies. 1. comparison of overfitting and overtraining, J. Chem. Inf. Comput. Sci. 35:826–833

Widrow B, Hoff ME (1960) Adaptive switching circuits. In: 1960 IRE WESCON Convention Record, New York IRE, pp. 96–104

Wasserman PD (1993) Advanced methods in neural computing. Van Nostrand Reinhold, New York

## Section II: Application Study

## 3.6 Load Forecasting

Load forecasting is necessary for knowing the electrical power demand beforehand. There are many benefits of such prediction, like, proper generation planning, efficient operation and distribution, interruption free power supply etc. Several techniques for short-, medium- and long-term load forecasting have been discussed, such as Kalman filters, regression algorithms and neural networks (IEEE 1980; Bunn and Farmer 1985). Among these techniques, load forecasting using ANN: linear network, multilayer perceptron, radial basis network has been particularly prominent with quite good results in terms of accuracy and low prediction error. In this section, we will discuss about an application study on short-term load forecasting to demonstrate how the different neural network technologies can be employed to forecast the hourly load for a particular day or 24 Hours.

There are different categories of load forecasting, as mentioned before, typically classified as short, medium and long term load forecasting. Short-term forecasting is typically oriented towards forecasting the load for one day (24 Hours) ahead. This is mainly used for unit commitment, energy transactions, system security analysis and optimum generation planning. Medium-term forecasting (from one day to several months) is utilized for medium term operation planning. Long-term forecasting (more than one year) is required for analyzing the demand growth (Novak 1995).

### *3.6.1 Data set for the Application Study*

For this application study on the short-term load forecasting, we consider load data for one month (30 days) for a medium sized city. This sort of load data can be acquired from load flow studies, load survey etc.

The data set starts on Friday 01:00 Hour and ends on the last Friday 24:00 Hour. The data is stored in an excel data file in the following format: in the first column of the data set are the hourly load data in mega watts, in the second column are the numbers for the day type (e.g., monday = 1, tuesday = 2, ..., sunday = 7) and in the third column the time from 01:00 Hour up to 24:00 Hour. We have 720 data (30 days × 24 Hours). From this data set, load-data for 29 days i.e., 696 data are used to train the neural networks, and the 30$^{\text{th}}$ day data (i.e., data from 697 to 720) data are used to verify the predicted result. The first twelve data from this data set of 720 data are shown as example in Fig. 3.40.

| | A | B | C |
|---|---|---|---|
| 1 | 247 | 5 | 1 |
| 2 | 236 | 5 | 2 |
| 3 | 231 | 5 | 3 |
| 4 | 213 | 5 | 4 |
| 5 | 213 | 5 | 5 |
| 6 | 224 | 5 | 6 |
| 7 | 264 | 5 | 7 |
| 8 | 264 | 5 | 8 |
| 9 | 321 | 5 | 9 |
| 10 | 324 | 5 | 10 |
| 11 | 329 | 5 | 11 |
| 12 | 332 | 5 | 12 |

**Fig. 3.40** Section of the data set for short-term load forecasting

### 3.6.2 Use of Neural Networks

For the implementation we will use three types of neural networks: linear network, backpropagation multilayer perceptron and radial basis network.

Linear network is the simplest one and uses basic neuron model with tapped delay line architecture (see Sect. 3.4.8.1). Multilayer backpropagation perceptron minimizes the output error by using the gradient descent algorithm (see Sect. 3.3.2) to backpropagate the estimated output error to the originating input node (see Sect. 3.3.4). Radial basis function (RBF) network uses a RBF multidimensional function which depends on the distance $r = \|x - c\|$, where $|| \bullet ||$ denotes a vector norm between the input vector $x$ and the centre $c$ (see Sect. 3.4.4). Matlab® neural network toolbox (Mathworks 2002) is used for the simulations for this application study.

To evaluate the performance of the different networks, we define a performance index, the Mean Absolute Prediction Error (MAPE):

$$MAPE = \frac{1}{N} \sum_{i=1}^{N} \frac{|t_i - p_i|}{t_i} \times 100, \tag{3.69}$$

where $t_i$ is the $i$-th sample of the true (measured) value of the load, $p_i$ is the predicted load value of the network, and $N$ is the total number of predicted samples.

### 3.6.3 Linear Network

The operation using the linear network is described below.

- The load data set is filtered to keep only the load data for fridays (5). These are used as the target values for training the linear network.
- An *input delay array* [number of delays from 1 to 3] is used as tapped delay line to format the input. Inputs with delayed samples are used for the betterment of the results.
- The linear network is implemented using the *newlind* (Mathworks 2002) command. The input array, input delay array and the training target values are converted to data cell format using *num2cell* (Mathworks 2002) command.
- Prediction is done by predicting the result for one to 24 values. As the linear network is trained only for the friday data with the first 696 data of the total 720 data excepting the last week's data, we expect the outcome for the last friday which is our desired date for prediction. For simulation we use the *sim* (Mathworks 2002) command.

Table 3.7 and Fig. 3.41 show the prediction result numerically and graphically respectively. The mean MAPE using (3.69) of the prediction results is $-17.4871\%$ which is relatively high.

It is evident from Fig. 3.41 that the predicted result (linear in nature) does not track the nonlinearity of the load values. This is expected from the simplistic linear network which is based on the least square problem solving algorithm. This is an example of underfitting (see Sect. 3.4.9.2).

**Table 3.7** Load forecasting result using the linear network

| Hour | Actual Value (MW) | Predicted Value (MW) |
|---|---|---|
| 1 | 224 | 247.0000 |
| 2 | 210 | 236.0000 |
| 3 | 203 | 273.8401 |
| 4 | 189 | 273.5312 |
| 5 | 194 | 273.2224 |
| 6 | 209 | 272.9135 |
| 7 | 234 | 272.6047 |
| 8 | 246 | 272.2959 |
| 9 | 311 | 271.9870 |
| 10 | 318 | 271.6782 |
| 11 | 321 | 271.3693 |
| 12 | 322 | 271.0605 |
| 13 | 310 | 270.7516 |
| 14 | 255 | 270.4428 |
| 15 | 252 | 270.1339 |
| 16 | 254 | 269.8251 |
| 17 | 254 | 269.5162 |
| 18 | 227 | 269.2074 |
| 19 | 272 | 268.8985 |
| 20 | 232 | 268.5897 |
| 21 | 237 | 268.2809 |
| 22 | 224 | 267.9720 |
| 23 | 218 | 267.6632 |
| 24 | 204 | 267.3543 |

The example computer program utilizing the Matlab® neural network toolbox (Mathworks 2002) for the linear network is shown below. However, it is to be noted that some of the functions might be different depending on the version of the Matlab®. Hence, the user is requested to check the compatibility of the functions with his/her Matlab® version.

*********************** **LinearNetwork.m** *********************

```
% Loadforecasting using linear network

clear;                          % Clears Workspace
data=load('loaddata.txt');      % Loads loaddata.txt file's data
                                  into 'data' variable

% ----------------- Preparing Target Data --------------------

 t1=data(1:600);      % Loading the first 600 data in 't1'
 day=data(1:600,2); % Loading Date information in 'day'
 c=(day==5);          % Getting 0-1 vector for Day 5 i.e. Friday
                        only
 tfinal=(t1.*c');     % All Friday-data are 1 other days are 0 in
                        'tfinal'
```

```
 tfinal=tfinal        % Filtering 'tfinal' to have only Friday
 (tfinal>0);            load data

% -------------------------------------------------------------

% --------------- Preparing Input and Input Delay -------------
P=1:96;              % 'P' as input corresponding to 'tfinal'
Pdelay=[1 3];        % Input Delay to implement 'Tapped Delay Line'
% -------------------------------------------------------------

% ------- Converting Array to Cell for Network Training -------

 P=num2cell(P);               % Converting Input Array to Cell
 Pdelay=num2cell(Pdelay); % Converting Input Delay Array to Cell
 target=num2cell(tfinal); % Converting Result Array to Cell
% -------------------------------------------------------------

% -------------- Implementing Network & Training ------------

 net=newlind(P,target,Pdelay);  % Creating and Training Network
% -------------------------------------------------------------

% -------------------------- Predicting -----------------------

 Predict=1:24;                  % Preparing for prediction 1 to
                                  24 for Friday
 Predict=num2cell(Predict);     % Convert Array to Cell for
                                  Network Training
 result=sim(net,Predict,Pdelay) % Prediction is done with 'sim'
                                  command
% -------------------------------------------------------------

*************************************************************
```

### *3.6.4 Backpropagation Network*

Load forecasting using the feedforward backpropagation multilayer perceptron network is discussed below.

- Training of the network is done with the first 600 data and the last 120 data (the last week) are used as the test data of which the last 24 values correspond to the 24 hours of the prediction date 5 (Friday).
- Number of delays ($d$) is set at 96 for most optimum result. This number 96 is the highest possible delay for predicting the last 24 values for the 24 hours of

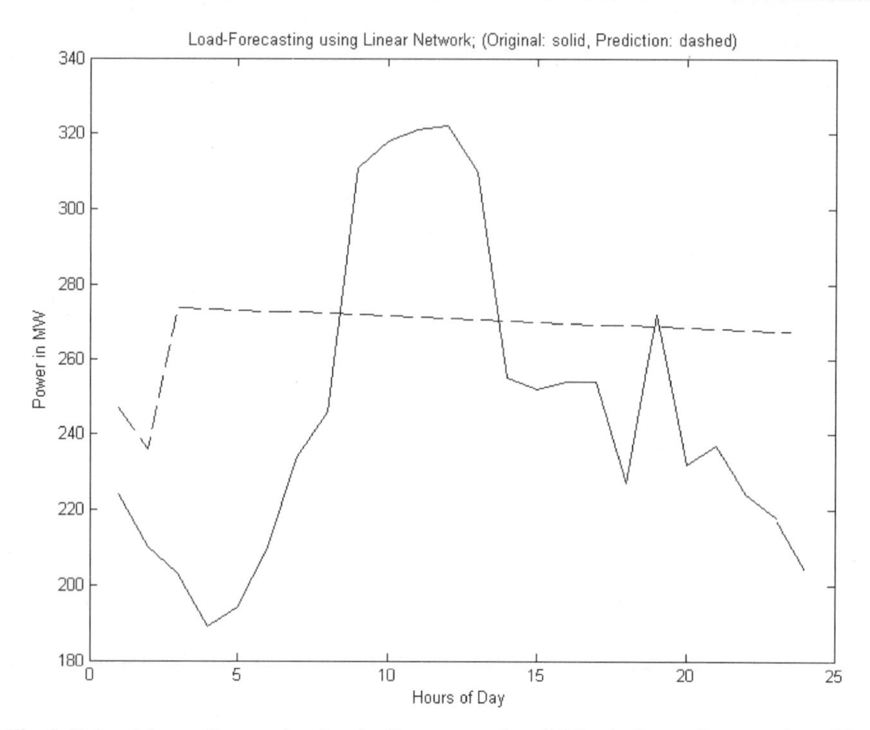


**Fig. 3.41** Load forecasting result using the linear network, solid line indicates the true value of the load and the dashed line indicates the predicted value

the day 5. After experimenting with delay ($d$) from 4 to 96, best result has been achieved with delay = 96.

- The input matrix is a kind of *toeplitz*[9] matrix, formed with the *toeplitz*[10] (Mathworks 2002) command by providing the first row and column value for the toeplitz matrix.
- The 600 training target data are normalized by dividing them with the highest load value in the range.
- We forecast last 120 values taking a delay of 96 i.e., the prediction window length is 120–96 = 24.

Matlab® program for the backpropagation network is shown below. However, it is to be noted that some of the functions might be different depending on the version of the Matlab®. Hence, the user is requested to check the compatibility of the functions with his/her Matlab® version.

The program generates the input toeplitz matrix and the target vectors for the training. Matlab® GUI (graphical user interface) for neural network toolbox (using

---

[9] A Toeplitz matrix is defined by one row and one column. A symmetric Toeplitz matrix is defined by just one row. For example, $a = \begin{bmatrix}1 & 2 & 3\end{bmatrix}$, Toeplitz matrix of $a = \begin{bmatrix}1 & 2 & 3\\ 2 & 1 & 2\\ 3 & 2 & 1\end{bmatrix}$.

[10] *toeplitz* command generates Toeplitz matrices given just the row or row and column description.

command *nntool*) (Mathworks 2002) has been used for these input and target data to train the network. The following parameters are chosen.

Network type was 'Feed-forward backprop' (Mathworks 2002), training function was 'TRAINLM' (Mathworks 2002), adaption learning function was 'LEARNGDM' (Mathworks 2002), performance function was 'MSE' (Mathworks 2002), number of layers (for best result) was 3, transfer function for the layer 1 & 2 was 'TANSIG' (Mathworks 2002) and that for layer 3 was 'PURELIN' (Mathworks 2002).

```
************************ BPNetwork.m ************************
% Loadforecasting using backpropagation network

clear;                      % Clears Workspace
data=load('loaddata.txt');  % Loads loaddata.txt file's data
                              into 'data' variable
target=data(1:720)';        % Preparing the target variable for
                              the Network
% -------- Preparing Training Data for the Network ----------

 total=720;              % Total number of data available
 ts=600;                 % Total Number of training samples
 tsignal=target(1:ts);   % Training signal
 d=96;                   % Number of delays (96 is Max to predict
                           last 24)
 w=ts-d;                 % Training Window length

 % Input matrix - Toepliz system
 InpTrain=toeplitz(tsignal(d:w+d-1),tsignal(d:-1:1))';

 TargetTrain=tsignal(d+1:d+w)';     % Target vector (Training)

 TargetTrain=TargetTrain./max(target); % Normalized target data

% ---------------------------------------------------------------
% -------- Preparing Training Data for the Network ----------

 tp = total-ts;                  % Total No. of prediction samples
 psignal = target(ts+1:ts+tp);   % Training signal for Prediction
 w = tp - d;                     % Prediction Window Length

 % Input matrix - Toepliz system
 InpPredict=toeplitz(psignal(d:w+d-1),psignal(d:-1:1))';

 % Target data for validating prediction
 TargetPredict=psignal(d+1:d+w)';
% ---------------------------------------------------------------

*****************************************************************
```

**Table 3.8** Load forecasting result using the backpropagation network

| Hour | Actual Value (MW) | Predicted Value (MW) |
|---|---|---|
| 1 | 224 | 229.8002 |
| 2 | 210 | 207.2383 |
| 3 | 203 | 209.0539 |
| 4 | 189 | 188.9999 |
| 5 | 194 | 195.9173 |
| 6 | 209 | 207.8305 |
| 7 | 234 | 237.0849 |
| 8 | 246 | 237.5848 |
| 9 | 311 | 310.2271 |
| 10 | 318 | 317.9722 |
| 11 | 321 | 315.0892 |
| 12 | 322 | 325.0992 |
| 13 | 310 | 315.8145 |
| 14 | 255 | 256.4545 |
| 15 | 252 | 247.9500 |
| 16 | 254 | 263.0439 |
| 17 | 254 | 241.7427 |
| 18 | 227 | 239.1299 |
| 19 | 272 | 282.8983 |
| 20 | 232 | 227.7221 |
| 21 | 237 | 241.9452 |
| 22 | 224 | 217.8235 |
| 23 | 218 | 211.0190 |
| 24 | 204 | 203.7538 |

Table 3.8 and Fig. 3.42 show the prediction result numerically and graphically respectively. The mean MAPE using (3.69) of the prediction results is $-2.0015\%$ which is acceptable, less than 3%. Experiments with different parameters that have been performed using the backpropagation network to achieve best result are as follows.

- Changing the number of delays (4 to 96, 96 gives best result)
- Changing the number of layers (2 to 4, 3 gives best result),
- Varying number of training data: all data to only weekdays data (all data training gives best result).

### *3.6.5 Radial Basis Function Network*

The operation using the radial basis function (RBF) network is described below.

- Training of the network is done with the first 600 data. The last 120 data (the last week) are used as the test data against which the predicted 24 values (24 hours) for the prediction date 5 (friday) are checked.
- The 600 training target data are normalized by dividing them with the maximum load value in the range.

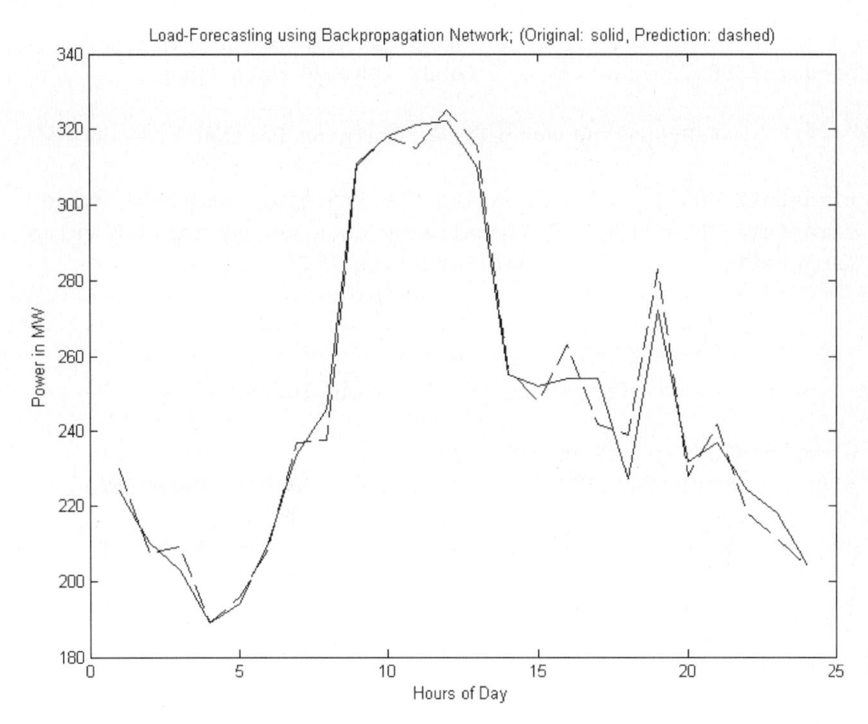


**Fig. 3.42** Load forecasting result using the backpropagation network, solid line indicates the true load value and the dashed line indicates the predicted value

- The input is given as concatenated array of the day-information (monday to friday, 1 to 5) and the hour-information (1 to 24) by the *cat* (Mathworks 2002) command. Output is also formed by concatenating the final Day (5) and the hour (1–24) information.
- Radial basis function network is implemented using the *newgrnn* (Mathworks 2002) command. This creates generalized regression neural network which is often used for function approximation with a 'SPREAD' (Mathworks 2002) of 0.58 instead of default 1.
- Prediction is done by *sim* (Mathworks 2002) command.

Matlab® program for the generalized regression RBF network is shown below. It is to be noted that some of the functions might be different depending on the version of the Matlab®. Hence, the user is requested to check the compatibility of the functions with his/her Matlab® version.

********************** **RBFNetwork.m** ************************

```
% Loadforecasting using radial basis function network

clear;                   % Clears Workspace
data=load('loaddata.txt'); % Loads loaddata.txt file's data into
```

```
                              'data' variable
Ref=data(696:720);          % Loads last 24 data into

% ---------- Preparing Normalized Training Dataset ------------

 t1=data(1:600);          % Loading the Training Load Data value
 m=max(t1); t1=t1./m;     % Normalizing data set by maximum value
 Target=t1;               % Set Target to 't1'

% ---------------------------------------------------------------

% --------------- Preparing the Training Input ----------------

 p1=ones(600,1);  p2=ones(600,1);
 p1(1:576,1)=data(1:576,2);  % Loading 'p1' with Date info
                               1-Mon,... ,7-Sun
 p2(1:576,1)=data(1:576,3);  % Loading 'p2' with Hour info
                               (1-24)
 Input=cat(1,p1',p2');       % 'Input' by concatenating Date
                               & Hour info
% ---------------------------------------------------------------

% -------------- Preparing the Prediction Input ---------------

 a1=ones(24,1);    a2=ones(24,1);
 a1=data(697:720,2);        % Loading 'a1' with Date info
                              1-Mon,...,7-Sun
 a2=data(697:720,3);        % Loading 'a2' with Hour info (1-24)
 Predict=cat(1,a1',a2');    % Prediction input, concatenating
                              Date & Hour info
% ---------------------------------------------------------------

% ----------------- Implementing RBF Network ------------------

 % Generalized regression network with SPREAD=0.58
 net=newgrnn(Input,Target,0.58);
 Result = sim(net,Predict); % Getting Normalized Result
 Result=Result.*m;          % Denormalizing result
% ---------------------------------------------------------------

***************************************************************
```

Table 3.9 and Fig. 3.43 show the prediction result using the RBF network numerically and graphically respectively. The mean MAPE using (3.69) of the prediction results is −4.8072% which is acceptable, being less than 5%.

Table 3.9 Load forecasting result using the radial basis function network

| Hour | Actual Value (MW) | Predicted Value (MW) |
|---|---|---|
| 1 | 224 | 227.3113 |
| 2 | 210 | 212.6371 |
| 3 | 203 | 217.0635 |
| 4 | 189 | 210.8347 |
| 5 | 194 | 211.0615 |
| 6 | 209 | 224.4144 |
| 7 | 234 | 252.5383 |
| 8 | 246 | 258.8428 |
| 9 | 311 | 302.6886 |
| 10 | 318 | 313.3246 |
| 11 | 321 | 316.6342 |
| 12 | 322 | 317.5345 |
| 13 | 310 | 309.8787 |
| 14 | 255 | 267.8440 |
| 15 | 252 | 251.0577 |
| 16 | 254 | 254.0187 |
| 17 | 254 | 251.4262 |
| 18 | 227 | 242.7677 |
| 19 | 272 | 289.5590 |
| 20 | 232 | 265.1567 |
| 21 | 237 | 251.7423 |
| 22 | 224 | 232.8657 |
| 23 | 218 | 230.3787 |
| 24 | 204 | 220.1883 |

Experiments with different parameters that have been performed using the generalized regression radial basis function network to achieve the best result are as follows.

- Changing the network type (generalized regression network gives best result).
- Changing the training data: without date information, with date information (with date information produces the best result).
- Varying the number of training data: all data to only friday data (all data training gives best result).
- Changing the spread (Mathworks 2002) for the regression RBF network (spread parameter range 0.5–2.0, 0.58 produces best result).

## References

Bunn DW, Farmer ED (1985) Comparative models for electrical load forecasting. John Wiley & Sons, New York

IEEE Committee (1980) Load forecasting bibliography phase I. IEEE Transactions on Power Applications and Systems 99:53–58

Mathworks Inc. (2002) MATLAB® Documentation–Neural Network Toolbox, Version 6.5.0.180913a, Release 13

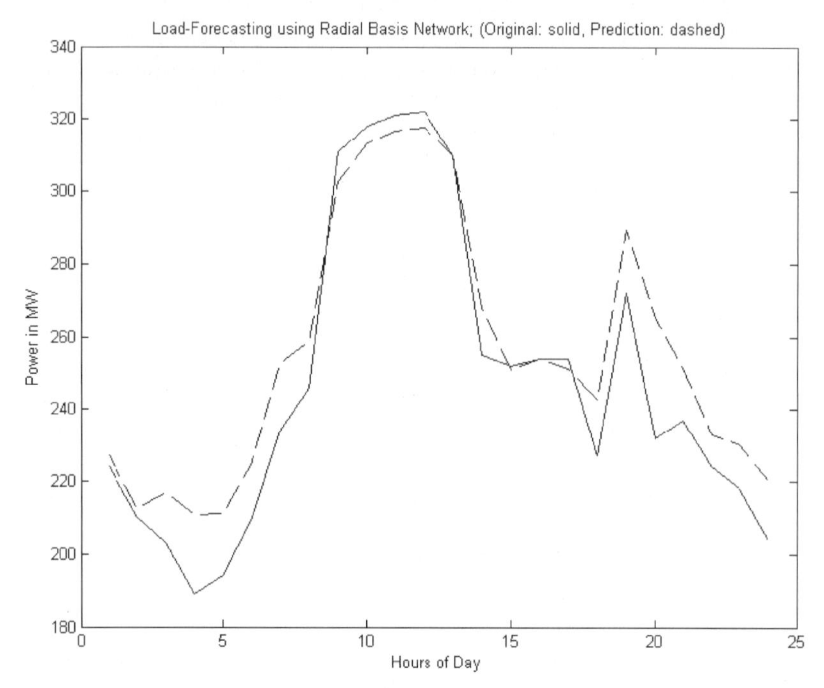


**Fig. 3.43** Load forecasting result using the radial basis function network, solid line indicates the true load value and the dashed line indicates the predicted value

Novak B (1995) Superfast autoconfiguring artificial neural networks and their applications to power systems. Electric Power Systems Research 35:11–16

## 3.7 Feeder Load Balancing

Feeder load balancing in the distribution system have different needs, from minimizing the losses in the system to relieving the transformer during the peak time and so forth. There are a number of normally closed and normally opened switches in a distribution system. By changing the open/close status of the feeder switches, load currents can be transferred from feeder to feeder, that is, from the heavily loaded to the less loaded feeders.

With the uses of the artificial intelligence, telecommunication and power electronics equipments in the power system, it is becoming easier to automate the phase balancing problem. The automation implementation will be technically advantageous as well as economical for the utilities and the customers, in terms of the variable costs reduction and better service quality, respectively (Ukil et al. 2006).

This application study shows how neural network can be used to switch on/off the different switches and keep the phases balanced. We assume that the loads are connected in a three-phase system. Each load will cater only one of the three phases following the constraint that for each load only one switch (to the phase) should be closed, while other two should remain open. For each loading condition, the neural network will be trained for the relevant minimum loss configuration. In this example we apply this on a small network of six houses as the unbalanced loads.

### 3.7.1 Phase Balancing Problem

To balance the three phase currents in every segment and then depressing the neutral line current is a very difficult task. We consider a distribution feeder as a three-phase, four-wire system which can be radial or open loop structure (Ukil et al. 2006). The example feeder shown in Fig. 3.44 has three phase conductors for the section between the main transformer and the different load points. We limit this application study to six load points, as shown in Fig. 3.44. To improve the system phase voltage and current unbalances, the connections between the specific feeders and the distribution transformers should be suitably arranged.

### 3.7.2 Feeder Reconfiguration Technique

In the case of a distribution system with some of the branches overloaded, and other branches lightly loaded, there is the need to reconfigure the networks such that loads are transferred from the heavily loaded feeder or transformers to the less loaded feeder (or transformers). The maximum load current the feeder conductor can take may be taken as the reference. The transfer of load must be done by satisfying the predefined objective to have minimum real power loss. Consequently, network reconfiguration may be redefined as the rearrangement of the network such as to minimize the total real power losses arising from the line branches. Mathematically, the total power loss can be expressed as follows (Chen and Cho 1992):

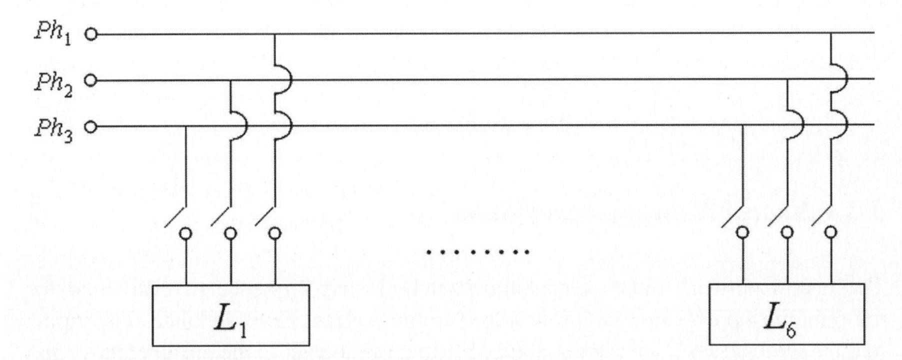


**Fig. 3.44** Example distribution feeder

$$\sum_{i=1}^{n} r_i \frac{P_i^2 + Q_i^2}{|V_i|^2}, \tag{3.70}$$

where, $r_i$, $P_i$, $Q_i$, $V_i$ are respectively the resistance, the active power, the reactive power and the voltage of the branch $i$, and $n$ is the total number of branches in the system. Due to some practical considerations, there could be a constraint on the number of switch–on and off. Given a distribution system as shown in Fig. 3.44, a network with three phases with a known structure, the problem consists of finding a condition of balancing. The mathematical model (Ukil et al. 2006) can be expressed as:

$$\mathbf{I}_{ph1k} = \sum_{i=1}^{3} sw_{k1i}\mathbf{I}_{ki} + \mathbf{I}_{ph1(k+1)}, \tag{3.71}$$

$$\mathbf{I}_{ph2k} = \sum_{i=1}^{3} sw_{k2i}\mathbf{I}_{ki} + \mathbf{I}_{ph2(k+1)}, \tag{3.72}$$

$$\mathbf{I}_{ph3k} = \sum_{i=1}^{3} sw_{k3i}\mathbf{I}_{ki} + \mathbf{I}_{ph3(k+1)}, \tag{3.73}$$

where, $\mathbf{I}_{ph1k}$, $\mathbf{I}_{ph2k}$ and $\mathbf{I}_{ph3k}$ represent the currents (phasors) per phase (1, 2 & 3) after the $k$ point of connection; $sw_{k11}, \ldots, sw_{k33}$ are different switches (the value of '1' means the switch is closed and '0' means it is open). Following the constraint of allowing only one breaker in each of the (3.71–3.73) to be closed, we can write the following set of modified constraints:

$$\sum_{i=1}^{3} sw_{k1i} - 1 = 0, \tag{3.74}$$

$$\sum_{i=1}^{3} sw_{k2i} - 1 = 0, \tag{3.75}$$

$$\sum_{i=1}^{3} sw_{k3i} - 1 = 0. \tag{3.76}$$

### 3.7.3 Neural Network-based Solution

The neural network must control the switch-closing sequence of each load for the minimum power loss which will lead to the optimal phase balance. The inputs to the neural network are the unbalanced load currents (six in the current study) and the outputs are the switch closing sequences for each load.

The input layer of the network has $N$ input neurons, $N$ being number of unbalanced load currents to be controlled. The following column vector has been assumed as the input.

$$C_{Sw} = [I_{L1}, I_{L2}, \ldots, I_{LN}]. \tag{3.77}$$

The output of the network is in the range {1, 2, 3} for each load, i.e., which switch (to the specific phase) should be closed for that specific load.

For this application, we will use the radial basis network (Ukil et al. 2006). Experimentations with the backpropagation and the radial basis network indicated faster training and better convergence for the latter. Radial basis networks may require more neurons than the standard feed-forward backpropagation networks, but often they can be designed in a fraction of the time needed to train the standard feed-forward networks. They work best when many training vectors are available. Matlab® neural network toolbox (Mathworks 2002) has been used for the implementation.

### 3.7.4 Network Training

The data set consisted of unbalanced load data from a load survey. The test data set had average load current values per houses for the different times of each day in a month. We randomly select six houses as our test data for each specific time, and we test our result on 500 data. First, we balance the loads heuristically (Ukil et al. 2006).

We consider the loads to be equally distributed per phase, i.e., we assume two loads to be connected per phase. So, the problem is to find the optimum three sets of two loads, with *minimum* differences among the individual sums of the three sets. To achieve this, first we calculate the ideal phase balance current value $I_{ideal}$, which is equal to the one-third of the sum of the all six load currents $I_L$.

$$I_{ideal} = \frac{1}{3}\sum_{j=1}^{6} I_{L_j}. \tag{3.78}$$

In the second step, we optimally select our 3 sets of currents for the three phase currents $I_{ph}$, each set comprising of two load currents $\{I_j, j = 1, 2\}$.

$$I_{Load} = \{I_{L_j}, j = 1, \ldots, 6\}, \tag{3.79}$$

$$I_{ph} = \{I_j, j = 1, 2\} \quad \text{where} \quad I_j \in I_{Load}. \tag{3.80}$$

Difference between the individual sum of these sets and the $I_{ideal}$ should be *minimum*, ideally 0 for the perfect phase balance. So, we need to find three sets of $\{I_j, j = 1, 2\}$, subject to the constraint,

$$\min\left|\sum_{j=1}^{2} I_j - I_{ideal}\right|, \text{ where } I_j \in I_{Load}. \qquad (3.81)$$

Following this, we obtain the output switching sequences as the target data set for training the networks. Using the output switching sequences and the input load currents, we calculate the balanced phase currents $I_{ph1}$, $I_{ph2}$ and $I_{ph3}$. For example, $I_{ph1}$ is calculated by adding the two load currents corresponding to the output switching sequences marked 1. Then we calculate the differences between $I_{ph1}$, $I_{ph2}$ and $I_{ph3}$, which ideally should be zero. The differences indicate the quality of the phase balance (Ukil et al. 2006).

Using the unbalanced load as the input vector, and the output switching sequences as the target vector, we trained the above-mentioned neural network. Then, we tested the network with different unbalanced load data set. The output was the optimal switching sequences of $\{1, 2, 3\}$ for the three-phases as explained above. Using the similar procedure as explained above, we computed the balanced phase currents and the differences between the phase currents, which indicate the quality of the balance.

### 3.7.5 Results

An Intel® Celeron® 1.9 GHz, 256 MB RAM computer was used for the simulation. Test results of the neural network-based approach for the simulated six load data format are shown in Table 3.10 to 3.13, for three different sample data. Table 3.10 shows the unbalanced load (current) data, Table 3.11 the output switching

**Table 3.10** Unbalanced load (current) data

| Data<br>$I_L$ (A) | 1 | 2 | 3 |
|---|---|---|---|
| 1 | 89 | 35 | 45 |
| 2 | 85 | 0 | 67 |
| 3 | 74 | 90 | 87 |
| 4 | 38 | 21 | 64 |
| 5 | 56 | 87 | 30 |
| 6 | 45 | 112 | 90 |

**Table 3.11** Output switching sequences

| Output Switching Sequences for 6 Loads | Data 1 | Data 2 | Data 3 |
|---|---|---|---|
| 1 | 1 | 1 | 1 |
| 2 | 2 | 2 | 2 |
| 3 | 3 | 1 | 3 |
| 4 | 1 | 3 | 2 |
| 5 | 3 | 3 | 3 |
| 6 | 2 | 2 | 1 |

**Table 3.12** Balanced phase currents

| $I_{Phase}$ (A) | Data 1 | Data 2 | Data 3 |
|---|---|---|---|
| Phase 1 | 127 | 125 | 135 |
| Phase 2 | 130 | 112 | 131 |
| Phase 3 | 130 | 108 | 117 |

**Table 3.13** Differences between the phase currents

| \|Difference\| (A) | Data 1 | Data 2 | Data 3 |
|---|---|---|---|
| Phase 1-2 | 3 | 13 | 4 |
| Phase 2-3 | 0 | 4 | 14 |
| Phase 3-1 | 3 | 17 | 18 |

sequences, Table 3.12 the balanced phase currents and Table 3.13 the absolute differences between the balanced phase currents.

## References

Chen, CS, Cho, MY (1992) Energy loss reduction by critical switches, of distribution feeders for loss minimization. IEEE Transactions on Power Delivery 4:

Mathworks Inc. (2002) MATLAB® documentation–neural network toolbox, version 6.5.0.180913a, release 13.

Ukil A, Siti W, Jordaan JA (2006) Feeder load balancing using neural network. Lecture Notes in Computer Science 3972:1311–1316

## Section III: Objective Projects

## 3.8 Fault Classification

The analysis of faults and disturbances (Nagrath and Kothari 1998) during abnormal conditions in power network can provide valuable information regarding the disturbances and the remedies. Fault classification is an important topic in disturbance and fault analysis to differentiate various types of faults. Typically faults are either symmetrical (three phase short-circuit or three phase-to-ground faults) or nonsymmetrical (single phase-to-ground or double phase short-circuit or double phase-to-ground faults). Typical probability of occurrence of different types of faults is shown in Fig. 3.45.

The faults considered in Fig. 3.45 are three-phase short-circuit (L-L-L), three phase-to-ground (L-L-L-G), line-line short-circuit (L-L), single line-to-ground (L-G) and double line-to-ground (L-L-G) faults, where the terms 'L' and 'G' refer

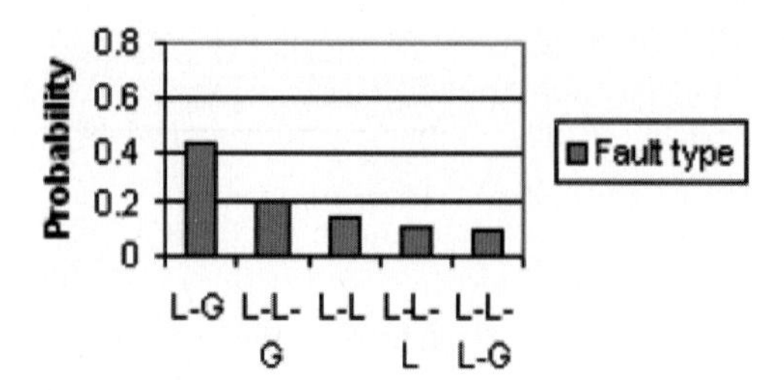


**Fig. 3.45** Probability of occurrence of different faults

to 'Line' and 'Ground' respectively. The probabilities associated with a fault type depend upon the operating voltage and can vary from system to system.

There have been many approaches to fault classification. One of the promising techniques is the neural networks. By nature, neural networks perform very well for pattern classification type problems. A brief outlook into designing a fault classifier using the neural networks is discussed below.

## 3.8.1 Simple Ground Fault Classifier

The input could be the three phase voltage and current recordings (pre- and post-fault). Advanced simulation could incorporate neutral voltage and current as well.

The output could be formatted as, 1 indicating the ground fault, and $-1$ indicating the non-ground fault and 0 indicating normal conditions.

This is a linear classification task. So, multilayer perceptron, feedforward back-propagation network are ideally suited for this type of task. However, radial basis function networks could also be tried.

The network training can be done by using the three-phase voltage and current recordings (pre- and post-fault or the neural voltage and current values. The target values should be 1 for the ground faults, $-1$ for the non-ground faults and 0 for normal conditions. A large training set should be chosen with many different cases of ground and non-ground faults to train the network.

Finally, the network should be tested by providing suitable input voltage and currents and it should identify the situation.

## 3.8.2 Advanced Fault Classifier

Once the simple fault classifier is developed, more advanced fault classifier could be designed. This can be a cascaded fault classifier which classifies the typical faults mentioned in Fig. 3.45.

We start with the simple ground fault classifier as discussed above. Once we get the decision whether the fault is ground or non-ground, we employ two more networks in cascade.

If the fault is non-ground in nature, one of the networks takes as input the three-phase voltages and currents to determine whether it is a three-phase or a single-phase fault. The output can be designed as 1 for the three-phase fault and 0 for the single-phase fault.

On the other hand, if the fault is a ground one, the third neural network can be designed to identify whether it is a single, double or three-phase line-to-ground fault. The network can take as input the three-phase voltages and currents along with neural voltage and current. The output can be designed as $-1$ for single line-to-ground fault, 0 for double line-to-ground fault and 1 for three-phase line-to-ground fault.

Feedforward multilayer perceptron, radial basis function networks should be the ideal networks.

Some hints to achieve better result:

- Use fully interconnected network.
- Number of neurons in the hidden layer should roughly be at least twice that of input layer.
- Use normalized input values.
- Use as many different data as possible to cover the whole spectrum.

## Reference

Nagrath J, Kothari DP (1998) Power system engineering. Tata McGraw Hill Publications, New delhi

## 3.9 Advanced Load Forecasting

We discussed about short-term load forecasting in Sect. 3.6. We can enhance that model for more advanced load forecasting (Bunn and Farmer 1985).

For the model in Sect. 3.6, we considered as input the hourly load values and the day information. We can incorporate additional parameters to improve the learning and prediction. Some of these parameters could be

- **Seasonal information**. Power consumption depends a lot on the seasonal variation, for example, during the summer or holidays (Aboul-Magd and Ahmed 2001) air-conditioners, fans are used more while in winter heater is used predominantly. Inclusion of a seasonal parameter should model better these seasonal load behaviors.
- **Weather information**. Load consumption varies a lot depending on the daily or weekly weather variability. A weather parameter (Taylor and Buizza 2003) in terms of temperature, relative humidity, rain could be incorporated.

- **Area specific modeling**. Even within a city load consumption varies from locality to locality due to economic and many other factors. An area-specific parameter dealing with the economic level, industry information etc could be added.
- **Domestic or industrial modeling**. The load forecasting for a city could be divided into domestic load forecasting and industrial load forecasting because there is a lot of difference between these two categories. The final load forecasting could be achieved by combining the domestic and the industrial load forecasting.
- **Tariff structure**. Tariff is an important factor for load consumption. Hence, a parameter providing updated tariff information should produce better prediction result.

All these parameters should be incorporated in the input. This would make the network more complex in nature. Hence, the network should be chosen accordingly. But feedforward backpropagation network, radial basis function network or combination of both should be the optimum choice.

## References

Aboul-Magd MA, Ahmed EEDES (2001) An artificial neural network model for electrical daily peak load forecasting with an adjustment for holidays. In Proc. Large Engg. Syst. Conf. on Power Engg., Halifax, Canada, pp. 105–113

Bunn DW, Farmer ED (1985) Comparative models for electrical load forecasting. John Wiley & Sons, New York

Taylor JW, Buizza R (2003) Using weather ensemble predictions in electricity demand forecasting. Int. J. Forecasting 19:57–70

## 3.10 Stability Analysis

This is related to the load forecasting. With the load forecasting it is expected that the actual load will differ from the predicted value (even if with very small error). This change in load (in terms of active and reactive power) would have an effect on the stability. This is depicted by the typical PV curve (Grisby et al. 2007) in Fig. 3.46. This is also referred to as the stability analysis (Grisby et al. 2007). The stability margin is defined as the distance between the current operating point and the maximum loading point. The latter is defined as the maximum load at which the system can operate without suffering any voltage stability loss. Voltage stability limit (Taylor 1993) can be defined as the difference between the maximum permissible power $P_m$ and the current operating point $P_0$.

$$\Delta P = P_m - P_0. \tag{3.82}$$

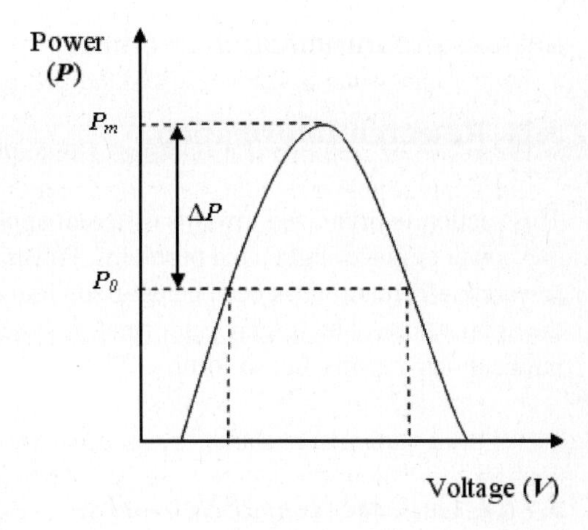


**Fig. 3.46** Typical power-voltage (PV) curve

An interesting application of neural network could be to estimate the stability limit corresponding to the current operating point. The steps of designing such a system could be as follows.

- Get the short-term (next day) load forecast value.
- Using the forecast value and the actual value the error is estimated.
- The forecast error and the load profile is used to train an ANN[11] to predict a random load variation.
- A second ANN[12] is trained with the load variation and the load forecast to calculate the stability margin.
- The stability margin is fed back to the first ANN to fine tune the prediction of the load variation.

## References

Grigsby LL (editor) (2007) The electric power engineering handbook, 2nd Ed., five Vol. CRC Publication, Florida

Taylor CW (1993) Power system voltage stability. McGraw Hill Inc., New York

[11] This ANN should be feedforward backpropagation type which extrapolates (correlated computation) random prediction from the deterministic observations.

[12] The second ANN should be RBF network as it measures the distance between two objectives.

# Section IV: Information Section

## 3.11 Research Information

This section is organized towards different applications of neural networks in various power engineering related problems. Power engineering specific neural network research information has been categorized into different sub-sections depending on the applications. The research information is organized in terms of relevant books, publications, reports and so forth.

### *3.11.1 General Neural Networks*

The following resources provide good basic understandings of general neural network theories.

Anderson JA, editor (1995) An introduction to neural networks. MIT Press, Cambridgc, MA

Callan R (1999) The essence of neural networks. Prentice Hall, London

Fausett LV (1994) Fundamentals of neural networks. Prentice-Hall Inc., New York

Haykin S (1999) Neural networks: a comprehensive foundation. Prentice-Hall Inc., Englewood Cliffs, NJ

Hertz J, Krogh A, Palmer RG (1991) Introduction to theory of neural computation. Addison-Wesley, Reading

Vojislav K (2001) Learning and soft computing–support vector machines, neural networks and fuzzy logic models. The MIT Press, Cambridge

Wasserman PD (1993) Advanced methods in neural computing. Van Nostrand Reinhold, New York

### *3.11.2 Neural Network and Power Engineering*

CIGRE TF 38.06.06 (1995) Artificial neural networks for power systems. Electra, No. 159, pp. 77–101

Dillion TS, Niebur D (1996) Neural networks applications in power systems. CRL Publishing Company, New York

El-Sharkawi M, Neibur D (1996) Artificial neural networks with applications to power systems. IEEE-PES Special Publ. 96TP-112-0

Madan S, Bollinger KE (1997) Applications of artificial intelligence in power systems. Electric Power Systems Research 41:117–131

Niebur D (1993) Artificial neural networks for power systems: a literature survey. Engng. Int. Syst. 1:133–158

Novak B (1995) Superfast autoconfiguring artificial neural networks and their application to power systems. Electric Power Systems Research 35:11–16
Vankayala VSS, Rao ND (1993) Artificial networks and their applications to power systems–a bibliographical survey. Electric Power Systems Research 28:67–79
Wehenkel L (1998), Automatic learning techniques in power systems. Kluwer Academic, Boston

### *3.11.3 Electrical Load Forecasting*

#### 3.11.3.1 Short-term

Asbury CE (1975) Weather load model for electric demand and energy forecasting. IEEE Transactions on Power App. and Systems PAS-94: 1111–1116
Bakirtzis A, Petridis V, Klartzis S (1996) A neural network short term load forecasting model for the greek power system. IEEE Transactions on Power Systems 11:858–863
Charytoniuk W, Chen MS (2000) Very short-term forecasting using artificial neural networks. IEEE Transactions on Power System 15:263–8
Chen S, Yu DC, Mooghaddamjo AR (1992) Weather sensitive short-term load forecasting using nonfully connected artificial neural network. IEEE Transactions on Power System 7:1098–1105
Chow TWS, Leung CT (1996) Neural network based short-term load forecasting using weather compensation. IEEE Transactions on Power System 11: 1736–1742
Khotanzad A, Davis MH, Abaye A, Martukulam DJ (1996) An artificial neural network hourly temperature forecaster with applications in load forecasting. IEEE Transactions PWRS 11:870–876
Kodogiannis VS, Anagnostakis EM (1999) A study of advanced learning algorithms for short-term load forecasting. Engineering Applications of Artificial Intelligence 12:159–173
Lu CN, Wu HT, Vemuri S (1993) Neural network based short term load forecasting. IEEE Transactions on Power Systems 8:337–342
Moghram I, Rahman S (1989) Analysis and evaluation of five short-term load forecasting techniques. IEEE Transactions on Power Systems 4:1484–1491
Msu YY, Yang CC (1991) Design of artificial neural networks for short-term load forecasting. Part I: self-organizing feature maps for day type identification peak load and valley load forecasting. IEE Proc. 138:407–413
Park JH, Park YM, Lee KY (1991) Composite modeling for adaptive short-term load forecasting. IEEE Transactions on Power System, 6:450–457
Peng TM, Huebele NF, Karady GG (1992) Advancement in the application of neural networks for short-term load forecasting. IEEE Transactions Power System 7:250–257

Ranaweera DK, Hubele NF, Papalexopoulos AD (1995) Application of radial basis function neural network model for short-term load forecasting. IEE Proceedings-Generation, Transmission, Distribution 1:45–50

Tamimi M, Egbert R (2000) Short term electric load forecasting via fuzzy neural collaboration. Electric Power Systems Research 56:243–248

#### 3.11.3.2 Medium- and Long-term

Abdel-Aal RE (2005) Improving electric load forecasts using network committees. Electric Power Systems Research 74:83–94

Barakat EH, Al-Rashed SA (1992) Long range peak demand forecasting under conditions of high growth. IEEE Trans. Power Systems 7:1483–1486

Bulsari AB, Saxen H (1993) A recurrent network for time-series modeling. In: Proc. Int. Conf. on Joint Neural Networks, Baltimore, USA, pp. 285–291

Carpinteiro OAS, Leme RC, de Souza ACZ, Pinheiro CAM, Moreira EM (2007) Long-term load forecasting via a hierarchical neural model with time integrators. Electric Power Systems Research 77:371–378

Gent MR (1992) Electric supply and demand in the United States: next 10 years. IEEE Power Eng. Rev., pp. 8–13

Ghiassi M, Zimbra DK, Saidane H (2006) Medium term system load forecasting with a dynamic artificial neural network model. Electric Power Systems Research 76:302–316

Kermanshahi B (1998) Recurrent neural network for forecasting next 10 years loads of nine Japanese utilities. Neurocomputing 23:125–33

Paarmann LD, Najar MD (1995) Adaptive online load forecasting via time series modeling, Electric Power Systems Research 32:219–225

Papalexopoulos A, How S, Peng TM (1994) An implementation of a neural network based load forecasting model for the EMS. In: IEEE PES 1994 Winter Meeting Paper 94, pp. 207–209

### *3.11.4 Fault Locator & Analysis*

Bi TS, Ni YX, Shen CM, Wu FF, Yang QX (2000) A novel radial basis function neural network for fault section estimation in transmission network. In: Proc $5^{th}$ Int. Conf. Power System Ctrl, Oprn and Mgmt, Hong Kong, 1:259–263

Dalsten T, Friedrick T, Kulicke B, Sobajic D (1996) Multi-neural network based fault area estimation for high speed protective relaying. IEEE Transactions on Power Delivery 11: (1996)

Dausten T, Kulieke B (1995) Neural network approach to fault classification for high speed protective relaying. IEEE Trans. Power Delivery 10:1002–1009

Jack LB, Nandi AK (2002) Fault detection using support vector machines and artificial neural networks: augmented by genetic algorithms. Mech. Syst. Signal Process 16:373–390

Kezunovic M, Rikalo I (1996) Detect and classify faults using neural nets. IEEE Trans. Comput. Appl. Power 9:42–47

Lu CN, Tsay MT, Hwang YJ (1994) An artificial neural network based trouble call analysis. IEEE Transactions on Power Delivery 9:1663–1668

Mahanty RN, Gupta PBD (2004) Application of rbf neural network to fault classification and location in transmission lines. Proc. Inst. Elect. Eng. 151

Oleskovicz M, Coury DV, Aggarwal RK (2001) A complete scheme for fault detection, classification and location in transmission lines using neural networks. In Proc. Inst. Elect. Eng., $7^{th}$ Conf. Power Syst. Protect., pp. 335–338

Purushothama GK, Narendranath AU, Thukaram D, Parthasarathy K (2001) ANN applications in fault locators. Electrical Power Energy Systems 23:491–506

Samantaray SR, Dash PK, Panda G (2006) Fault classification and location using HS-transform and radial basis function neural network. Electric Power Systems Research 76:897–905

Silva KM, Souza BA, Brito NSD (2006) Fault detection and classification I transmission lines based on wavelet transform and ANN. IEEE Transactions on Power Delivery 21:2058–2063

Souza BA, Brito NSD, Neves WA, Silva KM, Lima RV, Silva SSB (2004) Comparison between backpropagation and rprop algorithms to fault classification in transmission lines. In Proc. Int. Joint. Conf. N. Networks, Budapest, Hungary, pp. 2913–2918

Thukaram D, Shenoy UJ, Ashageetha H (2006) Neural network approach for fault location in unbalanced distribution networks with limited measurements. In: 2006 IEEE Power India Conference, India

Vasilic S, Kezunovic M (2005) Fuzzy art neural network algorithm for classifying the power system faults. IEEE Transactions on Power Delivery 20:1306–1314

Vazquez EE, Altuve H, Chacon O (1996) Neural network approach to fault detection in electric power systems. IEEE Trans. Power Delivery 11:2090–2095

Yang HT, Chang WY, Huang CL (1994) A new neural networks approach to on-line fault section estimation using information of protective relays and circuit breakers. IEEE Transactions on Power Delivery 9:220–230

Zhu J, Lubkeman DL, Girgis AA (1997) Automated fault location and diagnosis on electric power distribution feeders. IEEE Transactions on Power Delivery 12: 801–809

### *3.11.5 Power Systems Protection*

Cannas B, Celli G, Marchesi M (1998) Neural networks for power system condition monitoring and protection. Neurocomputing 23:111–123

Coury DV, Jorge DC (1998) Artificial neural network approach to distance protection. IEEE Transactions on Power Delivery 13:102–108

Dalstein T, Kulicke B (1995) Neural network approach to fault classification for high speed protective relaying. IEEE Trans. on Power Delivery 10:1002–1011

Dash K, Pradhan AK, Panda G (2000) A novel fuzzy neural network based distance relaying scheme. IEEE Transactions on Power Delivery 15:902–907

Kezunovic M (1997) A survey of neural net application to protective relaying and fault analysis. Eng. Int. Sys. 5:185–192

Novosel D, Backmann B, Hart D, Hu Y, Saha M (1996) Algorithms for locating faults on series compensated lines using neural network and deterministic methods. IEEE Transactions on Power Delivery 11:1728–1736

Sanaye-Psand M, Malik OP (1998) High speed transmission system directional protection using an Elman network. IEEE Transactions on Power Delivery 13: 1040–1045

Sanaye-Psand M, Korashadi-Zadeh H (2003) Transmission line fault detection & phase selection using ANN. In: Proc. IPST Conf., New Orleans, USA

Song YH, Johns AT, Yuan QY (1996) Artificial neural network-based protection scheme for controllable series-compensated EHV transmission lines. IEE Proc. on Generation, Transmission and Distribution 143:2032–2040

### *3.11.6 Harmonic Analysis*

Gongbao W, Dongyang X, Weiming M (2005) An improved algorithm with high accuracy for non-integer harmonics analysis based on FFT algorithm and neural network. In: Proc. PSCC Conf., Liege, Belgium

Lai LL, Chan WL, Tse CT, So ATP (1999) Real-time frequency and harmonic evaluation using artificial neural network. IEEE Transactions on Power Delivery 14:52–59

Renyong W, Zhiyong L (1999) Measurement of harmonics in power system based on artificial neural network. Power System Technology 23:20–23

Srinivasan D, Ng WS, Liew AC (2006) Neural-network-based signature recognition for harmonic source identification. IEEE Trans. Power Deliv. 21:398–405

Temurtas F, Gunturkun R, Yumusak N, Temurtas H (2004) Harmonic detection using feed forward and recurrent neural networks for active filters. Electric Power Systems Research 72:33–40

### *3.11.7 Transient Analysis*

Ferreira WP, Silveira MDCG, Lotufo ADP, Minussi CR (2006) Transient stability analysis of electric energy systems via a fuzzy ART-ARTMAP neural network. Electric Power Systems Research 76:466–475

Qian A, Shrestha GB (2006) An ANN-based load model for fast transient stability calculations. Electric Power Systems Research 76:217–227

Sawhney H, Jeyasurya B (2006) A feed-forward artificial neural network with enhanced feature selection for power system transient stability assessment. Electric Power Systems Research 76:1047–1054

Thukaram D, Khincha HP, Khandelwal S (2006) Estimation of switching transient peak overvoltages during transmission line energization using artificial neural network. Electric Power Systems Research 76:259–269

### 3.11.8 Power Flow Analysis

Gavoyiannis AE, Vlassis NA, Hatziargyriou ND (1999) Probabilistic neural networks for power flow analysis. In: Proc. ACAI'99, Machine Learning and Applications Conf., Crete, GREECE, pp. 44–51

Kalyani RP, Venayagamoorthy GK (2003) A continually online trained neural controller for the series branch control of the UPFC. In: Int. Joint Conf. on Neural Networks

Luo X, Patton AD, Singh C (2000) Real power transfer capability calculations using multi-layer feed-forward neural networks. IEEE Transactions on Power System 15:903–908

Narendra KG, Sood VK, Khorasani K, Patel RV (1997) Investigation into an artificial neural network based on-line current controller for an HVDC transmission link. IEEE Transactions on Power Systems 12:1425–1431

Park JH, Kim YS, Eom IK, Lee KY (1993) Economic load dispatch for piecewise quadratic cost function using Hopfield neural network. IEEE Transactions on Power Systems 8:1030–1038

Ranjan R, Venkatesh B, Chaturvedi A (2004) Power flow solution of three-phase unbalanced radial distribution network. Electric Power Components and Systems 32:421–433

### 3.11.9 Power Systems Equipments & Control

Bakhshai A, Espinoza J, Joos G, Jin H (1996) A combined ANN and DSP approach to the implementation of space vector modulation techniques. In: Conf. Rec.IEEE–IAS Annu. Meeting, pp. 934–940

Bansal RC, Bhatti TS, Kothari DP (2002) Artificial intelligence techniques for reactive power/voltage control in power systems: a review. In Proc. Applicat. Evol. Strategies to Power, Signal Processing Contr., pp. 57–63

Ekwue A, Cheng DTY, Macqueen JF (1997) Artificial intelligence techniques for voltage control. Artificial Int. Techniques in Power Systems, pp. 109–122

Pandit M, Srivastava L, Sharma J (2001) Voltage contingency ranking using fuzzified multilayer perceptron. Electric Power Systems Research 59:65–73

Ramakrishna G, Rao ND (1999) Adaptive neuro-fuzzy inference system for volt/var control in distribution systems. Elec. Power Sys. Research 49: 87–97

Shayeghi H, Shayanfar HA, Malik OP (2007) Robust decentralized neural networks based LFC in a deregulated power system. Electric Power Systems Research 77:241–251

### 3.11.10 Power Systems Operation

Dieu VN, Ongsakul W (2005) Hopfield Lagrange for short-term hydrothermal scheduling. In: Proc. 2005 IEEE PowerTech Conf., St. Petersburg, Russia
Dillon JD, Walsh MP, O'Malley MJ (2002) Initialization of the augmented Hopfield network for improved generator scheduling. IEE Proc. Generation, Transmission, Distribution 149:593–599
Ekonomou L, Gonos IF, Iracleous DP, Stathopulos IA (2007) Application of artificial neural network methods for the lightning performance evaluation of Hellenic high voltage transmission lines. Elec. Power Sys. Research 77:55–63
Hong YY, Wang CW (2005) Switching detection/classification using discrete wavelet transform and self-organizing mapping network. IEEE Transactions on Power Delivery 20:1662–1668
Jain T, Srivastava L, Singh SN (2003) Fast voltage contingency using radial basis neural networks. IEEE Transactions on Power Systems 18:1359–1366
Kottick D (1996) Neural-networks for predicting the operation of an underfrequency load shedding system. IEEE Trans. Power Systems, 10:1350–1370
Rajan CCA, Mohan MR, Manivannan K (2003) Neural-based tabu search method for solving unit commitment problem. IEE Proc. Generation, Transmission, Distribution 150:469–474
Ranaweera DK, Karady GG (1994) Active power contingency ranking using a radial basis function network. Int. J. Eng. Int. Syst. Elect. Eng. Comm. 2: 201–206.
Sasaki H, Watanabe M, Yokoyama R (1992) A solution method of unit commitment by artificial neural networks. IEEE Trans. Power Systems, 7:974–981
Thalassinakis EJ, Dialynas EN, Agoris D (2006) Method combining ANNs and monte carlo simulation for the selection of load shedding protection strategies in autonomous power systems. IEEE Trans. Power Systems, 21:1574–1582
Yang Y, Tai NL, Yu WY (2006) ART artificial neural networks based adaptive phase selector. Electric Power Systems Research 76:115–120

### 3.11.11 Power Systems Security

Chauhan S (1998) Artificial neural network application to power system security assessment. Ph.D. dissertation, Dept. of Electrical Engineering, University of Roorkee, Roorkee, India
Chauhan S (2005) Fast real power contingency ranking using counter propagation network: feature selection by neuro-fuzzy model. Electric Power Systems Research 73:343–352
Chauhan S, Dave MP (2002) ANN for transmission system static security assessment. Electric Power Energy Syst. 24:867–873
El-Keib AA, Ma X (1995) Application of artificial neural networks in voltage stability assessment. IEEE Transactions on Power Systems 10:1890–1896

Jensen CA, El-Sharkawi MA, Marks RJ (2001) Power system security assessment using neural networks: feature selection using Fisher discrimination. IEEE Transactions on Power Systems 16:757–763

Kim H, Singh C (2005) Power system probabilistic security assessment using Bayes classifier. Electric Power Systems Research 74:157–165

Neibur D, Germond A (1991) Power system static security assessment using kohonen neural network classifier. IEEE Trans. Power Systems 7:865–872

Sidhu TS, Cui L (2000) Contingency screening for steady-state security analysis by using FFT and artificial neural network. IEEE Transactions on Power Sytems 15:421–426

Sobajic DJ, Pao YH (1989) Artificial neural-net based dynamic security assessment for electric power systems. IEEE Trans. Power Systems 4:220–228

### *3.11.12 Power Systems Reliability*

Yu DC, Nguyen TC, Haddawy P (1999) Bayesian network model for reliability assessment of power systems. IEEE Trans. Power Syst. 14:426–432

### *3.11.13 Stability Analysis*

Aghamohammadi M, Mohammadian M, Saitoh H (2002) Sensitivity characteristic of neural network as a tool for analyzing and improving voltage stability. In: Proc. Transm. and Distrib. Conf., 2:1128–1132

Arya LD, Titare LS, Kothari DP (2006) Determination of probabilistic risk of voltage collapse using radial basis function (RBF) network. Electric Power Systems Research 76:426–434

Chauhan S, Dave MP (1997) Kohonen neural network classifier for voltage collapse margin estimation. Electric Machines Power Syst. 25:607–619

El-Keib AA, Ma X (1995) Application of artificial neural networks in voltage stability assessment. IEEE Transactions on Power Systems 10:1890–1896

Hsu YY, Chen CR (1991) Tuning of power system stabilizers using an artificial neural network. IEEE Trans. on Energy Conversion 6:612–619

Jeyasurya B (2000) Artificial neural networks for online voltage stability assessment. In: IEEE Power Engg. Society Summer Meeting, 4:2014–2018

Repo S (1999) General framework for neural network based real-time voltage stability assessment of electric power system. In: Proc. 1999 IEEE Midnight-Sun Workshop on Soft Computing methods on Ind. App., pp. 91–96

Sawhney H, Jeyasurya B (2006) A feed-forward artificial neural network with enhanced feature selection for power system transient stability assessment. Electric Power Systems Research 76:1047–1054

### 3.11.14 Renewable Energy

Al-Alawi A, Al-Alawi SM, Islam SM (2006) Predictive control of an integrated PV-diesel water and power supply system using an artificial neural network. Renewable Energy (In Press)

Bahgat ABG, Helwa NH, Ahamd GE, El Shenawy ET (2004) Estimation of the maximum power and normal operating power of a photovoltaic module by neural networks. Renewable Energy 29:443–457

Dutton AG, Kariniotakis G, Halliday JA, Nogaret E (1999) Load and wind power forecasting methods for the optimal management of isolated power systems with high wind penetration. Wind Engineering 23:69–87

Hontoria L, Aguilera J, Zufiria P (2005) A new approach for sizing stand alone photovoltaic systems based in neural networks. Solar Energy 78:313–319

Kalogirou SA (2001) Artificial neural networks in renewable energy systems applications: a review. Renewable and Sustainable Energy Reviews 5:373–401

Karatepe E, Boztepe M, Colak M (2006) Neural network based solar cell model. Energy Conversion and Management 47:1159–1178

Shen WX, Chan CC, Lo EWC, Chau KT (2002) A new battery available capacity indicator for electric vehicles using neural network. Energy Conversion & Management 43:817–826

Zhang L, Bai YF (2005) Genetic algorithm-trained radial basis function neural networks for modelling photovoltaic panels. Engineering Applications of Artificial Intelligence 18:833–844

### 3.11.15 Transformers

Booth C, McDonald JR (1998) The use of artificial neural networks for condition monitoring of electrical power transformers. Neurocomputing 23:97–109

Bastard P, Meunier M, Regal H (1995) Neural network-based algorithm for power transformer differential relays. Proc. Inst. Elect. Eng. 142:386–392

De A, Chatterjee N (2002) Recognition of impulse fault patterns in transformers using Kohonen's self-organizing feature map. IEEE Transactions on Power Delivery 17:489–494

Khalil ALON, Valencia JAV (1999) A transformer differential protection based on finite impulse response artificial neural network. Computers & Industrial Engineering 37:399–402

Mao PL, Aggarwal RK (2001) A novel approach to the classification of the transient phenomena in power transformers using combined wavelet transform and neural network. IEEE Transaction on Power Delivery 16:654–660

Mohamed EA, Abdelaziz AY, Mostafa AS (2005) A neural network-based scheme

for fault diagnosis of power transformers. Electric Power Systems Research 75:29–39
Perez LG, Flechsig AJ, Meador JL, Obradovic Z (1994) Training an artificial neural network to discriminate between magnetizing inrush and internal faults. IEEE Transactions on Power Delivery 9:434–441.
Thang KF, Aggarwal RK, McGrail AJ, Esp DG (2003) Analysis of power transformer dissolved gas data using the self-organizing map. IEEE Trans. Power Delivery 18:1241–1248
Wang MH (2003) Extension neural network for power transformer incipient fault diagnosis. IEE Proc. Generation, Transmission, Distribution 150: 679–685
Wang ZY, Liu YL, Griffin PJ (1998) A combined ANN and expert system tool for transformer fault diagnosis. IEEE Trans. Power Delivery 13:1224–1229
Yann CH (2003) Evolving neural nets for fault diagnosis of power transformers. IEEE Transactions on Power Delivery 18:843–848
Zaman MR, Rahman MA (1998) Experimental testing of an artificial neural network based protection of power transformer. IEEE Transactions on Power Delivery 13:510–517
Zhang Y, Ding X, Liu Y, Griffin PJ (1996) An artificial neural network approach to transformer fault diagnosis. IEEE Trans. Power Delivery 11:1836–1841

### *3.11.16 Rotating Machines*

Buhl MR, Lorenz RD (1991) Design and implementation of neural networks for digital current regulation of inverter drives. In: Conf. Rec.IEEE–IAS Annu. Meeting, pp. 415–423
Goedtel A, da Silva IN, Serni PJA (2007) Load torque identification in induction motor using neural networks technique. Electric Power Systems Research 77:35–45
Pinto JOP, Bose BK, Silva LEB, Karmierkowski MP (2000) A neural network based space vector PWM controller for voltage-fed inverter induction motor drive. IEEE Trans. on Ind. Appl. 36:1628–1636
Pryymak B, Moreno-Eguilaz JM, Peracaula J (2006) Neural network flux optimization using a model of losses in induction motor drives. Mathematics and Computers in Simulation 71:290–298
Ren TJ, Chen TC (2006) Robust speed-controlled induction motor drive based on recurrent neural network. Electric Power Systems Research 76:1064–1074
Song JW, Lee KC, Cho KB, Won JS (1991) An adaptive learning current controller for field–oriented controlled induction motor by neural network. In: Proc. IEEE–IECON'91, pp. 469–474
Yang BS, Han T, An JL (2004) ART-KOHONEN neural network for faults diagnosis of rotating machinery. Mech. System and Signal Process 18: 645–657
Ye Z, Sadeghian A, Wu B (2006) Mechanical fault diagnostics for induction motor

with variable speed drives using adaptive neuro-fuzzy inference system. Electric Power Systems Research 76:742–752

### 3.11.17 Power Quality

Abdel-Galil TK, El-Saadany EF, Salama MMA (2003) Power quality event detection using Adaline. Electric Power Systems Research 64:137–144

Ai Q, Zhou Y, Xu W (2006) Adaline and its application in power quality disturbances detection and frequency tracking. Elec. Power Sys Research (In Press)

Anis Ibrahim WR, Morcos MM (2002) Artificial intelligence and advanced mathematical tools for power quality applications: a survey. IEEE Transactions on Power Delivery 17:668–673

Cannas B, Celli G, Marchesi M, Pilo F (1998) Neural networks for power system condition monitoring and protection. Neurocomputing 23:111–123

Dash PK, Panda SK, Liew AC, Mishra B, Jena RK (1998) A new approach to monitoring electric power quality. Electric Power Systems Research 46:11–20

Qian A, Yuguang Z, Weihua X (2007) Adaline and its application in power quality disturbances detection and frequency tracking. Electric Power Systems Research 77:462–469

### 3.11.18 State Estimation

Schweppe FC, Wildes J, Rom DB (1970) Power system static-state estimation, parts I, II and III. IEEE Transactions on Power Apparatus and Systems PAS-89: 120–135

### 3.11.19 Energy Market

Conejo AJ, Contreras J, Espínola R, Plazas MA (2005) Forecasting electricity prices for a day-ahead pool-based electric energy market. International Journal of Forecasting 21:435–462

Gao F, Guan X, Cao XR, Papalexopoulos A (2000) Forecasting power market clearing price and quantity using a neural network. IEEE PES Summer Meeting, Seattle

Gareta R, Romeo LM, Gil A (2006) Forecasting of electricity prices with neural networks. Energy Conversion and Management 47:1770–1778

Hong YY, Lee CF (2005) A neuro-fuzzy price forecasting approach in deregulated electricity markets. Electric Power Systems Research 73:151–157

Szkuta BR, Sanabria LA, Dillon TS (1999) Electricity price short-term forecasting using artificial neural networks. IEEE Trans. Power Systems 14:2116–2120

Zhang L, Luh PB, Kasiviswanathan K (2003) Energy clearing price predication and confidence interval estimation with cascaded neural networks. IEEE Transactions on Power Systems 18:99–105

### 3.11.20 Power Electronics

Bose BK (1994) Expert system, fuzzy logic, and neural network applications in power electronics and motion control. Proceedings of the IEEE 82:1303-1323
Bose BK (2000) Fuzzy logic and neural networks in power electronics and drives. IEEE Industry Applications Magazine 6:57–63
Lin BR, Hoft RG (1994) Neural networks and fuzzy logic in power electronics. Control Engineering Practice 2:113–121
Lin BR (1997) Analysis of neural and fuzzy-power electronic control. IEE Proceedings: Science, Measurement and Technology 144:25–33
Tey LH, So PL, Chu YC (2002) Neural network-controlled unified power quality conditioner for system harmonics compensation. In Proc. IEEE/PES Transm. Distrib. Conf., Japan
Tey LH, So PL, Chu YC (2005) Adaptive neural network control of active filters. Electric Power Systems Research 74:37–56
Vazquez JR, Salmeron PR (2001) Neural network application to control an active power filter. In Proc. IEEE Power Electronics Applications Conf. Austria

# Chapter 4
# Support Vector Machine

## Section I: Theory

## 4.1 Introduction

The classical regression and Bayesian classification statistical techniques stand upon a strict assumption that the underlying probability distribution is known. However, in real life, oftentimes we are confronted with distribution-free regression or classification tasks with only recorded training patterns which are high-dimensional and empty in nature.

Support vector machine (SVM) is one of the relatively new and promising methods for learning separating functions in pattern recognition (classification) tasks, or for performing function estimation in regression problems. SVMs were originated from the statistical learning theory (SLT) by Vapnik (Vapnik 1995) for 'distribution-free learning from data'.

### *4.1.1 History and Background*

Vladimir Vapnik and Alexey Chervonenkis developed the Vapnik-Chervonenkis theory (also known as VC theory) during 1960–1990 (Vapnik and Chervonenkis 1971). VC theory is a computational learning theory related to the statistical learning theory from distribution-free data. This was the basis and starting point for the support vector machine (SVM), a supervised learning technique for classification and regression. Although Vapnik (Vapnik 1995, 1998) introduced the subjects of linear classifier and optimal separating hyperplanes in the 1960s, it did not receive much attention until the nineties. Bernhard Boser, Isabelle Guyon and Vapnik (Boser et al. 1992) introduced a way to create nonlinear classifiers by applying the kernel trick to maximum-margin hyperplanes. The kernel trick is a method for converting a linear classifier algorithm into a nonlinear one by using a nonlinear function to map the original observations into a higher-dimensional space, originally proposed by Aizerman (Aizerman et al. 1964). SVMs have been successfully applied in various classification and machine learning applications. In 1996, Vapnik, Harris Drucker,

Chris Burges, Linda Kaufman and Alex Smola proposed a version of SVM for utilization in the regression tasks (Drucker et al. 1996). This is popularly known as support vector regression (SVR). SVM is an expanding field in the machine learning field. SVM incorporates different ideas like the VC-theory, statistical learning, maximum margin optimal hyperplane, kernel tricks and so on. This makes it particularly powerful over the traditional empirical risk minimization-based methods like the neural networks, etc, as we shall see later.

### *4.1.2 Applications*

Since the nineties, support vector machines have been widely used in different machine learning, particularly classification as well as regression tasks. SVM is a growing field with more and more different kinds of application being explored. Some of these are mentioned below.

- Image processing
  - Object recognition
  - 3D object recognition
  - Face, facial expression recognition
  - Hand-written character recognition
  - Content-based image retrieval
  - Image clustering
- Speech processing
  - Speaker identification/verification
  - Speech recognition, synthesis
- Time-series prediction
  - Financial/currency prediction
  - Weather forecast
  - Load forecast
  - Marketing database
- Data mining
  - Data classification
  - Decision tree modeling
  - Text classification
  - Feature extraction
  - Noise classification and reduction
  - Signal conditioning
- Power systems
  - Fault classification
  - Disturbance analysis
  - Load forecasting

- Process Control
  - Density estimation
  - Generalized predictive control
  - Model predictive control
  - ANOVA decomposition
- Automotive
  - Combustion engine knock detection
  - Engine misfire detection
- Security
  - Intrusion detection
- Bioinformatics
  - Protein structure prediction
  - Breast cancer prognosis
  - Detection of remote protein homologies
  - Classification of microarray gene expression data
- Environmental sciences
  - Geo(spatial) and spatio-temporal environmental data analysis
  - Seismology: modeling of seismic liquefaction potential
  - Determining layered structure of earth
- High-energy Physics
  - Particle and quark flavor identification
  - Charmed quark detection
- Chaos Theory
  - Dynamic reconstruction of chaotic systems
  - Forecasting using chaos theory-based model.

### *4.1.3 Pros and Cons*

Any new field comes with certain advantages and disadvantages. We highlight some specific advantages and disadvantages associated with the SVMs.

- Pros
  - Solution for nonlinear and unknown systems
  - Distribution-free learning from data
  - No overfitting, robust to noise
  - Better convergence guarantee
  - Very effective for sparse, high-dimensional data.

- Cons
  - Binary classification. For multi-class classifications, pair-wise classifications can be used (one class against all others, for all classes) which might not be optimum.
  - Computationally expensive, large memory requirement, runs slow
  - Training might be slow.

## 4.2 Basics about Statistical Learning Theory

Support vector machines are learning algorithm based on the statistical learning theory, overcoming certain typical problems like the overfitting (see Sect. 3.4.9.1) in machine learning. Therefore, before going into the details of support vector machine, it will be worthwhile to have a short look at the statistical learning theory and the associated topics.

### *4.2.1 Machine Learning & Associated Problem*

Machine learning, in its simplest meaning, means mapping the input-output relationship from the observed data pairs (input-output) with the hope that this learned mapping would deduce the system response for unknown condition (unknown input data). However, the biggest challenge in such a training operation is the trade off between the accuracy and the generalization. This can be seen as having a learning machine which attains a particular *accuracy* level for particular training sets, with certain *capacity* to be able to learn any other (probably unseen) training set without error. Thus, the two main things to notice are the accuracy and the capacity.

Traditional neural network approaches have had difficulties in attaining appropriate generalization capacity in spite of achieving accuracy. Later, we will discuss in details about the difficulties of neural networks in a comparative manner over other approaches. Nevertheless, the goal in machine learning is to choose a model (input-output mapping) from the hypothesis space, which is closest (with respect to some error measure) to the underlying function in the target space (Gunn 1998). A model with too much capacity is overtrained (also referred as overfitted), while a model with very less capacity is undertrained or underfitted (see Chap. 3, Sect. 3.4.9 in context to the neural network). An overtrained model is like a hypothetical dog trainer who trained so many German shepherds (alsatian) that when presented with a greyhound, he concluded that it was not a dog because it did not match his known mapping for canine characteristics! In other way, an undertrained dog trainer when presented with a cat accepts it as a dog because it has four legs! This is the striking point of balancing the accuracy and the generalization in a learning process. Figure 4.1 depicts the over- and underfitting characteristics. Different fittings are shown in Fig. 4.1 for the two classes of square and circular points. A test square

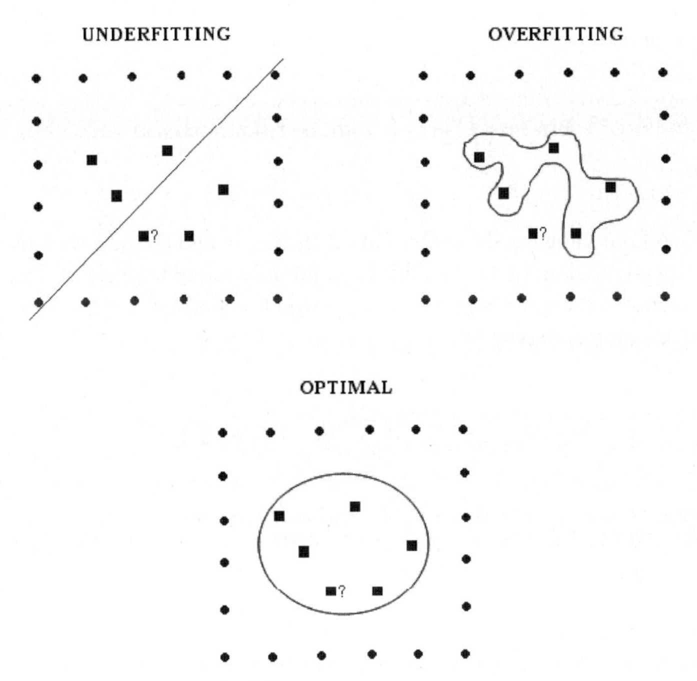


**Fig. 4.1** Example of over- and underfitting

point (with a question mark) is checked for correct classification. For the top two plots (under- and overfitting case), the test point is incorrectly classified while the optimal fitting gives the correct classification. This indicates that neither the under- nor overfitting is a desired option. The theory of statistical learning takes on from here.

## *4.2.2 Statistical Learning Theory*

Statistical learning theory (also known as SLT) is primarily concerned about the relation between the capacity of a learning machine and its performance (Burges 1998). Suppose we are given $n$ observations, each consisting of an input-output data pairs $\{x_i, y_i, i = 1, 2, \ldots, n\}$, where $x_i$'s are the input data and $y_i$'s are the corresponding 'true' output data (from a trusted source). In a deterministic learning process, the machine can be defined to achieve the functional mapping $x_i \mapsto f(x_i, \alpha)$, governed by the adjustable learning parameter(s) $\alpha$. For example, in case of a neural network, $\alpha$ corresponds to the weights and biases, for fuzzy logic-based modeling $\alpha$ corresponds to the fuzzy variables and rules, and so on.

Regarding these input-output data pairs, we also assume that there exists some unknown probability distribution $P(x, y)$, with probability density function (pdf) $p(x, y)$, from which these data are drawn independently and identically distributed (i.i.d). Then, the expectation of the test error or risk ($R$) can be cited in terms of the learning parameter $\alpha$ as

$$R(\alpha) = \int_{x,y} \tfrac{1}{2}|y_i - f(x_i, \alpha)|p(x, y)dxdy. \tag{4.1}$$

However, without knowing the probability distribution and the pdf, we cannot compute the expected risk in (4.1). And this is commonly found in practice. The alternative is to compute the empirical risk ($R_{emp}$) which is nothing but the average error rate on the training data pairs,

$$R_{emp}(\alpha) = \frac{1}{2n}\sum_{i=1}^{n}|y_i - f(x_i, \alpha)|. \tag{4.2}$$

It is evident from (4.2) that $R_{emp}$ does not depend on the underlying probability distribution, and is a fixed number for a particular choice of $\alpha$ and training set $\{x_i, y_i, i = 1, 2, \ldots, n\}$ (Burges 1998). The quantity on the right hand side of (4.2) is referred to as *loss* which depends on the range of the output.

Let us choose some parameter $\eta$ such that $0 \leq \eta \leq 1$. It is to be noted that with proper scaling it is easily possible to associate the parameter $\eta$ with the loss (in ideal case i.e., for no error $\eta = 0$, while in worst case, $\eta = 1$). For losses following the parameter $\eta$, Vapnik proved that the following bound (Vapnik 1995) holds with probability $1 - \eta$.

$$R(\alpha) \leq R_{emp}(\alpha) + \sqrt{\left(\frac{h(\log(2n/h) + 1) - \log(\eta/4)}{n}\right)}, \tag{4.3}$$

where, $h$ a nonnegative integer, is called the Vapnik Chervonenkis (VC) dimension (Vapnik and Chervonenkis 1971). VC dimension is a quantitative measure of the capacity of the learning machine.

It can be easily noted from the expression of the VC bound in (4.3) that it does not depend on the probability distribution $P(x, y)$. Also, it is noticeable that knowing the VC dimension $h$, we can compute the second term in the right hand side of (4.3), enabling us to compute the expected risk of the left hand side, which otherwise is not usually computable.

From (4.3), we can intuitively see that we would like to have a learning process with the minimal expected risk (left hand side of (4.3)). From the right hand side of (4.3), we can interpret this as choosing a learning machine (having particular empirical risk) which gives the lowest upper bound (Burges 1998). This way, we have a structure for the risk and we try to minimize it without knowing the underlying probability distribution. This refers to the structural risk minimization (SRM) principle. Before going into the details of SRM, let's see something about the all important VC dimension.

### 4.2.3 Vapnik Chervonenkis (VC) Dimension

As mentioned before, the VC dimension is a scalar value that measures the capacity of a set of functions. Before discussing about the VC dimension, it is necessary to understand the concept of shattering of points. If a given set of $n$ points can be categorized in all $2^n$ possible ways, and for each categorization, we can find a member of the set of functions which perfectly separates the set of points, we say that the set of points is shattered by the set of functions.

We see an example in Fig. 4.2. In Fig. 4.2, we see a set of three points in two dimensional space ($\mathbf{R}^2$). The separating function is a straight line. For all the different combinations, we see that any member of the set of function can separate out 2 points (say class 1) from the remaining 1 point.

The VC dimension $h$ of a set of functions $\{f(x, \alpha)\}$ is defined as the maximum number of training points that can be shattered by the members of the set of functions (Vapnik 1995; Burges 1998). Therefore, for the example shown in Fig. 4.2, the VC dimension of the set of oriented lines in $\mathbf{R}^2$ is $3(= 2 + 1)$. The separating function in $\mathbf{R}^2$ is the set of oriented lines. In general, in an $n$-dimensional space $\mathbf{R}^n$, the separating functions would be set of oriented hyperplanes. And it can be shown that the VC dimension of the set of the hyperplanes in $\mathbf{R}^n$ is $n + 1$ (Burges 1998). Please note that for the $\mathbf{R}^2$ example above, we wrote the VC dimension as $2 + 1$, conforming to this general relation.

The VC dimension is an indicator of the (shattering) capacity of the set of functions, not only for the indicator functions like the one shown in Fig. 4.2, but also for real-valued functions. The VC dimension of a class of real-valued functions $\{Q(x, \alpha)\}$ is defined to be the VC dimension of the indicator class $\{I(Q(x, \alpha) - \beta > 0)\}$, where $\beta$ takes values over the range of $Q$ (Hastie et al. 2001). Figure 4.3 shows an example of the VC dimension for real-valued function.

From the definition of the VC dimension, it might appear that the VC dimension for learning machines (functions) with many parameters is higher than learning machines with few parameters. However, there are examples of univariate functions with single parameter having infinite VC dimension, i.e., the family of classifiers capable of shattering $l$ points, no matter how large $l$ (Vapnik 1995; Burges 1998). One classic example is the sinusoid, $f(x, \omega) = I(\sin \omega x > 0)$, the corresponding VC dimension $h = \infty$. Figure 4.4 shows the example of the infinite VC dimension

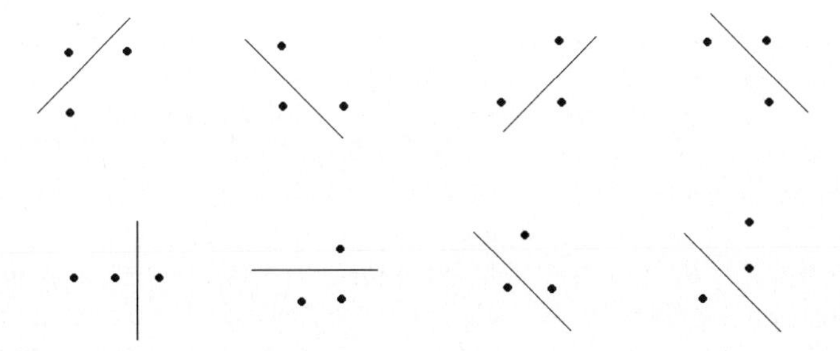

**Fig. 4.2** Shattering of points in a two-dimensional space

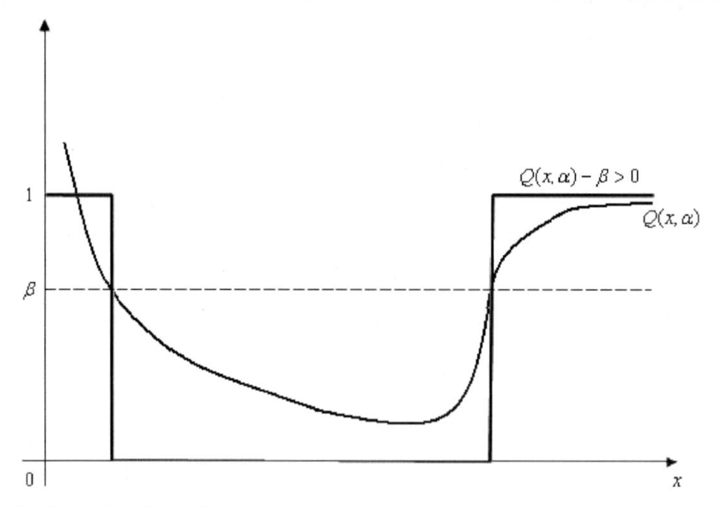


**Fig. 4.3** VC dimension for real-valued functions

of the sinusoidal function which can shatter an arbitrarily large number of points by choosing an appropriately high frequency $\omega$, the only parameter.

We can compute the VC bound or confidence (second term on the right hand side of (4.3)) knowing the VC dimension $h$. Figure 4.5 shows the plot of the VC confidence against the VC dimension $h$, which is a monotonically increasing function (Burges 1998). From (4.3) and the nature of the curve in Fig. 4.5, the following inference can be cited. For any nonzero empirical risk, the optimal learning machine would have associated set of functions with minimal VC dimension. This ensures minimization of the right hand side of (4.3), leading to a better upper bound on the actual error (Burges 1998).

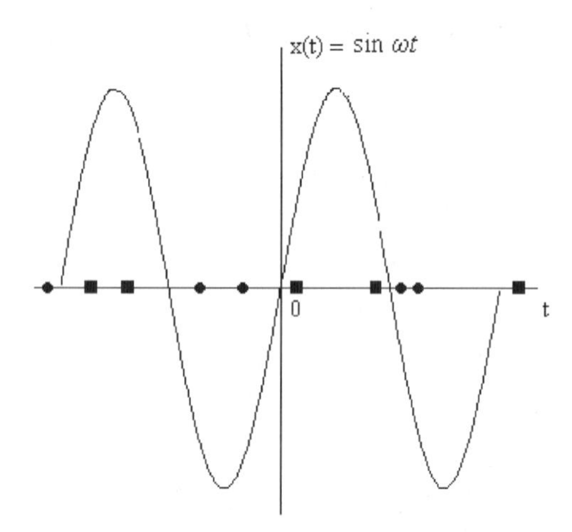


**Fig. 4.4** Sinusoidal function with an infinite VC dimension

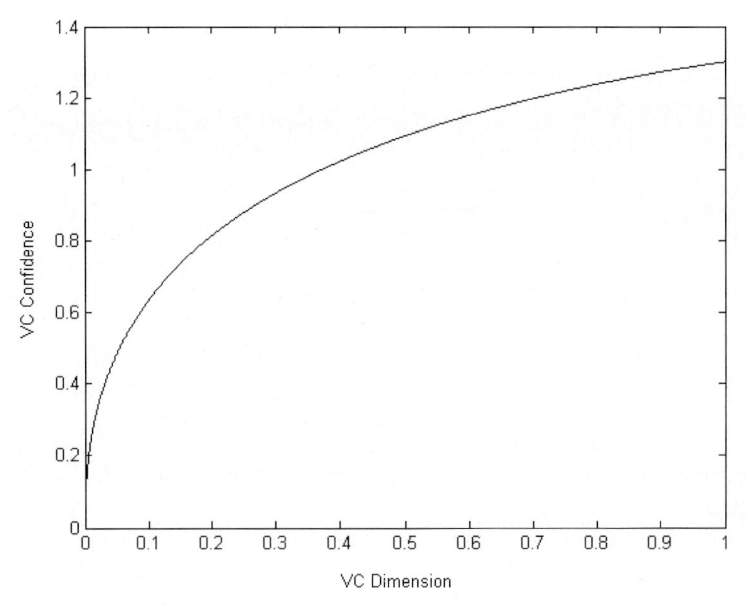


**Fig. 4.5** VC confidence is monotonically increasing for the VC dimension $h$

## *4.2.4 Structural Risk Minimization*

The main objective of the structural risk minimization (SRM) principle is to choose the model complexity optimally for a given training sample. For determining the optimal complexity, we note that the first term in the right hand side of (4.3) depends on a particular function from the set of functions. The second term in the right hand side of (4.3) depends mainly on the VC dimension of the set of functions. So, the goal is to find the subset of the chosen set of functions such that the risk bound for that subset is minimized. As the VC dimension $h$ is an integer, we cannot arrange things so that $h$ varies smoothly. Under SRM, the set $S$ of loss functions $\{Q(x, \alpha)\}$ has a structure (Vapnik 1995). As shown in Fig. 4.6, the structure consists of nested subsets $S_k$ such that

$$S_1 \subset S_2 \subset \ldots \subset S_k \subset \ldots, \text{ where } h_1 < h_2 < \ldots < h_k < \ldots. \tag{4.4}$$

From (4.3), it is to be noted that the VC confidence can be seen as the (estimated) difference between the true risk and the empirical risk. With SRM and the nested subset structure, we try to find the model with the right balance in order to avoid the over- and underfitting. Figure 4.7 shows the connection between the SRM-nested subset structure and how it is used to choose the optimal model with the balance of accuracy and generalization capacity.

The SRM principle can be implemented by training a series of learning machines, one for each subset, where for a given subset the goal of training is to minimize the empirical risk (Burges 1998). From the series of trained machines, the optimal one

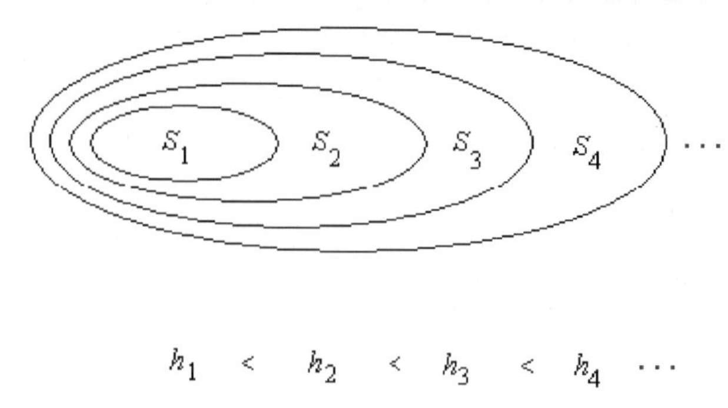


**Fig. 4.6** Nested subsets of functions at par with the VC dimension

can be found for which the sum of the empirical risk and the VC confidence is minimum.

Multiple output problems can usually be reduced to a set of independent single output problems. Hence, irrespective of the number of outputs, the SRM principle can, in general, be applied to deduce the optimal model from multiple inputs.

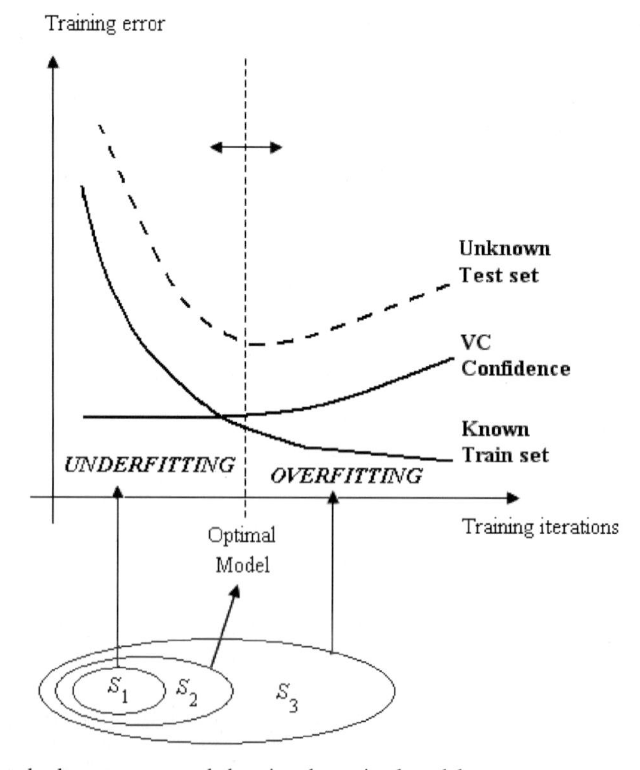


**Fig. 4.7** Nested subset structure and choosing the optimal model

## 4.3 Support Vector Machine

### 4.3.1 Linear Classification

The aim of linear classification task is to construct linear decision boundaries that explicitly try to separate data into different classes. Figure 4.8 shows an example of mixture of two different data classes represented using the plus and the minus signs.

As shown in Fig. 4.8, in two-dimensional space, the separating hyperplane would be a straight line which can be defined, in general, by the weight vector $w$ and the bias vector $b$. So, it can be generally represented as

$$w^T x + b = 0, \tag{4.5}$$

where, the superscript $T$ indicates the transpose. In the classified data classes, the plus-sign data class can be represented as class 1 while the minus-sign data class can be represented as class $-1$. Hence, class 1 and $-1$ can be described by (4.6) and (4.7) respectively.

$$w^T x + b = 1. \tag{4.6}$$

$$w^T x + b = -1. \tag{4.7}$$

Concentrating on the example shown in Fig. 4.8, and the general hyperplane, and the two classes described by (4.5–4.7), we can observe the following equations corresponding to the five data points.

$$w_1.1 + w_2.1 + b = -1, \tag{4.8}$$

$$w_1.2 + w_2.1 + b = -1, \tag{4.9}$$

$$w_1.2 + w_2.2 + b = -1, \tag{4.10}$$

$$w_1.4 + w_2.4 + b = 1, \tag{4.11}$$

$$w_1.5 + w_2.5 + b = 1. \tag{4.12}$$

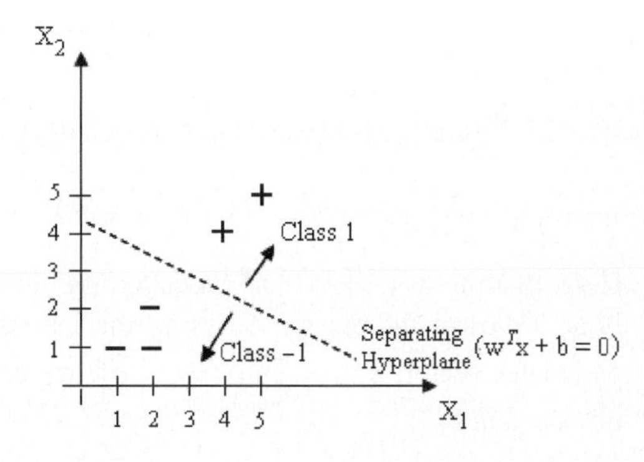


**Fig. 4.8** Example of linear data classification

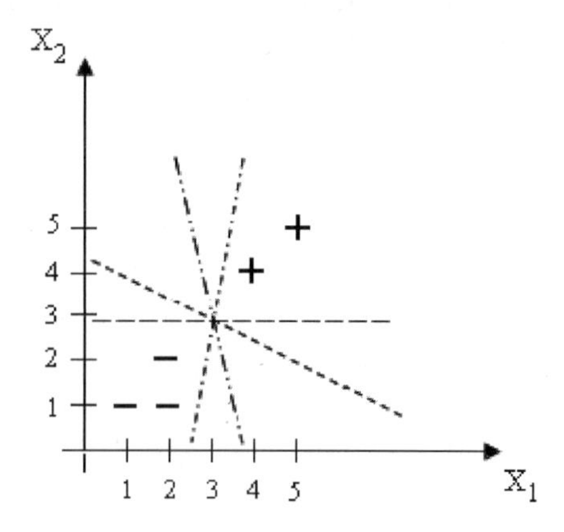


**Fig. 4.9** Infinite solution to the classification problem

So, we have three unknown variables, $w_1, w_2, b$, and five equations. Hence, we cannot have a unique solution. In fact, for the data classes shown in Fig. 4.8, we have infinite number of classification solutions, i.e., possible separating hyperplanes as shown in Fig. 4.9. So, how to choose the optimal one? This is the question that support vector tries to answer as we shall see later.

#### 4.3.1.1 Least Square Solution

We can represent (4.8–4.12) in the matrix form as

$$\begin{bmatrix} 1 & 1 & 1 \\ 2 & 1 & 1 \\ 2 & 2 & 1 \\ 4 & 4 & 1 \\ 5 & 5 & 1 \end{bmatrix} . \begin{bmatrix} w_1 \\ w_2 \\ b \end{bmatrix} = \begin{bmatrix} -1 \\ -1 \\ -1 \\ 1 \\ 1 \end{bmatrix}. \tag{4.13}$$

Equation (4.13) can be represented as

$$A.\beta = y. \tag{4.14}$$

Therefore, the least square estimate of the parameter vector $\beta$ would be

$$\hat{\beta} = A^{-1}y. \tag{4.15}$$

Here, $A^{-1}$ represents the matrix inversion operation.

$$A^{-1}y = (A^T)^{-1}A^{-1}A^Ty = (A^TA)^{-1}A^Ty = pinv(A).y. \tag{4.16}$$

Here, there term $(A^TA)^{-1}A^T$ is called the *pseudo-inverse* (*pinv*) (Golub and Van Loan 1996) of the matrix $A$. Using the least square estimate (4.15–4.16) of the parameter vector, $\hat{\beta}^T = \begin{bmatrix} \hat{w}_1 & \hat{w}_2 & \hat{b} \end{bmatrix}$, we can obtain the classification for any unseen data as

$$y = \text{sign}(w^T x + b). \tag{4.17}$$

However, we have a three-dimensional hyperplane for a two-dimensional classification problem.

#### 4.3.1.2 Perceptron Learning

Perceptron learning (Schölkopf and Smola 2001) is oriented towards finding a separating hyperplane by minimizing the distances of the misclassified points to the decision boundary. Misclassified points mean the points which are categorized into wrong classes. This is shown in Fig. 4.10, by the circled data points which lie in the wrong data class. So, intuitively we can imagine that minimizing the distances of such misclassified points brings us close to the minimum error condition, i.e., towards the optimal solution.

Continuing further the description of the two-dimensional classification problem from the previous section, now we additionally introduce these misclassified points. So, we have two misclassification cases.

Case 1: If a response $y_i = 1$ is misclassified, then $w^T x + b < 0$,
Case 2: If a response $y_i = -1$ is misclassified, then $w^T x + b > 0$.

Combining the two cases, the goal is

$$\min \left[ -\sum_{i \in D} y_i (w^T x_i + b) \right], \text{ with respect to } w, b, \tag{4.18}$$

where, $D$ is the set of misclassified points. We can define the entity to minimize as

$$J = -\sum_{i \in D} y_i (w^T .x_i + b) = -\sum_{i \in D} y_i (x_i^T .w + b). \tag{4.19}$$

In (4.19), as $w$ and $x$ are operated using a scalar dot product, hence they are interchangeable. $J$ is non-negative with respect to misclassification and proportional

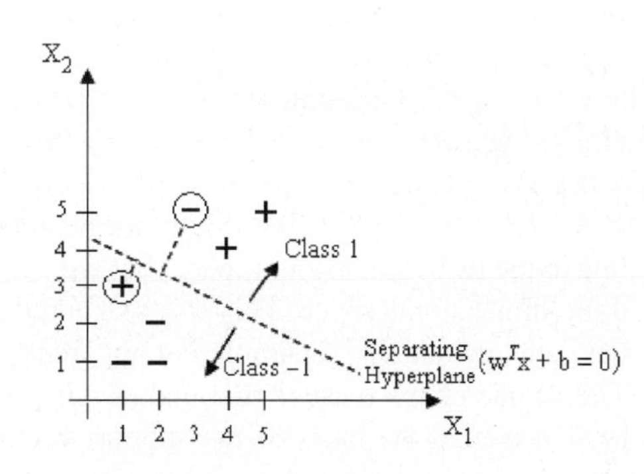


**Fig. 4.10** Misclassified points

to the distance (perpendicular) of the misclassified points (shown by the dashed lines in Fig. 4.10) to the decision boundary which is defined by (4.5).

Using the gradient descent algorithm (see Chap. 3, Sect. 3.3.2), we get

$$\frac{\partial J}{\partial w} = -\sum_{i \in D} y_i x_i, \tag{4.20}$$

$$\frac{\partial J}{\partial b} = -\sum_{i \in D} y_i. \tag{4.21}$$

So, following the update rule (see (3.5) in Chap. 3, Sect. 3.3.1) we have

$$w_{i_{new}} = w_i + (-\alpha).\frac{\partial J}{\partial w} = w_i + \sum_{i \in D} \alpha y_i x_i. \tag{4.22}$$

$$b_{new} = b + (-\alpha).\frac{\partial J}{\partial b} = b + \sum_{i \in D} \alpha y_i. \tag{4.23}$$

Here, $\alpha$ is the learning rate.

### 4.3.2 *Optimal Separating Hyperplane*

From the classification example in Sect. 4.3.1, we've seen that the separating hyperplane is not unique in nature. There are many solutions when data are separable. There are three associated problems.

1. If data are separable, the solution found depends on the starting values, as can be seen in the cases of the least square solution and the perceptron learning.
2. Smaller the gap between the separable data classes, larger the time to find the separating hyperplane.
3. If data are not separable, the algorithm does not converge and cycles develop.

For the separable data, problem 1 can be eliminated by adding additional constraints to the separating hyperplane. Problem 2 can be eliminated by seeking a hyperplane not in the original space, but in a much enlarged spaced obtained by transformation. For example, we change the geometry of the original given two-dimensional input data set $x_i = \{x_1, x_2\}$ by projecting it into a higher dimensional space using a transformation to have new three-dimensional input data set, say, $X_i = \{X_1, X_2, X_1X_2\}$. We might not be able to find a suitable separating straight line in the two-dimensional space, but we could easily find a separating plane in the three-dimensional space. This way, we could also sometimes solve the problem 3 if the data are actually separable but apparently non-separable in a lower dimension. This is one of the reasons that makes this projecting and transformation technique (which is also the basis of the support vector machines) more robust in terms of convergence than the neural network-based techniques. We will discuss more about this later.

#### 4.3.2.1 Margin

For tackling the problem of choosing the optimal hyperplane from the many possible options one key parameter is the margin. In other words, the optimal separating hyperplane separates the classes (two in case of the problem in Sect. 4.3.1) and maximizes the distance between the closest point from each class, i.e., maximizes the margin. Intuitively, this is obvious because we are more confident about an optimal separation if we can maintain a maximum safety gap (margin). The effect of the margin is shown in Fig. 4.11.

#### 4.3.2.2 Affine Set

Figure 4.12 shows a separating hyperplane and two data points on it.

From Fig. 4.12, we have the following description of the two data points.

$$W^T x_1 + b = 0. \tag{4.24}$$

$$W^T x_2 + b = 0. \tag{4.25}$$

Subtracting (4.25) from (4.24), we get

$$w^T (x_1 - x_2) = 0. \tag{4.26}$$

Hence, the difference vector $(x_1 - x_2)$ of the two data points is lying in the affine set (Vapnik 1998) or the hyperplane defined by (4.5).

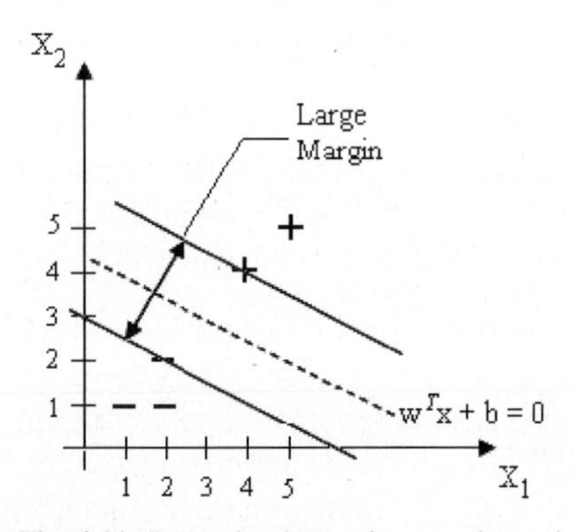


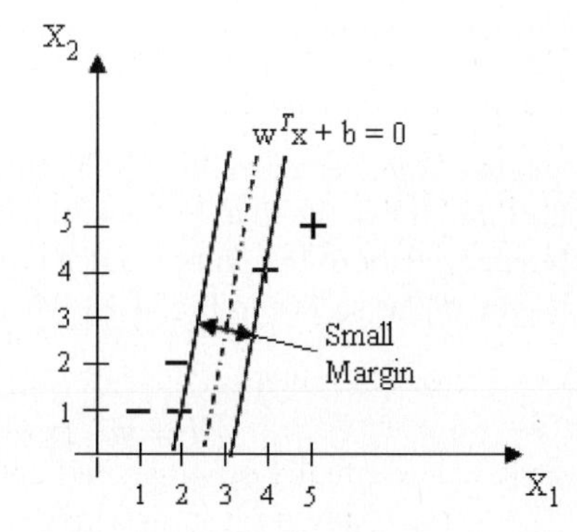


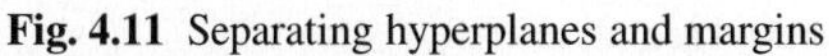

**Fig. 4.11** Separating hyperplanes and margins

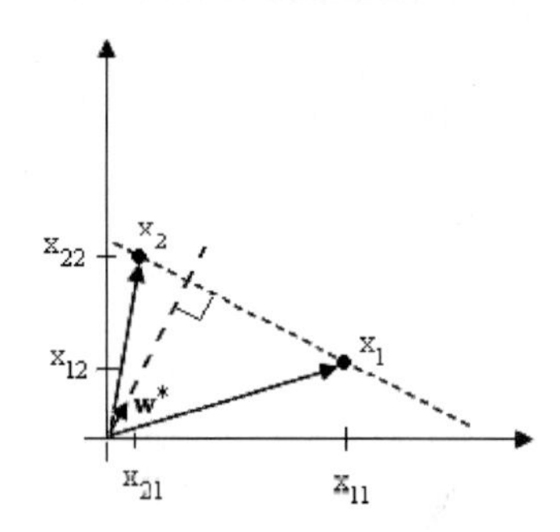


Fig. 4.12 Hyperplane with data points lying on it

#### 4.3.2.3 Orthogonal Vector

Two vectors (**a**, **b**) are said to be orthogonal if the scalar (dot) product is zero.

$$\mathbf{a}^T.\mathbf{b} = 0. \tag{4.27}$$

Following this, we can define a vector **w** which is normal to the surface $w^T x + b = 0$. This is the orthogonal vector to the separating hyperplane. The length of the orthogonal vector is

$$\|\mathbf{w}\| = \sqrt{\mathbf{w}^T\mathbf{w}} = \sqrt{\mathbf{w}^2}. \tag{4.28}$$

We can also define a unit vector $\mathbf{w}^*$ as

$$\mathbf{w}^* = \frac{\mathbf{w}}{\|\mathbf{w}\|}. \tag{4.29}$$

#### 4.3.2.4 Distance of a Point

Now, we want to deduce the distance of a point from the optimal separating hyperplane. Figure 4.13 shows a general point $x$ at a distance $d$ from the surface $w^T x + b = 0$.

The distance between two vectors is given by the scalar (dot) product equal to the length of one vector multiplied by the algebraic projection of the another vector on the direction of the first. For determining the distance of the point $x$, we utilize a difference vector of the point $x$ and a point $x_0$ on the affine set and the unit orthogonal vector $\mathbf{w}^*$, as shown in Fig. 4.13. Then, we have,

$$d = \mathbf{w}^{*^T}(\mathbf{x} - \mathbf{x}_0) = \frac{\mathbf{w}^T}{\|\mathbf{w}\|}(\mathbf{x} - \mathbf{x}_0), \tag{4.30}$$

as $x_0$ is a point on the affine set, $\mathbf{w}^T\mathbf{x}_0 + b = 0$. Hence,

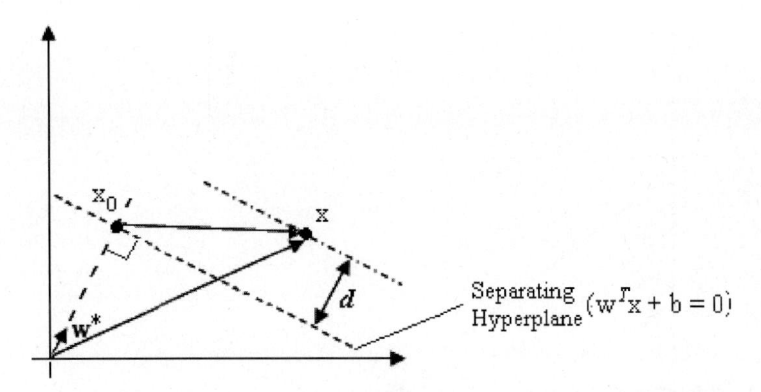


**Fig. 4.13** Distance of a point

$$d = \frac{\mathbf{w}^T\mathbf{x} - \mathbf{w}^T\mathbf{x}_0}{\|\mathbf{w}\|} = \frac{\mathbf{w}^T\mathbf{x} + b}{\|\mathbf{w}\|}. \tag{4.31}$$

We can generalize (4.31) as follows. Let there be a given point $p(x_{1p}, x_{2p}, \ldots, x_{np})$ and a hyperplane $k(x, w, b) = 0$ defined by

$$w_1x_1 + w_2x_2 + \ldots + w_nx_n \pm b = 0. \tag{4.32}$$

The distance $d_p$ from point $p$ to the hyperplane $k$ is given as

$$d_p = \frac{|(\mathbf{w}^T.\mathbf{x}_p) \pm b|}{\|\mathbf{w}\|} = \frac{|(\mathbf{w}.\mathbf{x}_p^T) \pm b|}{\|\mathbf{w}\|} = \frac{\|w_1x_{1p} + w_2x_{2p} + \ldots + w_nx_{np} \pm b|}{\sqrt{w_1^2 + w_2^2 + \ldots + w_n^2}}. \tag{4.33}$$

Let's see an example. We want to calculate the distance $d$ between the point (1,1,1,1,1) and the 5-dimensional hyperplane defined as $x_1 + 2x_2 + 3x_3 + 4x_4 + 5x_5 + 7 = 0$. Using (4.33),

$$d = \frac{|[1 \;\; 2 \;\; 3 \;\; 4 \;\; 5][1 \;\; 1 \;\; 1 \;\; 1 \;\; 1]^T + 7|}{\sqrt{1^2 + 2^2 + 3^2 + 4^2 + 5^2}} = \frac{22}{\sqrt{55}}.$$

#### 4.3.2.5 Distance of Margin

Figure 4.14 shows the optimal separating hyperplane separating the two classes $(1, -1)$ and the maximal margin. We define two data points $x_1$ and $x_2$ from the two classes on the two margins represented by the vectors $\mathbf{x}_1$ and $\mathbf{x}_2$. Figure 4.14 also shows the orthogonal vector $\mathbf{w}$ to the hyperplane. The distance between the margins is $d$. And the vectors $\mathbf{x}_1$ and $\mathbf{x}_2$ have angles of $\theta_1$ and $\theta_2$ with the orthogonal vector $\mathbf{w}$ respectively.

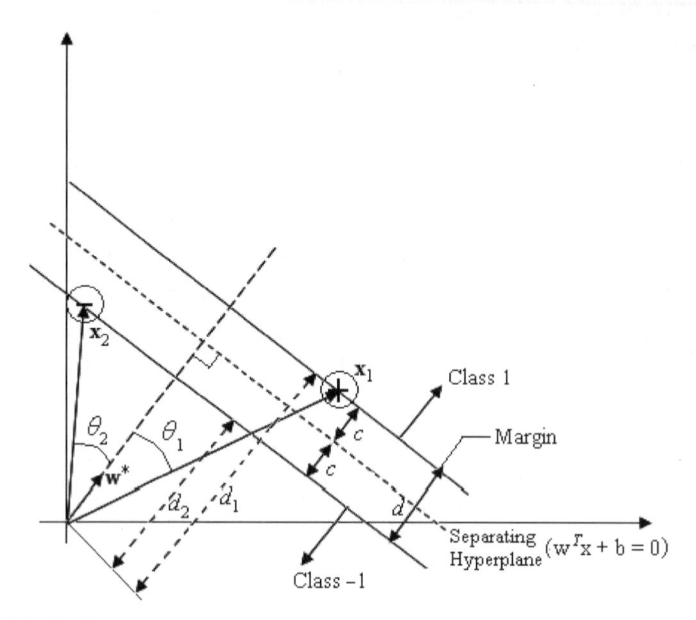


**Fig. 4.14** Margin distance

Then, the distance between the margins is given by

$$d = d_1 - d_2. \tag{4.34}$$

$$d_1 = |\mathbf{x}_1| \cos\theta_1. \tag{4.35}$$

$$d_2 = |\mathbf{x}_2| \cos\theta_2. \tag{4.36}$$

So,

$$d = |\mathbf{x}_1| \cos\theta_1 - |\mathbf{x}_2| \cos\theta_2. \tag{4.37}$$

We have

$$\cos\theta_1 = \frac{\mathbf{w}.\mathbf{x}_1^T}{|\mathbf{w}||\mathbf{x}_1|}, \tag{4.38}$$

$$\cos\theta_2 = \frac{\mathbf{w}.\mathbf{x}_2^T}{|\mathbf{w}||\mathbf{x}_2|}. \tag{4.39}$$

Hence,

$$d = \frac{\mathbf{w}.\mathbf{x}_1^T}{|\mathbf{w}|} - \frac{\mathbf{w}.\mathbf{x}_2^T}{|\mathbf{w}|} = \frac{\mathbf{w}.\mathbf{x}_1^T - \mathbf{w}.\mathbf{x}_2^T}{|\mathbf{w}|} \tag{4.40}$$

For class 1 we have

$$\mathbf{w}.\mathbf{x}_1^T + b = 1, \tag{4.41}$$

$$\therefore \mathbf{w}.\mathbf{x}_1^T = 1 - b.$$

For class $-1$ we have

$$\mathbf{w}.\mathbf{x}_2^T + b = -1, \tag{4.42}$$
$$\therefore \mathbf{w}.\mathbf{x}_2^T = -1 - b.$$

Using (4.40–4.42), we get

$$d = \frac{1 - b - (-1 - b)}{|\mathbf{w}|} = \frac{2}{|\mathbf{w}|} \tag{4.43}$$

From Fig. 4.14, we have the half-distance $c = d/2$. Hence, from (4.43)

$$c = \frac{d}{2} = \frac{1}{|\mathbf{w}|}. \tag{4.44}$$

### *4.3.3 Support Vectors*

SVMs can be seen as a new method for training classifiers, regressors, like polynomial models, neural networks, radial basis function, fuzzy models and so on. SVMs are based on the structural risk minimization (SRM) principle, whereas neural networks (refer to Chap. 3) are based on the empirical risk minimization (ERM) principle (that is why SVMs usually generalize better than the NNs). Unlike the classical adaptation algorithm that work on $L_1$ or $L_2$ norms, i.e., by minimizing an absolute error or an error-square respectively, the SVMs minimize the Vapnik Chervonenkis (VC) bounds (Vapnik and Chervonenkis 1971; Vapnik 1995). According to Vapnik's theory, minimizing the VC bounds means the expected probability of error is low, i.e., good generalization, and that too with unknown probability distribution.

For a separable classification task, the idea is to map the training data into a higher dimensional feature space using a kernel function where an optimal separating hyperplane (defined by the $w$: weight vector, $b$: bias) can be found that maximizes margin or the distance between the closet data points as discussed in Sect. 4.3.2.

For the two-class classification problem (Fig. 4.14) we have the indicator function defined by (4.17). For the class 1,

$$y = 1; \quad \mathbf{w}^T.\mathbf{x} + b \geq 0. \tag{4.45}$$

For the class $-1$,

$$y = -1; \quad \mathbf{w}^T.\mathbf{x} + b < 0. \tag{4.46}$$

So, correctly classified points are represented as

$$y\left(\mathbf{w}^T.\mathbf{x} + b\right) \geq 0. \tag{4.47}$$

Distance $d$ (of the points in either class) is defined by (4.31). So, for the correctly classified points we have

$$yd \geq 0. \tag{4.48}$$

For the optimal hyperplane, we need to find the maximum margin (defined by the half-distance $c = \frac{1}{|\mathbf{w}|}$ as in Sect. 4.3.2.5). So, in general, for $n$ data points, the optimization problem can be described as

$$\max c, \text{ with respect to} (\mathbf{w}, b). \tag{4.49}$$

Subject to the constraint

$$y_i d_i = \frac{y_i \left(\mathbf{w}^T.\mathbf{x}_i + b\right)}{|\mathbf{w}|} \geq c, \quad i = 1, 2, \ldots, n. \tag{4.50}$$

Substituting $c = \frac{1}{|\mathbf{w}|}$ in (4.50), the constraint becomes

$$y_i \left(\mathbf{w}^T.\mathbf{x}_i + b\right) \geq 1, \quad i = 1, 2, \ldots, n. \tag{4.51}$$

If we consider, in general, a monotonic function $\sqrt{f}$, then minimization of $\sqrt{f}$ is equivalent to the minimization of $f$. Consequently, a minimization of the norm $|\mathbf{w}|$ equals minimization of $\mathbf{w}^T\mathbf{w} = (\mathbf{ww}) = \sum_{i=1}^{n} w_i^2$ and this leads to a maximization of margin $d (= 2c)$. So, the transformed optimization problem (for finding the optimal hyperplane) is

$$\min \frac{1}{2}|\mathbf{w}|^2, \text{ with respect to } (\mathbf{w}, b), \tag{4.52}$$

subject to the constraint described by (4.51).

The $m$ number of points closest to the optimal separating hyperplane with the largest margin $2c = \frac{2}{|\mathbf{w}|}$, are specified as the *support vectors* (SVs) defined by

$$y_i \left(\mathbf{w}^T.\mathbf{x}_i + b\right) = 1, \quad i = 1, 2, \ldots, m. \tag{4.53}$$

#### 4.3.3.1 Interpretation of SV in Higher Dimension

In the higher dimension, the cost function (analogous to (4.52)) will be quadratic and the constraints (analogous to (4.51)) linear, i.e., it would a quadratic programming (Vapnik 1995) problem. The optimum separating hyperplane can be represented based on the kernel function (Schölkopf and Smola 2001) as

$$f(x) = sign\left[\sum_{i=1}^{n} \alpha_i y_i \psi(x, x_i) + b\right], \tag{4.54}$$

where, $n$ is the number of training samples, $y_i$ is the label value of the example $i$, $\psi$ represents the kernel function, and $\alpha_i$ coefficients must be found in a way to maximize a particular Lagrangian[13] representation. Subject to the constraints $\alpha_i \geq 0$ and $\sum \alpha_i y_i = 0 \; \forall i$, there is a Lagrange multiplier ($\alpha_i$) for each training point and only those training points that lie close to the decision boundary have non-zero $\alpha_i$. These are called the support vectors (SVs).

In real world problems, data are noisy and no linear separations are possible in the feature space. The hyperplane margin can be made more relaxed by penalizing the training points that the system misclassifies. Hence, the optimum separating hyperplane equation is defined as

$$y_i \left( \mathbf{w}^T \mathbf{x}_i + b \right) \geq 1 - \beta_i, \beta_i \geq 0, \tag{4.55}$$

where, $\beta_i$ introduces a positive *slack* variable that measure the amount of violation from the constraints.

So, the optimization criterion to obtain the (optimum) separating hyperplane is

$$\min \left[ \frac{1}{2} \|\mathbf{w}\|^2 + \mu \sum_{i=1}^{n} \beta_i \right]. \tag{4.56}$$

The penalty parameter $\mu$ is a regularization parameter that controls the trade off between maximizing the margin and minimizing the training error. Hence, the new learning machine utilizing the SVs, is called the support vector machine (SVM). SVM has two associated parameters:

1. the kernel function
2. the penalty parameter $\mu$.

This is called the *soft-margin* (Burges 1998) approach.

### *4.3.4 Convex Optimization Problem*

A convex optimization problem consists of a quadratic criterion with linear inequality constraints. So, for the support vector machines, we introduce a Lagrangian primal

---

[13] Lagrangian is a function of the dynamic variables of a system. In mathematical optimization problems, Lagrange multipliers are a method for dealing with constraints. We often encounter problem of extremizing a function of several variables subject to one or more constraints given by further functions of the variables to given values. The method introduces a new unknown scalar variable, the Lagrange multiplier, for each constraint; and forms a linear combination involving the multipliers as coefficients. This reduces the constrained problem to an unconstrained problem. It may then be solved, for example by the usual gradient method. For general description on Lagrangian see (Fletcher 1987).

$$L_P = \frac{1}{2}\mathbf{w}^T\mathbf{w} - \sum_{i=1}^{n} \alpha_i [y_i(\mathbf{w}^T\mathbf{x}_i + b) - 1], \tag{4.57}$$

where, $\alpha_i$'s are the Lagrange multipliers. So, the complete optimization problem for SVMs in terms of the Lagrangian function is given as follows.

$$\text{Objective function } L(\mathbf{w}, b, \alpha) = \frac{1}{2}\mathbf{w}^T\mathbf{w} - \sum_{i=1}^{n} \alpha_i \left[y_i\left(\mathbf{w}^T\mathbf{x}_i + b\right) - 1\right], \tag{4.58}$$

$$\text{Inequality constraints } y_i\left(\mathbf{w}^T.\mathbf{x}_i + b\right) \geq 1, \quad i = 1, 2, \ldots, n.$$

The Lagrangian $L$ in (4.58) must be minimized with respect to $(\mathbf{w}, b)$, and has to be maximized with respect to nonnegative $\alpha_i\,(\alpha_i \geq 0)$.

The solution of the problem in the primal space (space described by $\mathbf{w}$ and $b$) is as follows. We search for the saddle point (i.e., $(\mathbf{w}_0, b_0, \alpha_0)$) using (4.57). At the saddle point,

$$\frac{\partial L_P}{\partial \mathbf{w}} = \mathbf{w} - \sum_{i=1}^{n} \alpha_i y_i \mathbf{x}_i = 0. \tag{4.59}$$

$$\frac{\partial L_P}{\partial b} = -\sum_{i=1}^{n} \alpha_i y_i = 0. \tag{4.60}$$

Equation (4.59) gives

$$\mathbf{w} = \sum_{i=1}^{n} \alpha_i y_i \mathbf{x}_i. \tag{4.61}$$

And (4.60) gives

$$\sum_{i=1}^{n} \alpha_i y_i = 0. \tag{4.62}$$

Substituting (4.61–4.62) in (4.57), we transform the Lagrangian primal to dual variable Lagrangian ($L_D$).

$$L_D = \sum_{i=1}^{n} \alpha_i - \frac{1}{2}\sum_{i=1}^{n}\sum_{j=1}^{n} \alpha_i \alpha_j y_i y_j {\mathbf{x}_i}^T \mathbf{x}_j. \tag{4.63}$$

It is to be noted that $L_D$ is expressed in terms of the given data points and depend only on the scalar products of the input vectors $(\mathbf{x}_i, \mathbf{x}_j)$. In the dual space (space described by $\alpha_i$), we have to maximize $L_D$ subject to the following conditions

$$\alpha_i \geq 0,$$

$$\sum_{i=1}^{n} \alpha_i y_i = 0. \tag{4.64}$$

For finding the optimum of the constraint function in the dual space, we have to use the Karush-Kuhn-Tucker (KKT) (Karush 1939; Kuhn and Tucker 1951; Fletcher 1987) conditions. According to the KKT conditions, the solution must also satisfy

$$\alpha_i \left[ y_i \left( \mathbf{w}^T \mathbf{x}_i + b \right) - 1 \right] = 0 \quad \forall i. \tag{4.65}$$

We have two cases for the (4.65).

Case 1:
If $\alpha_i > 0$, then $y_i \left( \mathbf{w}^T \mathbf{x}_i + b \right) - 1 = 0 \Rightarrow y_i \left( \mathbf{w}^T \mathbf{x}_i + b \right) = 1$. This indicates that the input vector $\mathbf{x}_i$ is on the boundary of the margin, i.e., they are support vectors.

Case 2:
If $\alpha_i = 0$, then $y_i \left( \mathbf{w}^T \mathbf{x}_i + b \right) - 1 > 0 \Rightarrow y_i \left( \mathbf{w}^T \mathbf{x}_i + b \right) > 1$. This indicates that the input vector $\mathbf{x}_i$ is not on the boundary of the margin.

After solving the dual problem for the two cases of $\alpha_i, i = 1, 2, \ldots, n$, the solution of the primal problem is obtained using the relation specified by (4.61). Solution vector $\mathbf{w}$ is defined in terms of a linear combination of the support points. $b$ is calculated using $y_i \left( \mathbf{w}^T \mathbf{x}_i + b \right) = 1$, for any support points. So, the optimal separating hyperplane produces a function

$$f(\mathbf{x}) = \mathbf{w}^T \mathbf{x} + b. \tag{4.66}$$

The classification (indicator) output is then given by $sign\ f(x)$. If the data classes are Gaussian in nature, then least square is the optimal solution (see Sect. 4.3.1.1). However, if data classes are non-Gaussian (which is more realistic in real life situation), the support vectors are more robust solutions.

### *4.3.5 Overlapping Classes*

Non-overlapping data sets maintain the margin properly, i.e., no data points from any class leak into the margin region. However, such a case is rare in practice. Instead, overlapping classes consist of data points getting into the margin region and misclassified data points, i.e., points which should be confined to any specific class, gets into the other class(es). Figure 4.15 shows an example of overlapping class.

In Fig. 4.15, the optimal margin algorithm is generalized to the non-separable problems by the introduction of the nonnegative slack variables $\beta_i$ in the statement of the optimization problem. We cannot use the quadratic programming solutions in the case of an overlapping class as the inequality constraints $y_i \left( \mathbf{w}^T \mathbf{x}_i + b \right) \geq 1, i = 1, 2, \ldots, n$ are not satisfied. So, the transformed optimization algorithm is

$$\max\ c, \text{ with respect to } (\mathbf{w}, b), \tag{4.67}$$

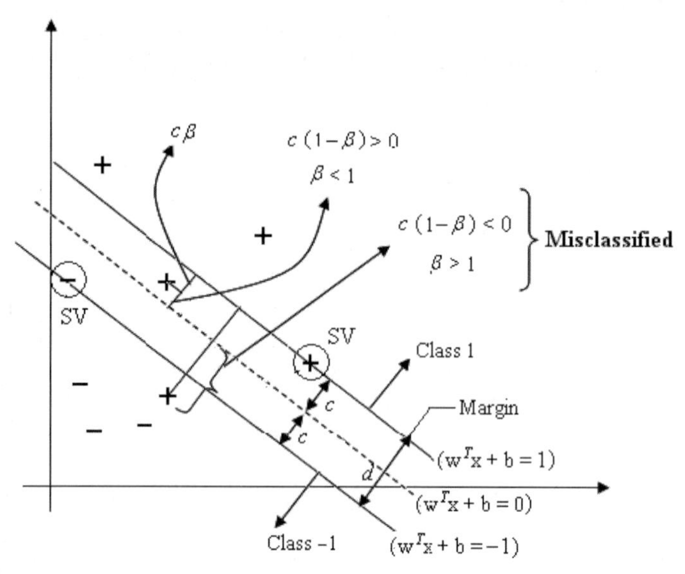


**Fig. 4.15** Overlapping class

subject to,

$$y_i \frac{\left(\mathbf{w}^T.\mathbf{x}_i + b\right)}{|\mathbf{w}|} \geq c(1 - \beta_i), \quad \forall i \tag{4.68}$$

where,

$$\beta_i \geq 0, \tag{4.69}$$

$$\sum_i \beta_i \leq k. \tag{4.70}$$

Equation (4.70) bounds the total number of misclassifications at $k$. We have $c = \frac{1}{|\mathbf{w}|}$, therefore, the equivalent form of the optimization problem is given as follows.

$$\min |\mathbf{w}|, \text{ with respect to } (\mathbf{w}, \beta), \tag{4.71}$$

subject to,

$$y_i \left(\mathbf{w}^T \mathbf{x}_i + b\right) \geq 1 - \beta_i, \ \forall \beta_i \geq 0. \tag{4.72}$$

The bound of misclassification is given by (4.70).

For the generalized optimal separating hyperplane in the overlapping case, we introduce in the minimization functional an extra term accounting for the cost of the overlapping errors. The changed objective function with penalty parameter $\mu$ is

$$\min \left[\frac{1}{2}\mathbf{w}^T \mathbf{w} + \mu \sum_{i=1}^{n} \beta_i\right], \text{ with respect to } (\mathbf{w}, \beta), \tag{4.73}$$

subject to the inequality constraints given by (4.69 & 4.72).

Increasing the penalty parameter $\mu$ corresponds to assigning a higher penalty to the misclassification errors, side by side resulting in larger weights. This makes it a convex optimization (Vapnik 1995, 1998; Schölkopf and Smola 2001) problem.

### *4.3.6 Nonlinear Classifier*

We made the linear classification example (Sect. 4.3.1) closer to the practical case by introducing the overlapping classes in the previous section. We can make it one step closer to the practical scenario by considering a nonlinear separating hypersurface (decision boundary) rather than the linear separation lines. This could be achieved by considering a linear classifier in so-called feature space (higher dimensional). Figure 4.16 shows an example of the nonlinear classifier.

In plot (a) of Fig. 4.16, the plus and minus data classes are separable using a linear separation line. However, if we interchange the positions of the circled plus and minus data points, we get a new class of data as shown in plot (b) of Fig. 4.16. We can still utilize the linear separation line, but this will result in two misclassifications. Instead, with the nonlinear decision boundary (plot (c) of Fig. 4.16), we would be able to separate the two classes without any error.

We can design the nonlinear SVM by mapping the input vectors $\mathbf{x} \in \mathbf{R}^n$, into vectors $\mathbf{z}$ of a higher dimensional feature space ($\mathbf{z} \in \psi(\mathbf{x})$), i.e., enlarging the input space using mapping (basis expansions).

$$\psi(\mathbf{x}_i) = (\psi_1(\mathbf{x}_i), \psi_2(\mathbf{x}_i), \ldots, \psi_M(\mathbf{x}_i)) . \tag{4.74}$$

We fit the SV classifier using input features $\psi_1(\mathbf{x}_i), \psi_2(\mathbf{x}_i), \ldots, \psi_M(\mathbf{x}_i); i = 1, 2, \ldots, n$, and produce the nonlinear boundary function $f(\mathbf{x}) = \mathbf{w}^T \psi(\mathbf{x}) + b$. By performing such a mapping, we hope that in a higher dimensional $\psi$-space, our learning algorithm will be able to linearly separate the projected images of the (original) input vector $\mathbf{x}$ by applying the linear SVM formulations.

As an example, we consider the following function

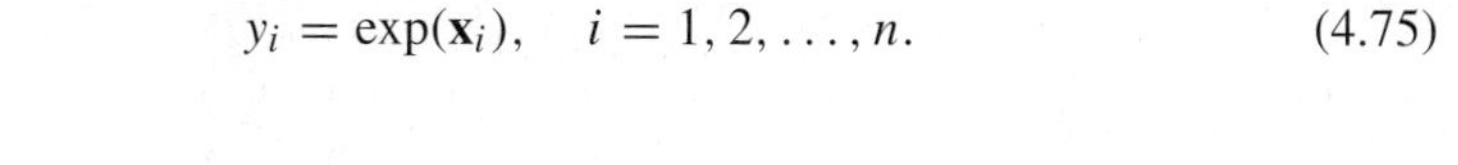

$$y_i = \exp(\mathbf{x}_i), \quad i = 1, 2, \ldots, n. \tag{4.75}$$

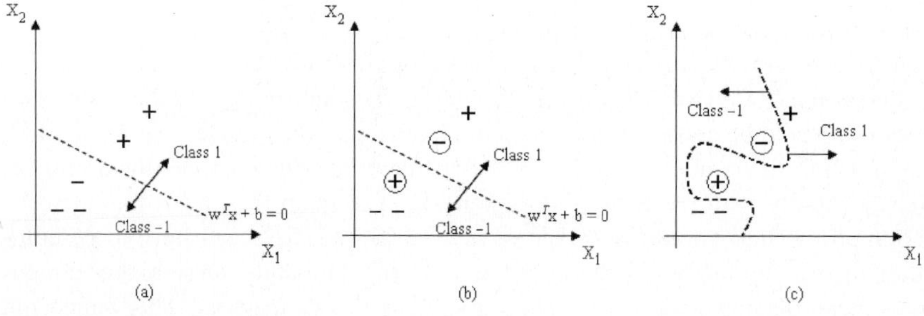


**Fig. 4.16** Nonlinear classifier example

$y_i$ is exponentially linearized with the following $\psi$-space mapping,

$$\psi(\mathbf{x}_i) = \left(\mathbf{x}_i, \mathbf{x}_i^2, \mathbf{x}_i^3, \ldots, \mathbf{x}_i^M\right). \quad (4.76)$$

And, linear coefficients $\mathbf{w}$ is equal to the Taylor series as given below.

$$y_i = \sum_{n=0}^{\infty} \frac{1}{n} \mathbf{x}_i^n \approx \sum_{n=1}^{M} \mathbf{w}_n \psi_n(\mathbf{x}_i) + \mathbf{w}_0. \quad (4.77)$$

## *4.3.7 Kernel Method*

### 4.3.7.1 Introduction

Kernel method implicitly introduces a notion of similarity between data. Kernels indicate the choice of similarity patterns, implying the choice of features. Let's see an analogy for kernel. We consider two books with same number of pages but of entirely different subjects. Although these two books are different subject-wise, yet we see somewhat similarity based on the feature of number of pages. So, now we can think of a book with same number of blank pages as a kernel. Then, taking this blank book we fill up with the appropriate subject to arrive at the individual pattern.

Mathematically, kernel methods exploit information about the inner products between data items. Hence, kernel functions can be thought of as the inner products in some feature space. Once the kernel function for the data is mentioned, we no longer require any information about the features of the data. As we saw in the previous book example, once we have the blank book, we do not need to specify that we are working on the basis number of pages, we just use that blank book to get the individual books!

We've seen in the previous section that linear classifiers cannot particularly deal with nonlinearly separable data and noisy data. In those cases, we require a nonlinear classifier. Again, instead of using a nonlinear classifier in lower dimension, we can intelligently map the data into a richer feature space (higher dimension) which would include the problematic nonlinear features as well. Then, in that transformed higher dimensional feature space, we might have a linear separation. Actually from optimization point of view, we continue to search for this nice feature space where we can find a linear separation. This is shown in Fig. 4.17.

Oftentimes in this sort of classification tasks, the feature space or the basis functions ($\mu$) has to be chosen high. For example, the classification task, $y = sign(x)$, requires infinite number of sinusoidal basis functions. The computational cost increases rapidly with the increasing dimension. The main reason for this increase is that all the inner products $\psi(\mathbf{x}_i)^T.\psi(\mathbf{x}_j),\ i, j = 1, 2, \ldots, n$, have to be computed for solving the optimization problem. Computationally, these higher dimensions mean dealing with large vectors, requiring lots of memory. This dimension explosion and the associated computational demands can be effectively lowered by

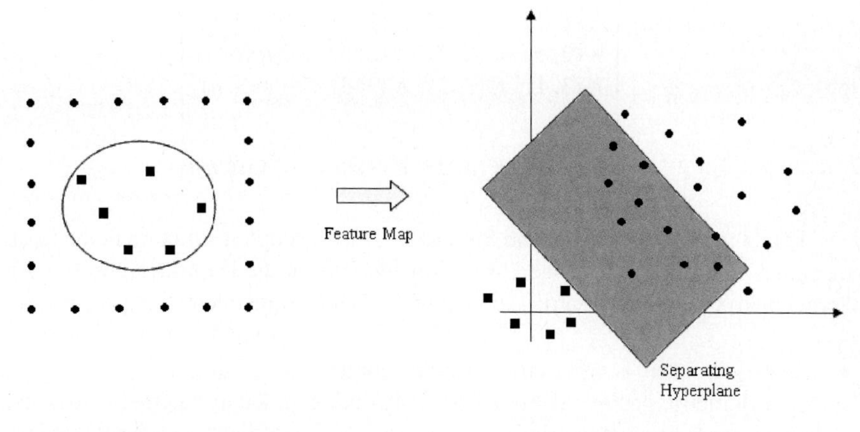


**Fig. 4.17** Linear classification in higher dimension

using the kernel functions (Burges 1998; Schölkopf and Smola 2001). The kernel methods can be used to replace the higher dimension feature space (mathematically, by replacing the exploding inner products). Hence, also from the practical implementation point of view, kernel methods are important to tackle infinite dimensions efficiently in time and space.

So, the 4-step synopsis for the kernel methods are given below.

1. We've nonlinear, noisy data which we want to classify.
2. One way is to construct a nonlinear classifier in the given space. But this is complex!
3. We project the data into a higher dimensional feature space where linear classification is possible. But this results in huge dimensions, computation problem!
4. We continue to use the excellent concept in point 3 by transforming the data into an easier feature space where we can perform linear classification. But we get rid of the higher dimension by performing this transformation with the aid of kernel functions.

#### 4.3.7.2 Definition of a Kernel

A kernel function is a function that returns the value of the dot product between the images of the two arguments.

$$K(x_1, x_2) = \langle \phi(x_1), \phi(x_2) \rangle. \tag{4.78}$$

An important concept associated with the kernel functions is the dot product which can be represented using the kernel matrix also known as the Gram matrix (Schölkopf and Smola 2001).

$$K = \begin{bmatrix} K(1,1) & K(1,2) & K(1,3) & \dots & K(1,n) \\ K(2,1) & K(2,2) & K(2,3) & \dots & K(2,n) \\ K(3,1) & K(3,2) & K(3,3) & \dots & k(3,n) \\ \vdots & \vdots & \vdots & \ddots & \vdots \\ K(n,1) & K(n,2) & K(n,3) & \dots & K(n,n) \end{bmatrix}. \quad (4.79)$$

The kernel matrix (Schölkopf and Smola 2001) is the central structure in the kernel machines, containing all the necessary information for the learning algorithm. The kernel matrix (aka gram matrix) shown in (4.79) has some interesting properties.

- The kernel matrix is symmetric positive definite.
- Any symmetric positive definite matrix can be regarded as a kernel matrix, that is as an inner product matrix in some space.

These points are formally known as Mercer's theorem (Mercer 1909; Burges 1998; Schölkopf and Smola 2001). According to Mercer's theorem, for a continuous symmetric nonnegative definite kernel $K$, there is an orthonormal basis for the mapping, consisting of the eigenfunctions such that the corresponding sequence of eigenvalues ($\lambda_i$) is nonnegative. Using this, the kernel function in (4.78) can be written as

$$K(x_1, x_2) = \sum_i \lambda_i \phi_i(x_1)\phi_i(x_2). \quad (4.80)$$

This implies that the eigenfunctions act as the features.

#### 4.3.7.3 Properties of Kernel

We cite some useful properties of the kernel function below. The set of kernels is closed under some operations. If $K$, $K'$ are kernels, then:

- $K + K'$ is a kernel.
- $aK$ is a kernel, if $a > 0$.
- $aK + bK'$ is a kernel, if $a, b > 0$.

Using these properties, we can construct different complex kernels from simple ones. However, to utilize the power of kernels, we must pay attention whether the kernels that we can construct are good or bad ones. A bad kernel can be identified from the kernel or gram matrix. The gram matrix of a bad kernel would be mostly diagonal as shown in (4.81). This indicates that all points are orthogonal to each other, indicating no clusters or structures. In other words, if we try to map in a space with too many irrelevant features, the kernel matrix becomes diagonal. Therefore, to choose a good kernel, we always need some prior knowledge of the target.

$$K_{bad} = \begin{bmatrix} 1 & 0 & 0 & \dots & 0 \\ 0 & 1 & 0 & \dots & 0 \\ 0 & 0 & 1 & \dots & 0 \\ \vdots & \vdots & \vdots & \ddots & \vdots \\ 0 & 0 & 0 & \dots & 1 \end{bmatrix}. \tag{4.81}$$

#### 4.3.7.4 Typical Kernel Functions

Some typical kernel functions are mentioned below.

- Linear kernel: $\psi(\mathbf{x}, \mathbf{x}_k) = \mathbf{x}_k^T \mathbf{x}$.
- Polynomial kernel of degree $p$: $\psi(\mathbf{x}, \mathbf{x}_k) = (\mathbf{x}_k^T \mathbf{x} + 1)^p$.
- Radial basis function (RBF) kernel:

$$\psi(\mathbf{x}, \mathbf{x}_k) = \exp\left[\frac{-\|\mathbf{x} - \mathbf{x}_k\|^2}{2\sigma^2}\right].$$

- Two layer neural kernel: $\psi(\mathbf{x}, \mathbf{x}_k) = \tanh\left[a\mathbf{x}_k^T \mathbf{x} + b\right]$.

A kernel function is defined in the input space and no explicit mapping into the higher dimensional feature space is required by using it. Using a kernel function $K(\mathbf{x}_i, \mathbf{x}_j)$ in the input space, we can avoid the mapping in to a higher dimensional feature space. For any given training data vectors, the required (scalar) inner products are calculated by computing the kernel.

We see an example of mapping into the polynomial feature space defined as

$$\psi\colon \mathbf{x} = (\mathbf{x}_1, \mathbf{x}_2) \to \tilde{\mathbf{x}} = (\mathbf{x}_1^2, \sqrt{2}\mathbf{x}_1\mathbf{x}_2, \mathbf{x}_2^2). \tag{4.82}$$

In this feature space, the inner products of two vectors $\tilde{\mathbf{x}}$ and $\tilde{\mathbf{x}}_k$ are given as

$$\tilde{\mathbf{x}}^T \tilde{\mathbf{x}}_k = \mathbf{x}_1^2\mathbf{x}_{k1}^2 + 2\mathbf{x}_1\mathbf{x}_2\mathbf{x}_{k1}\mathbf{x}_{k2} + \mathbf{x}_2^2\mathbf{x}_{k2}^2 = (\mathbf{x}^T\mathbf{x}_k)^2 = \psi(\mathbf{x}, \mathbf{x}_k). \tag{4.83}$$

In general, the mapping of an input vector $\mathbf{x} \in \mathbf{R}^d$ into a polynomial space with degree $p$ implies the dimension $\binom{p+d-1}{d}$ of the feature space.

Any symmetric function $K(\mathbf{x}, \mathbf{y})$ in input space can represent a scalar product in feature space if

$$\int\int K(\mathbf{x}, \mathbf{y}) f(\mathbf{x}) f(\mathbf{y}) d\mathbf{x} d\mathbf{y} > 0, \tag{4.84}$$

where, $f(\mathbf{x})$ or $f(\mathbf{y})$ are any function with a finite $L_2$ norm in input space, i.e., a function for which

$$\int f^2(\mathbf{x}) d\mathbf{x} < \infty. \tag{4.85}$$

In previous sections, we discussed about the computation of the optimal boundary functions. We see them in light of the kernel functions below.

Linear case:

$$f(\mathbf{x}) = \mathbf{w}^T \mathbf{x} + b,$$

$$\mathbf{w} = \sum_{k=1}^{n} \alpha_k y_k \mathbf{x}_k,$$

$$\Rightarrow f(\mathbf{x}) = \sum_{k=1}^{n} \alpha_k y_k \mathbf{x}_k^T \mathbf{x} + b. \tag{4.86}$$

Nonlinear case:

$$f(\mathbf{x}) = \mathbf{w}^T \mathbf{x} + b,$$

$$\mathbf{w} = \sum_{k=1}^{n} \alpha_k y_k \psi(\mathbf{x}_k),$$

$$\Rightarrow f(\mathbf{x}) = \sum_{k=1}^{n} \alpha_k y_k \psi^T(\mathbf{x}_k) \psi(\mathbf{x}) + b. \tag{4.87}$$

The term $\mathbf{x}_k^T \mathbf{x}$ in (4.86) is an example of a linear kernel. The term $\psi^T(\mathbf{x}_k)\psi(\mathbf{x})$ in (4.87) is an example of a nonlinear kernel.

#### 4.3.7.5 Kernel Hilbert Space

In the SVM literature, the aforesaid feature space (higher dimensional) is usually referred to as the kernel Hilbert space, named after David Hilbert. We will briefly review about the vector spaces and the kernel Hilbert space in the following section.

A norm represents the distance of a vector in $\mathbf{R}^2$ or $\mathbf{R}^3$, as shown in Fig. 4.18. Therefore, for $x = (x_1, x_2) \in \mathbf{R}^2$, the norm is $\|x\| = \sqrt{x_1^2 + x_2^2}$. For,

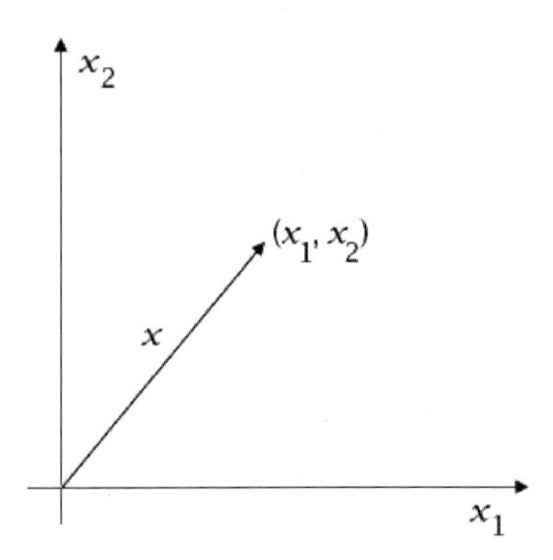


**Fig. 4.18** Norm in $\mathbf{R}^2$

$x = (x_1, x_2, x_3) \in \mathbf{R}^3$, the norm is $\|x\| = \sqrt{x_1^2 + x_2^2 + x_3^2}$. In general, for $x = (x_1, x_2, \ldots, x_n) \in \mathbf{R}^n$, the norm is $\|x\| = \sqrt{x_1^2 + x_2^2 + \ldots + x_n^2}$.

However, the norm is not linear on $\mathbf{R}^n$. To impose linearity, we need to use the dot product. For $x, y \in \mathbf{R}^n$, the dot product is defined as

$$x.y = x_1 y_1 + x_2 y_2 + \ldots + x_n y_n. \tag{4.88}$$

The dot product is a number in $\mathbf{R}^n$, not a vector. Clearly, $x.x = \|x\|^2, \forall x \in \mathbf{R}^n$. Instead of being real, if $x$ is complex, we can define

$$x = a + jb. \tag{4.89}$$

The magnitude $|x|$, and the complex conjugate $x^*$ are defined as

$$\|x\| = \sqrt{a^2 + b^2}, \tag{4.90}$$

$$x^* = a - jb. \tag{4.91}$$

Equations (4.89–4.91) are related as

$$xx^* = \|x\|^2. \tag{4.92}$$

For $x = (x_1, x_2, \ldots, x_n) \in \mathbf{C}^n$, the norm is given as

$$\|x\| = \sqrt{|x_1|^2 + |x_2|^2 + \ldots + |x_n|^2}. \tag{4.93}$$

As $x$ is complex, and we need a nonnegative number, we can think of the norm $\|x\|^2$. Using (4.92–4.93),

$$\|x\|^2 = |x_1|^2 + |x_2|^2 + \ldots + |x_n|^2 = x_1 x_1^* + x_2 x_2^* + \ldots + x_n x_n^*. \tag{4.94}$$

Similarly to the real space, we can think of $\|x\|^2$ as the inner product of $\|x\|$ with itself. Hence, the inner product of $y = (y_1, y_2, \ldots, y_n) \in \mathbf{C}^n$ with x will be

$$\langle x.y \rangle = y_1 x_1^* + y_2 x_2^* + \ldots + y_n x_n^*. \tag{4.95}$$

Comparing (4.88) with (4.95), we get that the inner product is a generalization of the dot product.

Thus, for a real or complex vector space $H$, every inner products produce a norm like

$$\|x\| = \langle x, x \rangle. \tag{4.96}$$

According to the Cauchy criterion, any sequence $\{x_n\}$ in this space is a Cauchy sequence if for every positive real number $\varepsilon$, there exists a natural number $N$ so that

$$\|x_n - x_m\| < \varepsilon, \ \forall m, n > N. \tag{4.97}$$

If $H$ is a Hilbert space, it is complete with respect to the norm in (4.96), i.e., every Cauchy sequence (defined by (4.97)) in $H$ converges to some point. In other words, Hilbert space is a space of infinite dimensions in which distance is preserved by making the sum of squares of coordinates a convergent Cauchy sequence.

Two (or more) Hilbert spaces can be combined into a single Hilbert space by taking their direct sum or their tensor product. Example of Hilbert space is the Euclidian space in $L_2$. For SVM purpose, the higher dimensional feature space replaced by and represented as the kernel Hilbert space, can be thought of as a generalized Euclidian space that behaves in a gentlemanly fashion (Burges 1998).

#### 4.3.7.6 Kernels & SVM Classifier Optimization

We review the SVM classifier optimization problem in terms of kernel functions in this section. The SVM classifier satisfies

$$\begin{aligned} \mathbf{w}^T \psi(\mathbf{x}_k) + b &\geq 1, \quad if \ y_k = 1, \\ \mathbf{w}^T \psi(\mathbf{x}_k) + b &< 1, \quad if \ y_k = -1. \end{aligned} \tag{4.98}$$

This is equivalent to

$$y_k \left[ \mathbf{w}^T \psi(\mathbf{x}_k) + b \right] \geq 1, \quad k = 1, 2, \ldots, n. \tag{4.99}$$

For overlapping classes,

$$y_k \left[ \mathbf{w}^T \psi(\mathbf{x}_k) + b \right] \geq 1 - \beta_k, \quad k = 1, 2, \ldots, n, \tag{4.100}$$

where, $\beta_k$'s are nonnegative slack variables taking into account the overlaps. Involving the penalty parameter $\mu$, the optimization problem becomes

$$\min \left[ \frac{1}{2} \mathbf{w}^T \mathbf{w} + \mu \sum_{k=1}^{n} \beta_k \right], \text{ with respect to } (\mathbf{w}, \beta), \tag{4.101}$$

subject to the inequality specified by (4.100). Here, large $\mu$ implies discouragement of overlapping, i.e., overfitting, while small $\mu$ implies small $\|\mathbf{w}\|$, i.e., smooth boundary.

Introducing the Lagrange multipliers $\alpha_k \geq 0$, $\nu_k \geq 0$ for $k = 1, 2, \ldots, n$, input data points, we get the Lagrangian functional as follows.

$$L = \frac{1}{2} \mathbf{w}^T \mathbf{w} + \mu \sum_{k=1}^{n} \beta_k - \sum_{k=1}^{n} \alpha_k \left[ y_k \left( \mathbf{w}^T \psi(\mathbf{x}_k) + b \right) - 1 + \beta_k \right] - \sum_{k=1}^{n} \nu_k \beta_k. \tag{4.102}$$

In the primal space (i.e., large amount of input data sets), solving (4.102) for the saddle points, we get

$$\frac{\partial L}{\partial \mathbf{w}} = 0, \; k = 1, 2, \ldots, n, \tag{4.103}$$
$$\Rightarrow \mathbf{w} = \sum_{k=1}^{n} \alpha_k y_k \psi(\mathbf{x}_k).$$

$$\frac{\partial L}{\partial b} = 0, \; k = 1, 2, \ldots, n, \tag{4.104}$$
$$\Rightarrow \sum_{k=1}^{n} \alpha_k y_k = 0.$$

$$\frac{\partial L}{\partial \beta_k} = 0, \; k = 1, 2, \ldots, n, \tag{4.105}$$
$$\Rightarrow 0 \leq \alpha_k \leq \mu.$$

In the dual space (i.e., high dimensional feature space), the quadratic programming problem becomes

$$\max \sum_{k=1}^{n} \alpha_k - \frac{1}{2} \sum_{i=1}^{n} \sum_{j=1}^{n} \alpha_i \alpha_j y_i y_j \psi(\mathbf{x}_i, \mathbf{x}_j), \text{ with respect to } \alpha_k, \tag{4.106}$$

subject to,

$$\sum_{k=1}^{n} \alpha_k y_k = 0, \tag{4.107}$$
$$0 \leq \alpha_k \leq \mu, \; k = 1, 2, \ldots, n.$$

The kernel function is

$$K(\mathbf{x}_i, \mathbf{x}_j) = \psi^T(\mathbf{x}_i) \psi(\mathbf{x}_j). \tag{4.108}$$

Hence, we do not need to compute the $\mathbf{w}$ or $\psi(\mathbf{x}_k)$ (see (4.103)). That is, we bypass the scalar inner products by computing the kernels. This is a global solution.

### *4.3.8 Support Vector Regression*

In regression, we typically use some measure or error of approximation instead of margin between an optimal separating hyperplane and the SVs. If $y$ and $f$ represent respectively the predicted and the measured values, the typical error functions are given as follows.

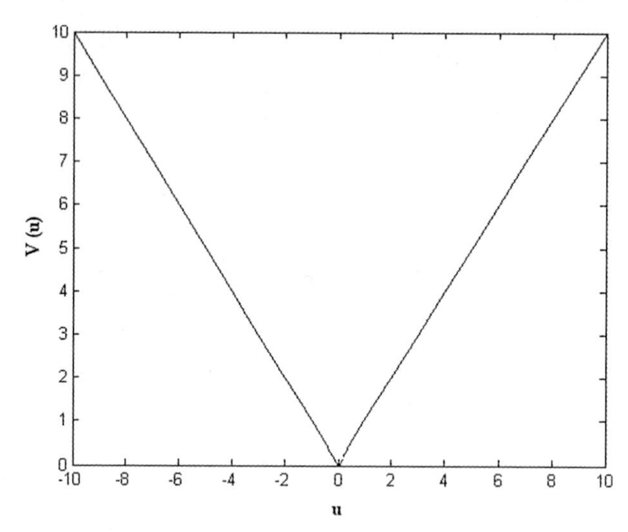


**Fig. 4.19** Error function in $L_1$ norm

- $L_1$ norm (least modules): $|y - f|$, shown in Fig. 4.19.
- $L_2$ norm (square errors): $(y - f)^2$, shown in Fig. 4.20.
- Huber's loss function:

$$\frac{1}{2}(y - f)^2, \quad for \quad |y - f| < \xi$$

$$\xi\,|y - f| - \frac{\xi^2}{2}, \quad otherwise.$$

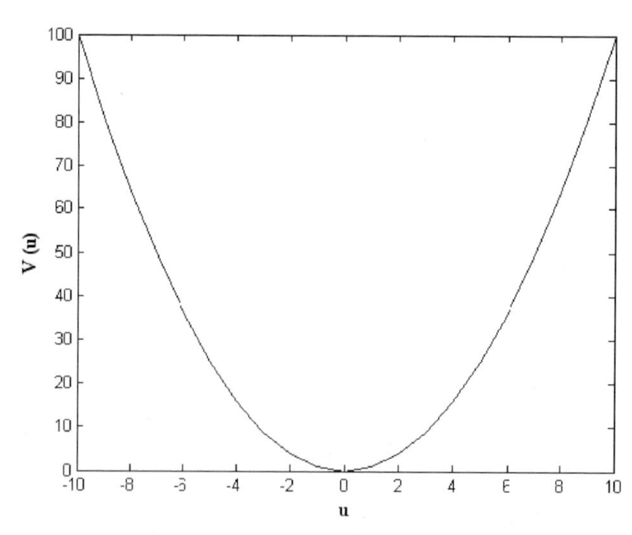


**Fig. 4.20** Error function in $L_2$ norm

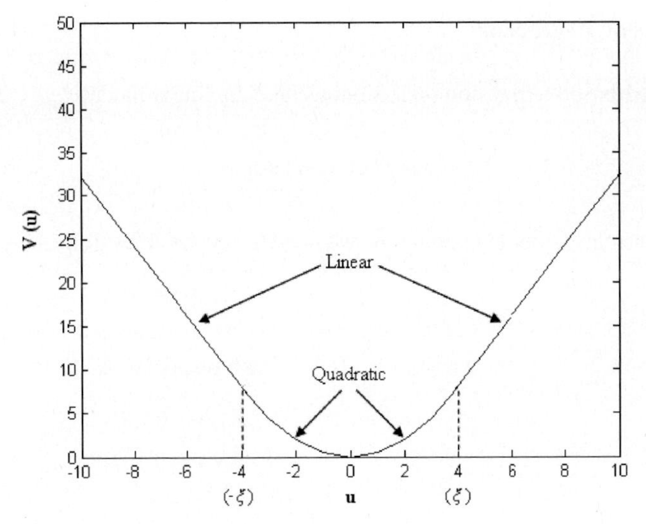


**Fig. 4.21** Huber's loss function for *robust regression*

Huber's loss function results in robust regression. It is to be noted that Huber's loss function, shown in Fig. 4.21, is related to the quadratic (square) loss function (Fig. 4.20). If the quadratic loss function (Fig. 4.20) becomes linear after $|\xi|$, it becomes Huber's loss function as in Fig. 4.21.

- Vapnik's linear loss function with $\varepsilon$-insensitivity zone, which is shown in Fig. 4.22.

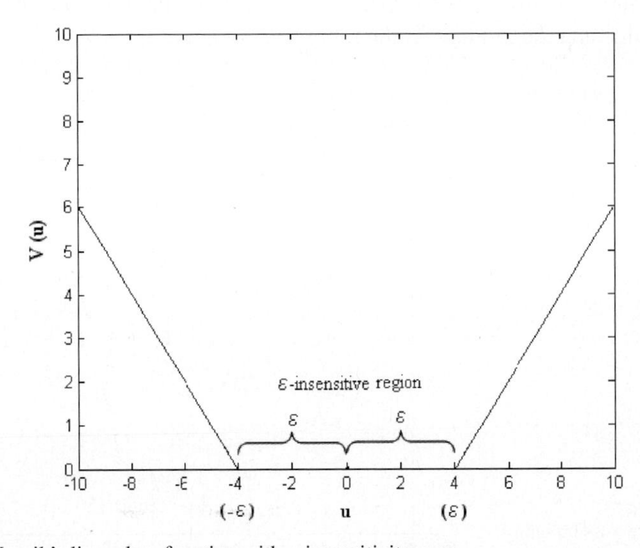


**Fig. 4.22** Vapnik's linear loss function with $\varepsilon$-insensitivity zone

#### 4.3.8.1 Linear Regression

For the SVM-based regression operation, we use the linear model

$$\mathbf{y} = f(\mathbf{x}) = \mathbf{w}^T\mathbf{x} + b. \tag{4.109}$$

For linear regression, to estimate **w**, we need to minimize the following objective function,

$$\min \sum_{i=1}^{n} V\left(y_i - f(x_i)\right) + \frac{\lambda}{2}\left\|\mathbf{w}\right\|^2, \text{ with respect to } \mathbf{w}, \tag{4.110}$$

where,

$$V(u) = \begin{cases} 0, & if\ |u| < \varepsilon, \\ |u| - \varepsilon, & otherwise. \end{cases} \tag{4.111}$$

In (4.110), $\lambda$ is a regularization parameter and can be estimated using cross-validation. In (4.111), $\varepsilon$ is the insensitive error measure.

In formulating an SV algorithm for regression, we will try to minimize both empirical risk and $\|\mathbf{w}\|^2$ simultaneously. Figure 4.23 shows the SV-based regression (function approximation). The outliers are marked with squares while the support vectors (denoting sparse fit) are marked as circled.

As shown in Fig. 4.23, Vapnik's $\varepsilon$-insensitive linear loss function defines an $\varepsilon$-tube. If any predicted value lies within the tube, the error is zero. For all other predicted points lying outside the tube, the loss equals the difference between the predicted value and the radius $\varepsilon$ of the tube.

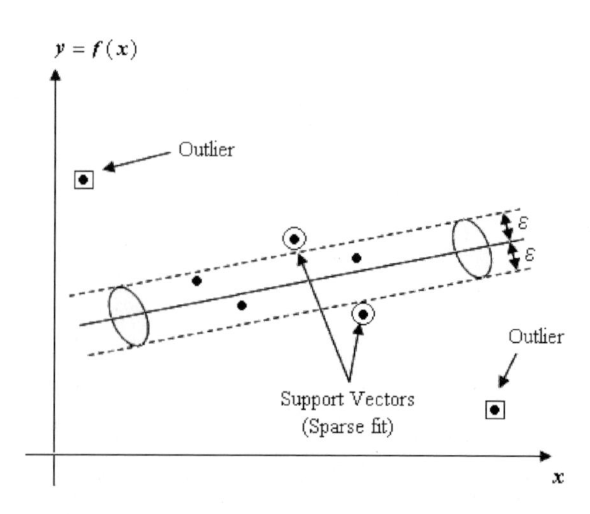


**Fig. 4.23** Support vector-based regression using an $\varepsilon$-tube

#### 4.3.8.2 Optimization Problem in Regression

Concerning (4.110–4.111) for SV-regression, the optimization problem can be formulated by introducing the Lagrange multipliers $(\alpha_i, \alpha_i^*)$ like the SV classifier. The two Lagrange multipliers correspond to the data points outside the aforesaid $\varepsilon$-tube. More specifically, $\alpha_i$ and $\alpha_i^*$ correspond to the data points above and below the $\varepsilon$-tube respectively (see Fig. 4.23). Hence, each data point can only lie either above or below $\varepsilon$-tube. Therefore, like SV classifier, the optimization problem in dual space becomes

$$\min \varepsilon \sum_{i=1}^{n}(\alpha_i^* + \alpha_i) - \sum_{i=1}^{n} y_i(\alpha_i^* - \alpha_i) + \frac{1}{2}\sum_{i,j=1}^{n}(\alpha_i^* - \alpha_i)(\alpha_j^* - \alpha_j)\mathbf{x}_i^T\mathbf{x}_j, \tag{4.112}$$

$$\text{with respect to } (\alpha_i, \alpha_i^*),$$

subject to,

$$\begin{aligned} \sum_{i=1}^{n}\alpha_i^* &= \sum_{i=1}^{n}\alpha_i, \\ \alpha_i^* &\le 1/\lambda, \\ 0 &\le \alpha_i. \end{aligned} \tag{4.113}$$

Solving (4.112) subject to the constraints given by (4.113) determines the Lagrange multipliers $\alpha_i, \alpha_i^*$. The regression function is given by (4.109), where

$$\mathbf{w} = \sum_{i=1}^{n}(\alpha_i^* - \alpha_i)\mathbf{x}_i, \tag{4.114}$$

$$b = \frac{1}{n}\sum_{i=1}^{n}\left(y_i - \mathbf{x}_i^T\mathbf{w}\right). \tag{4.115}$$

The support vectors are given by the solutions which satisfy the Karush-Kuhn-Tucker (KKT) condition (Karush 1939; Kuhn and Tucker 1951; Fletcher 1987). These are

$$\alpha_i\alpha_i^* = 0,\ i = 1, 2, \ldots, n. \tag{4.116}$$

That is, the support vectors are given by the data points where exactly one of the Lagrange multipliers is nonzero. After finding the Lagrange multipliers, the optimal weight vector $\mathbf{w}$ and the optimal bias term $b$ can be found using (4.114) and (4.115) respectively.

#### 4.3.8.3 Nonlinear Regression

In most practical cases, a nonlinear model performs better modeling than a linear one. In the same manner as the nonlinear support vector classification approach, a nonlinear mapping can be used to map the data into a high dimensional feature space where linear regression is performed. The kernel approach is again employed to address the curse of dimensionality.

Using the $\varepsilon$-insensitive linear loss function for nonlinear regression, the optimization problem becomes

$$\min \varepsilon \sum_{i=1}^{n}(\alpha_i^* + \alpha_i) - \sum_{i=1}^{n} y_i(\alpha_i^* - \alpha_i) + \frac{1}{2}\sum_{i,j=1}^{n}(\alpha_i^* - \alpha_i)(\alpha_j^* - \alpha_j)K(\mathbf{x}_i, \mathbf{x}_j), \tag{4.117}$$

$$\text{with respect to } (\alpha_i, \alpha_i^*),$$

subject to the constraints given by (4.113). In (4.117), $K(\mathbf{x}_i, \mathbf{x}_j)$ is the kernel function to replace the problematic inner products. Solving (4.117) subject to the constraints given by (4.113) determines the Lagrange multipliers. Then, the regression function is given by

$$f(\mathbf{x}) = \mathbf{w}^T\mathbf{x} + b = \sum_{SVs}(\alpha_i - \alpha_i^*)K(\mathbf{x}_i, \mathbf{x}) + b. \tag{4.118}$$

The optimal bias term $b$ in (4.118) can be estimated as

$$b = \frac{1}{2}\sum_{i=1}^{n}(\alpha_i^* - \alpha_i)\left[K(\mathbf{x}_i, \mathbf{x}_r) + K(\mathbf{x}_i, \mathbf{x}_s)\right]. \tag{4.119}$$

#### 4.3.8.4 Regression Example

We show an example of nonlinear regression below. The input (X) and the output (Y) data for the regression are shown in Table 4.1.

The nonlinear regression functions using different kernels to map the input and the output data are shown in Fig. 4.24 to 4.27.

For the simulations, the LS-SVM toolbox (Pelckmans et al. 2003), which is based on the least square support vector machine, has been used in MATLAB® environment.

**Table 4.1** Input and output data for nonlinear regression

| X | Y |
|---|---|
| 2 | 180 |
| 3 | 135 |
| 4 | 80 |
| 5 | 51 |
| 6 | 58 |
| 7 | 68 |
| 9 | 83 |
| 11 | 133 |
| 14 | 172 |
| 17 | 214 |

### *4.3.9 Procedure to use SVM*

We've seen so far the theories behind the statistical learning theory, support vector machines for classification and regression tasks. In this section, we briefly describe the procedures to apply the SVMs in practice for classification and regression task.

1. **Scaling**. The input and the output data need to be scaled properly for better performance. This is important as the kernel values depend on the inner products of the feature vectors. Hence, large attribute values might cause numerical problems (Hsu et al. 2003). Simple linear scaling, like the input and the output data range having bounds like $[-1, 1]$ or $[0, 1]$, provides good result.

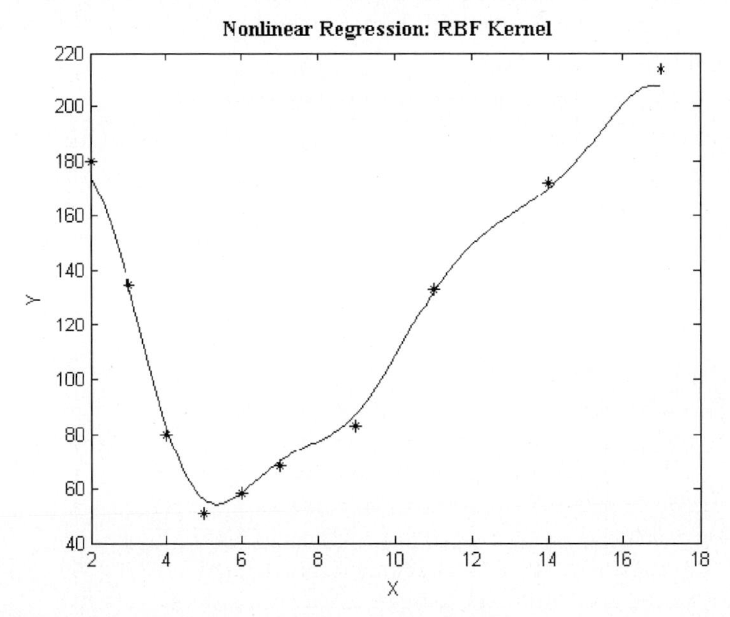


**Fig. 4.24** Nonlinear regression using radial basis function kernel

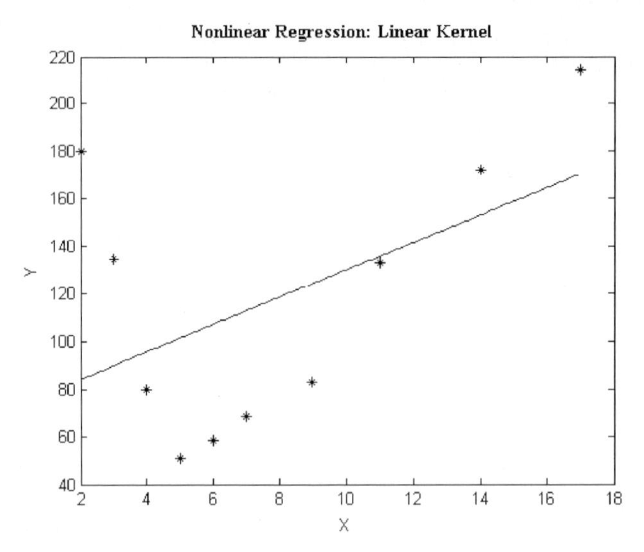


**Fig. 4.25** Nonlinear regression using linear kernel

2. **Choosing the kernel**. The next step is to choose an appropriate kernel function (see Sect. 4.3.7.4 for different kernels). The best starting point is to use the RBF kernel. Keerthi and Lin (Keerthi and Lin 2003) showed that if the RBF kernel is used as the basis kernel, there is no need to use the linear kernel. Moreover, the neural kernel involving the sigmoid function may not be positive definite and its accuracy is lower than the RBF kernel (Hsu et al. 2003).
3. **Adjusting the kernel parameters**. The next step is to adjust the kernel parameters, e.g., variance for the RBF kernel, polynomial order for the polynomial

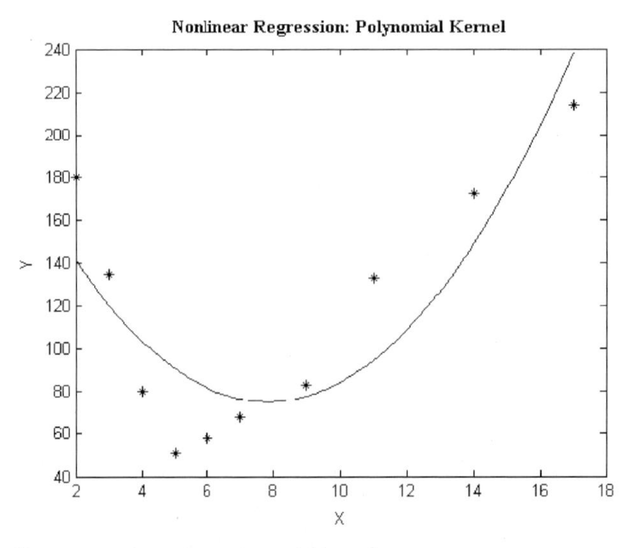


**Fig. 4.26** Nonlinear regression using polynomial kernel

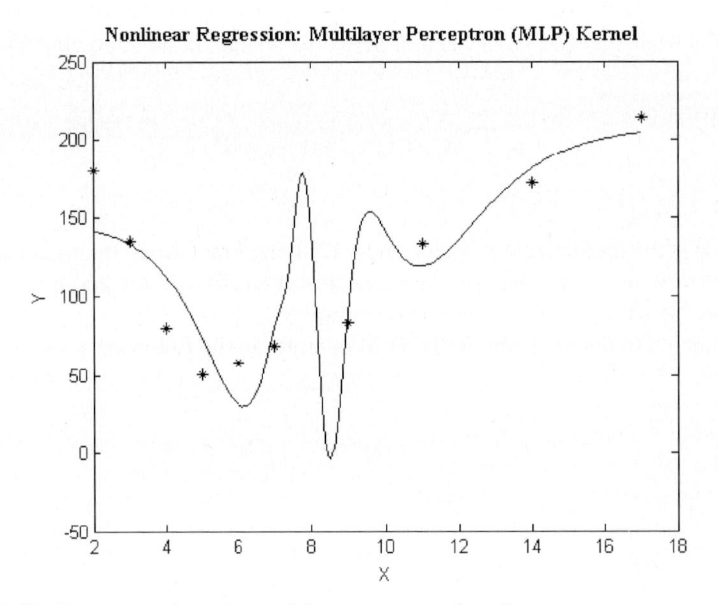


**Fig. 4.27** Nonlinear regression using multilayer perceptron kernel

kernel, etc. It is to be noted here that a very high order polynomial might cause computational problem. An optimum value for the kernel parameter of any specific kernel might be achieved using cross-validation.

4. **Training**. Train the SVM using the chosen kernel with optimal parameter, the input and the output (for regression) data. The choice of the appropriate kernel could be influenced by the number of training data. However, in practice, usually cross-validation technique is the best way to judge the appropriateness of the kernel.
5. **Testing**. After training the SVM, test it using the validation data.

### *4.3.10 SVMs and NNs*

Both the support vector machines (SVMs) and the neural networks (NNs) (Chap. 3) map the nonlinear relationship between the input ($x_i$) and the output ($y_i$). In this section, we will see a comparative discussion of the SVMs and the NNs.

Classical multilayer perceptron (MLP) operates like

$$e = \sum_{i=1}^{n} (y_i - f(\mathbf{x}_i, \mathbf{w}))^2, \qquad (4.120)$$

where, $e$ is the error measure. In (4.120), the right hand side term is a measure of the closeness to data.

Radial basis function (RBF) neural networks operate in the following manner.

$$e = \sum_{i=1}^{n} (y_i - f(\mathbf{x}_i, \mathbf{w}))^2 + \lambda \|\mathbf{P} f\|^2, \tag{4.121}$$

where, **P** is the regularization vector. In (4.121), the first term in the right hand side is a measure of the closeness to data similar to (4.120), and the additional second term indicates the smoothness of the operation.

Compared to the MLP and RBF, SVM operates in the following way.

$$e = \sum_{i=1}^{n} (y_i - f(\mathbf{x}_i, \mathbf{w}))^2 + \Omega(l, d, \eta). \tag{4.122}$$

As before, the first term in the right hand side of (4.122) is a measure of the closeness to data and the function $\Omega$ in the second term denotes the capacity of the machine. In (4.122), the structural risk minimization principle (SRM) uses the VC dimension $d$ as a controlling parameter for minimizing the generalization error $e$ (i.e., risk).

Vapnik (Vapnik 1995, 1998) basically introduced two basic, constructive approaches to the minimization of the right hand side of (4.122). These are as follows.

1. We choose an appropriate structure, for example, order of polynomials, number of hidden layer neurons in neural network, number of rules in fuzzy logic model etc. Keeping the confidence interval fixed in this way, we minimize the training error, i.e., we perform the empirical risk minimization (ERM). In simpler words, this way we concentrate on the structure hoping that the structure takes care of (minimizes) the associated risk (unseen!).
2. We keep the value of the training error fixed to zero or some acceptable level and minimize the confidence level. This way, we structure a model of the associated risk and try to minimize that (that's SRM!). The result is the optimal structure.

Classical NNs implement the first approach, while the SVMs follow the second strategy. In both cases, the resulting model should resolve the trade-off between under- and over-fitting. Nevertheless, the final model should ideally match the learning machines capacity with training data complexity.

It's evident from above discussion that, in the first ERM approach, as we start with an appropriate structure hoping to tackle the underlying risk of the input-output mapping, we might not be guaranteed that the structure achieves it. That's why NNs many times face the problem of convergence. Also, we do not have any idea about the underlying input-output mapping, hence, the structure, we started with, might get stuck at some local minima, as shown in Fig. 4.28. In comparison, in the second SRM approach, we start with the test error and arrive at an optimal structure. Hence, the SVMs, in general, generalize better. Also, surety of convergence is more for the

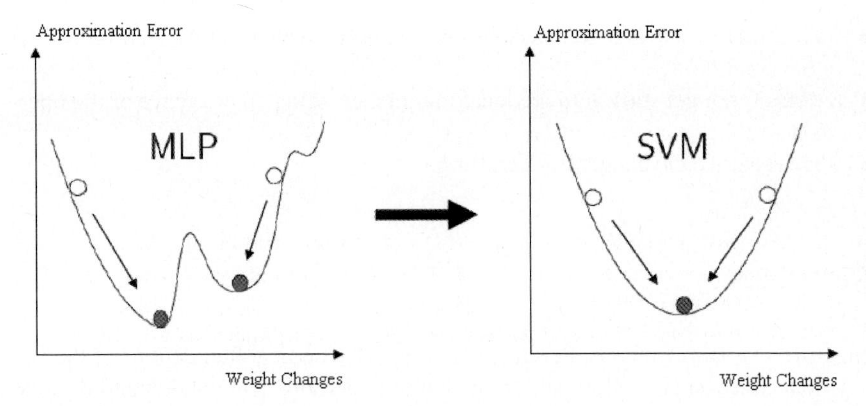


**Fig. 4.28** Local minima, convergence problem of NN

SVMs than the NNs. This is due to the fact that the former does not face the problem of getting stuck at local minima (hidden and unpredicted).

We summarize the comparison of the SVM and the NN below.

#### 4.3.10.1 Similarities between SVM and NN

- Both the SVMs and the NNs learn from experimental data, for which, more often in the practical scenario, the underlying probability distribution is not known.
- Struturally, the SVMs and the NNs are same.
- Both the SVMs and the NNs are universal approximators in the sense that they can approximate any function to any desired degree of accuracy.
- After the learning, they are given with the same mathematical model.

#### 4.3.10.2 Differences between SVM and NN

- NNs are based on the ERM principle, starting with an appropriate structure to minimize the error. SVMs are based on the SRM principle, starting with the error to achieve an optimal structure.
- SVMs and NNs follow different learning methods. NNs typically use the back-propagation algorithm (see Sects. 3.3, 3.3.4) or other algorithms like the gradient descent (see Sect. 3.3.2), other linear algebra based approach. The SVMs learn by solving the linear and quadratic programming problem.

#### 4.3.10.3 Advantages of SVM over NN

- SVMs can achieve a trade-off between the false positives and the false negatives using asymmetric soft margins.

- SVMs always converge to the same solution for a given data set regardless of the initial conditions.
- SVMs do not get stuck into the local minima, ensuring convergence of the operation.
- SVMs remove the danger of overfitting.

## References


Aizerman M, Braverman E, Rozonoer L (1964) Theoretical foundations of the potential function method in pattern recognition learning. Automation and Remote Control 25: 821–837

Boser BE, Guyon IM, Vapnik V (1992) A training algorithm for optimal margin classifiers. In Proc. 5th Annual ACM Workshop on COLT, D. Haussler, editor, pp. 144–152, ACM Press, Pittsburgh, PA

Burges CJC (1998) A tutorial on support vector machines for pattern recognition. Data Mining and Knowledge Discovery 2:121–167

Drucker H, Burges CJC, Kaufman L, Smola A, Vapnik V (1996) Support vector regression machines. Advances in Neural Information Processing Systems NIPS 1996, pp. 155–161, The MIT Press, Cambridge

Fletcher R (1987) Practical methods of optimization, 2nd edition. John Wiley and Sons, Inc., Chichester

Golub GH, Van Loan CF (1996) Matrix Computations, 3rd edition. Johns Hopkins University Press, Baltimore

Gunn SR (1998) Support vector machines for classification and regression. Technical report, University of Southampton, UK

Hastie T, Tibshirani R, Friedman J (2001) The elements of statistical learning. Springer, New York

Hsu CW, Chang CC, Lin CJ (2003). A practical guide to support vector classification. Technical report, Dept. of Computer Science, National University of Taiwan

Karush W (1939) Minima of functions of several variables with inequalities as side constraints. M.Sc. Dissertation. Dept. of Mathematics, Univ. of Chicago, Chicago, Illinois

Keerthi SS, Lin CJ (2003) Asymptotic behaviors of support vector machines with Gaussian kernel. Neural Computation 15:1667–1689

Kuhn HW, Tucker AW (1951) Nonlinear programming. In Proc. of 2nd Berkeley Symposium, pp. 481–492, University of California Press, Berkeley

Mercer J (1909) Functions of positive and negative type and their connection with the theory of integral equations. Philos. Trans. Royal Society, London

Pelckmans K, Suykens J, Van Gestel T, De Brabanter J, Lukas L, Hamers B, De Moor B, Vandewalle J (2003) LS-SVMlab toolbox user's guide, version 1.5.

Schölkopf B, Smola A (2001) Learning with kernels. The MIT Press, Cambridge

Vapnik V (1995) The nature of statistical learning theory. Springer, New York

Vapnik V (1998) Statistical learning theory. Springer, New York

Vapnik V, Chervonenkis A (1971) On the uniform convergence of relative frequencies of events to their probabilities. Theory of Probability and its Applications 16:264–280

# Section II: Application Study

## 4.4 Fault Classification

### *4.4.1 Introduction*

The analysis of faults and disturbances has always been a fundamental foundation for a secure and reliable electrical power supply. The introduction of digital recording technology, e.g. digital fault recorder (DFR), digital protective relay (DPR), etc, opened up a new dimension in the quantity and quality of fault and disturbance data acquisition, resulting in the availability of a huge amount of new information to power systems engineers. Information from the analysis of digital records, particularly the fault data can provide much-needed insight towards analyzing the disturbance as well as performance of protection equipments.

In power systems, faults are either symmetrical (e.g., three phase short-circuit, three phase-to-ground faults, etc) or nonsymmetrical (e.g., single phase-to-ground, double phase-to-ground, double phase short-circuit, etc). Typical probability of occurrence of different types of faults is shown in Fig. 4.29.

The faults considered in Fig. 4.29 are three phase short-circuit (L-L-L), three phase-to-ground (L-L-L-G), line-line short-circuit (L-L), single line-to-ground (L-G) and double line-to-ground (L-L-G) faults, where the terms ‘L’ and ‘G’ refer to ‘Line’ and ‘Ground’ respectively. The probabilities associated with a fault type depend upon the operating voltage and can vary from system to system.

### *4.4.2 Fault Classification*

We have already mentioned in the previous section the different types of faults. A particular important task is to perform the classification in an automated way. This

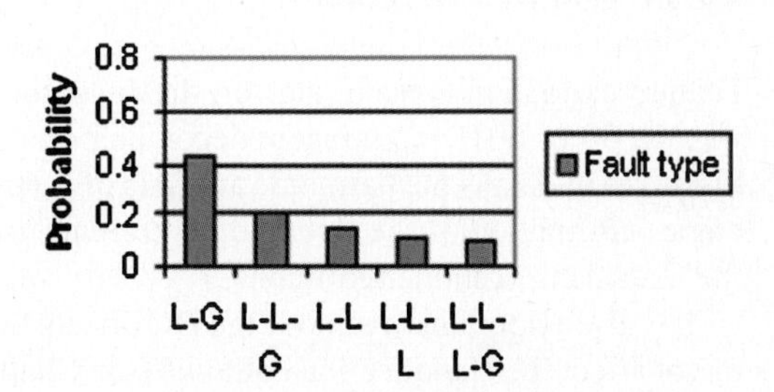


**Fig. 4.29** Probability of occurrence of different faults

is particularly important for automated fault analysis (Kezunovic et al. 2001) which supersedes time-consuming and cumbersome manual analysis. The ultimate aim of automatic fault classification is to analyze the fault signals (three-phase voltage and current as well as neural voltage and current signals) to retrieve the following important information (Keller et al. 2005).

- Faulted phase(s)
- Fault type
- Total fault duration
- Fault location
- Fault resistance
- DC offset
- Breaker operating time
- Auto reclose time.

These are typical requirements of automatic fault analyzer, however, additional features might be included. Nevertheless, a good starting point is to have the fault signals classified according to the fault type. As can be seen form the above list, fault classification also answers the first two requirements.

## *4.4.3 Fault Classifier*

Architecture of a fault classifier is shown in Fig. 4.30. The first module takes the fault signal as input. It is to be noted that a good practice is to use the general IEEE COMTRADE (**Com**mon Format for **Tra**nsient **D**ata **E**xchange) standard (IEEE 1991) for the input and output signal type. Use of COMTRADE allows flexible utilizations of variety of sources of transient data, such as DFRs, DPRs, transient simulation programmes from different manufacturers. The second module extracts essential features from the fault signals. These features are utilized in the third module by a SVM-based classifier to perform the automated fault classification task. The output of the classifier could be the information on the faulted phase and the type of the fault.

### 4.4.3.1 Feature Extraction

Feature extraction concentrates on the fundamental frequency of the fault signal, i.e., the 50 or 60 Hz component depending on the country of operation. Chapter 5 discusses in details the harmonic analysis of a signal. The magnitudes and the phase angle information of the three phase currents are particularly important to design the fault classification algorithm.

It is also important to remove the DC offsets from the fault signals, otherwise that might affect the classification outcome (see Chap. 5, Sect. 5.9.5.4 for more details).

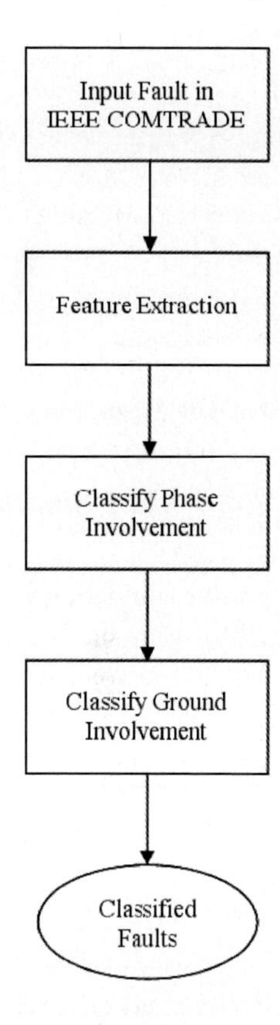


**Fig. 4.30** General architecture of the fault classifier

#### 4.4.3.2 Classification Algorithm

Here we focus on SVMs for two-class classification. First we distinguish between single-phase and multi-phase faults. After the determination of the number of phase involvement, we concentrate on whether the fault involved ground or not. So, we will use two-stage SVMs.

The inputs are the phase-currents. For this, we would utilize the sorted three-phase currents, i.e., we categorize the phase currents in a descending order (Henze 2005). This is depicted in Fig. 4.31. This is done to ensure the improved accuracy of the SVM-classifier.

The sufficient statistic for determining the number of phase involvement in the fault is the separating hyperplane between the plane of the maximum and the second highest phase current. The second classifier uses only the second highest and the lowest phase current. For three-phase faults, the lowest value will be also quiet high compared to the two-phase faults. Once we determine the phase involvement, we utilize the neutral currents for the different classes (maximum, second highest, lowest) to find out the involvement of the ground (Henze 2005; Keller et al. 2005). The flowchart of the classification algorithm is shown in Fig. 4.32.

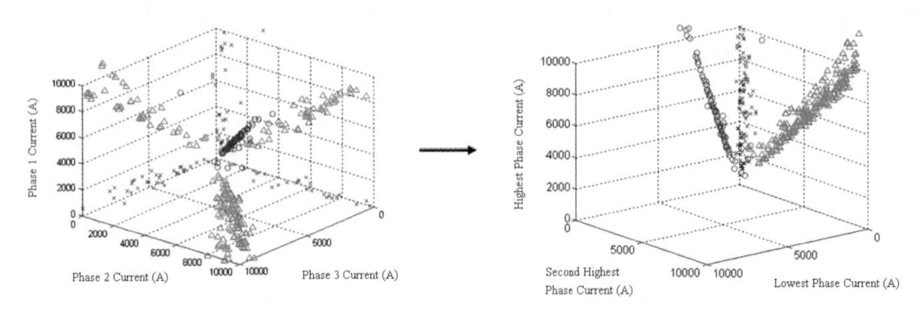


**Fig. 4.31** Inputs: sorted three-phase currents

## *4.4.4 SVM Simulation*

For the simulation we will use a least square support vector machine (Suykens et al. 2002) using the LSSVM toolbox (Pelckmans et al. 2003). A least square SVM is in general faster in speed. The modifications for the least square SVM with regard to the standard SVM classifiers are mentioned below.

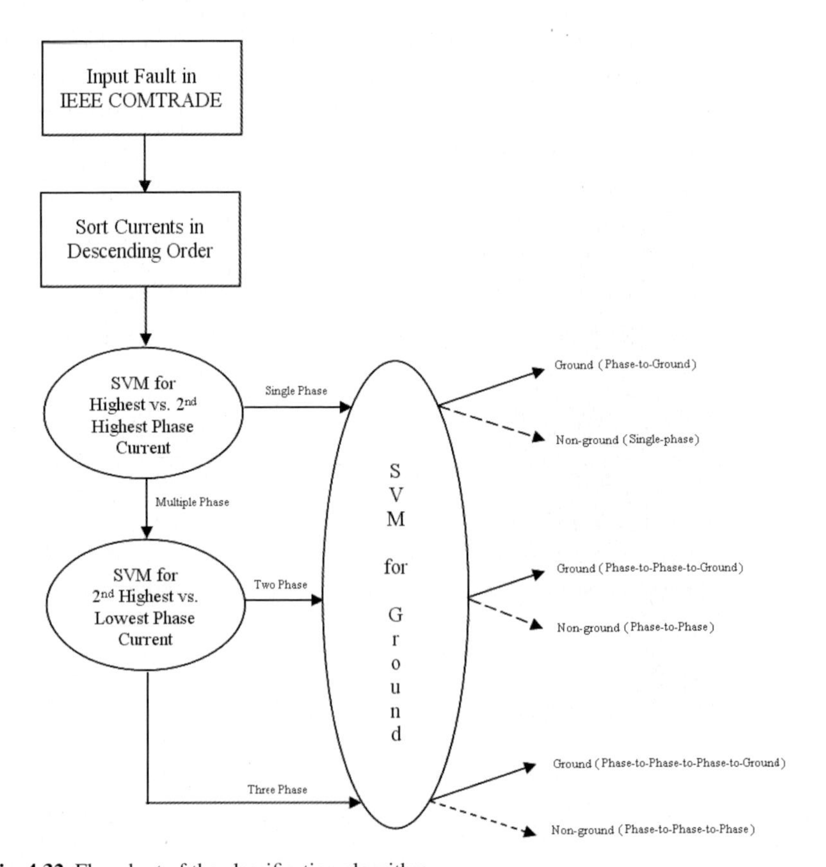


**Fig. 4.32** Flowchart of the classification algorithm

- Target values instead of threshold values are used in the constraints.
- Simplifies the classification problem via equality constraints and least squares.

The LSSVM would be trained using about 100 records for each fault type.

#### 4.4.4.1 Phase Involvement Determination

As per the number of phase involvement detection algorithm, we determine the separating hyperplane between the maximum and the second highest phase currents for bifurcating single and multiple-phase involvement. This is shown in Fig. 4.33. In Fig. 4.33, the pink points (closer to the Y-axis of highest phase current) indicate the single-phase involvement, while the white points (closer to the X-axis of second highest phase current) indicate the multiphase involvement.

We further take the multiphase faults (white points in Fig. 4.33) and plot the second highest ones against the lowest phase currents to bifurcate two-phase and

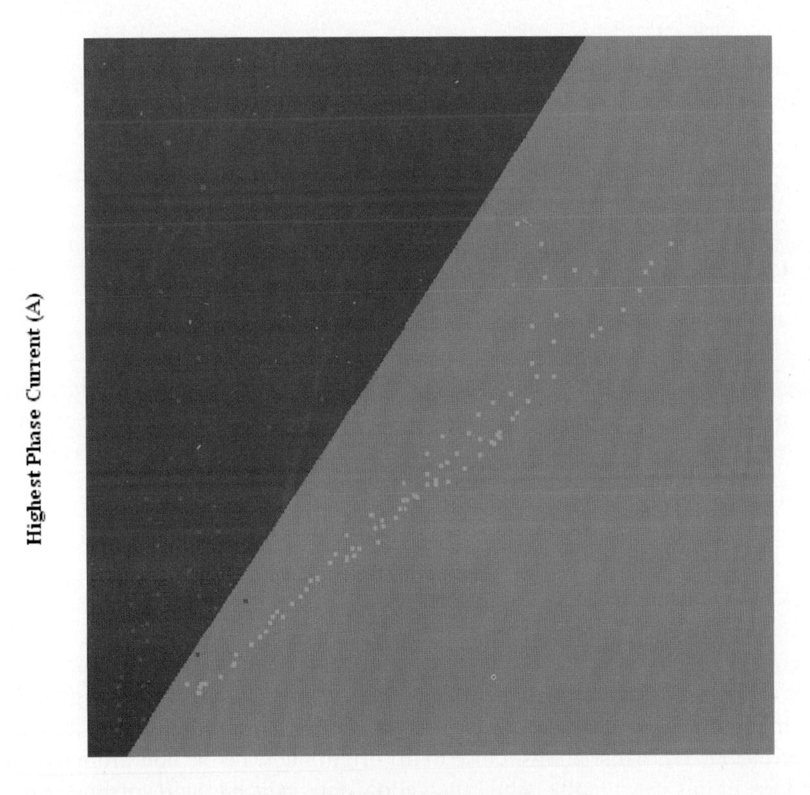


**Fig. 4.33** Determining single- or multi-phase involvement

**Fig. 4.34** Determining two- or three-phase involvement

three-phase involvement, as shown in Fig. 4.34. In Fig. 4.34, the white points (closer to the Y-axis of second highest phase current) indicate the two-phase involvement. Therefore, the rest of the points, i.e., the yellow points (closer to the X-axis of lowest phase current) indicate involvement of all the three-phases.

#### 4.4.4.2 Ground Fault Determination

After we classify the phase involvements, for each class we utilize the neural currents and seek the separating hyperplane. This is shown in Fig. 4.35. In Fig 4.35, we plot the neutral current against the highest phase current for the class with two-phase involvement. The white points (close to the origin) indicate the non-ground (phase-to-phase in this case) faults, while the yellow dots indicate the involvement of the ground, hence they can be classified as phase-to-phase-to-ground fault (L-L-G in Fig. 4.29).

**Fig. 4.35** Determining involvement of ground for two-phase faults

## References

Henze C (2005) Automatic fault classification using support vector machines. Technical report, Department of Electrical Engineering and Information Technology, Dresden University of Technology, Germany

IEEE standard C37.111–1991 (1991) IEEE standard common format for transient data exchange, version 1.8.

Keller P, Henze C, Zivanovic R (2005) Automated analysis of digital fault records-a utility perspective. In Proc. SAUPEC Conf., Jo'burg, South Africa

Kezunovic M, Chen-Ching L, McDonald JR, Smith L (2001) IEEE tutorial on automated fault analysis.

Pelckmans K, Suykens J, Van Gestel T, De Brabanter J, Lukas L, Hamers B, De Moor B, Vandewalle J (2003) LS-SVMlab toolbox user's guide, version 1.5.

Suykens J, Van Gestel T, De Brabanter J, De Moor B, Vandewalle J (2002) Least squares support vector machines. World Scientific Pub. Co., Singapore

# Section III: Objective Projects

## 4.5 Load Forecasting

Load forecasting is used to estimate the electrical power demand. There are many benefits of such prediction, like, proper generation planning, efficient operation and distribution, interruption-free power supply etc. Several techniques for short-, medium- and long-term load forecasting have been discussed, such as Kalman filters, regression algorithms and neural networks (IEEE 1980; Bunn and Farmer 1985).

### *4.5.1 Use of SVM in Load Forecasting*

SVM is a very useful technique for performing task like non-linear classification, function estimation and density estimation and so on. Load forecasting is an application in the area of function estimation (regression).

#### 4.5.1.1 Input and Output

The input and the output depends on the type of load forecasting. Nevertheless, the input and the output are, in general, load consumption in terms of megawatts (MW).

For short-term load forecasting, the output could be the load prediction for the next hour, and load values of the previous hours can be used as the inputs. We can additionally also use the day and hour information.

For medium- and long-term load forecasting, instead of hourly load input values, we could use monthly, yearly, or multiple year load consumption values as the input and the output in similar ranges. Additional inputs could be seasons, demand profile, industrial growth and so on.

Normalization of the input values often works better for the training purpose.

#### 4.5.1.2 SVM Type

Different linear and nonlinear SVM regression could be utilized. One good candidate is nonlinear radial basis function (RBF) kernel SVM because the inputs and the outputs follow a nonlinear relationship in load forecasting.

#### 4.5.1.3 SVM Tool

Several SVM-libraries for Matlab® are available for function approximation and regression. One good tool is the Least Squares SVM (LS-SVM) toolbox (Pelkmans et al. 2003).

### 4.5.2 Additional Task

Neural network is a traditional tool for load forecasting. Load forecasting using different types of neural networks have been discussed in details in Chap. 3, Sect. 3.6.

One interesting task would be to compare the performances of the neural networks with the SVM for different categories of load forecasting. Theoretical differences of the NN and the SVM have been addressed in Chap. 4.3.10.

## References

Bunn DW, Farmer ED (1985) Comparative models for electrical load forecasting. John Wiley & Sons, New York

IEEE Committee (1980) Load forecasting bibliography phase I. IEEE Transactions on Power Applications and Systems 99:53–58

Pelckmans K, Suykens J, Van Gestel T, De Brabanter J, Lukas L, Hamers B, De Moor B, Vandewalle J (2003) LS-SVMlab toolbox user's guide, version 1.5. Catholic University Leuven, Belgium

## 4.6 Differentiating Various Disturbances

### 4.6.1 Magnetizing Inrush Currents

Energizing of a transformer often goes hand in hand with high magnetizing inrush currents. Transformer protection must be set so that the transformer does not trip for this inrush current. This high current is the result of the remnant flux in the transformer core when it was switched out, and depends where on the sine wave the transformer is switched back in (Ukil and Zivanovic 2007). A typical example of magnetizing inrush current is shown in Fig. 4.36.

An interesting application involving the SVM could be to differentiate the magnetizing inrush currents from the normal fault condition (Faiz and Lotfi-Fard 2006). This could be a linear or nonlinear classification task. The two-dimensional feature space could be formed using the magnetizing inrush current and the fault current. Then, linear, Gaussian or RBF kernel SVMs could be used for the classification task.

### 4.6.2 Power Swing

Power swing is an important phenomenon in the power systems disturbance analysis. Power systems under steady-state conditions operate typically close to their

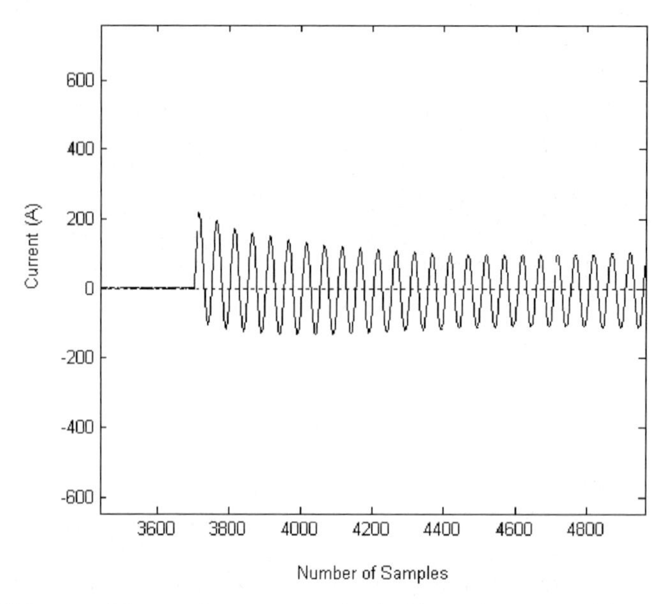


**Fig. 4.36** High magnetizing inrush currents associated with transformer energizing

nominal frequency. A balance between generated and consumed active and reactive powers exists during steady-state operating conditions and the sending and the receiving end voltages are within limit (typically 5%). Power system transient events, like, faults, line switching, generator disconnection, and the large load change result in sudden changes to electrical power, whereas the mechanical power input to generators remains relatively constant. These system disturbances cause oscillations in machine rotor angles and can result in severe power flow swings. Depending on the severity of the disturbance and the actions of power system controls, the system may remain stable and return to a new equilibrium state experiencing what is referred to as a stable power swing. Severe system disturbances, on the other hand, could cause large separation of generator rotor angles, large swings of power flows, large fluctuations of voltages and currents, and eventual loss of synchronism between groups of generators or between neighboring utility systems. Large power swings, stable or unstable, can cause unwanted relay operations at different network locations, which can aggravate further the power-system disturbance and possibly lead to cascading outages and power blackouts (Kundur 1994; Taylor 1993).

Under fault condition also the voltage levels are decreased. Figure 4.37 depicts the decrease in the voltage level during power swing (plot i) and under fault condition (plot ii). Differentiation of the power swing from the normal fault condition is usually a difficult task.

SVM-based classification technique could be utilized to differentiate the power swings from the normal fault condition. The classification could be performed using

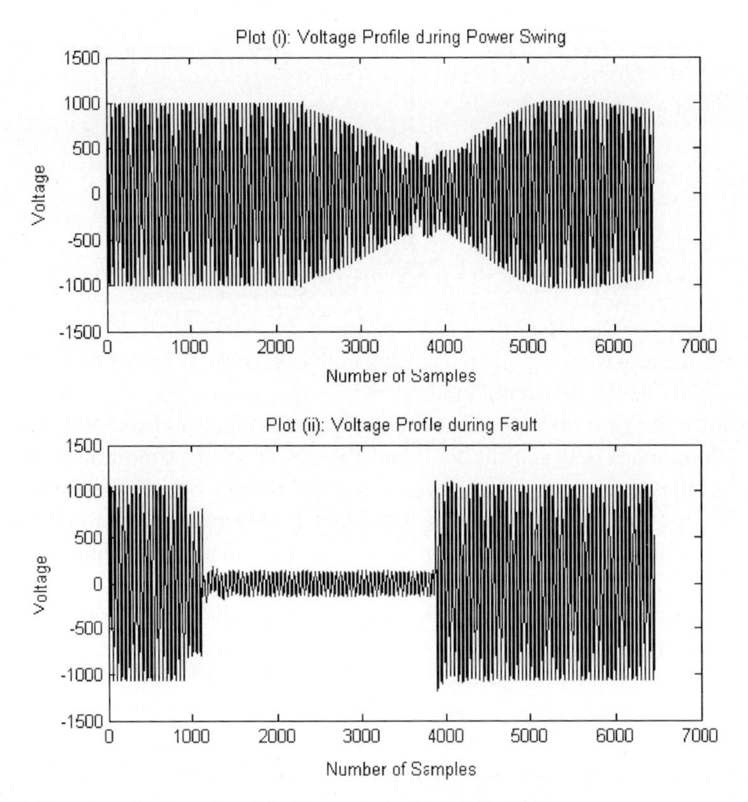


**Fig. 4.37** Decrease of voltage level during power swing and fault

Gaussian or nonlinear RBF kernel SVMs in a feature space comprising of the voltage profiles under fault condition and power swing.

### *4.6.3 Reactor Ring Down*

When a line reactor is connected to a line and the circuit-breakers are opened at both ends, the voltage does not disappear. Instead, an oscillating voltage waveform can be found which slowly reduces in magnitude. This phenomenon is called reactor ring down (Ukil and Zivanovic 2007). It is a result of the interaction between the reactor and the capacitance of the line. This forms an oscillatory circuit, a schematic being depicted in Fig. 4.38.

The voltage profile during the reactor ring down phenomenon is shown in Fig. 4.39.

From the discussion of the power swing in the previous section, it is evident that reactor ring down phenomenon adds more complexity to the disturbance differentiation problem. Hence, a composite SVM-based classifier, in conjunction with that from the Sect. 4.6.2, could be designed to differentiate from each other, the reactor

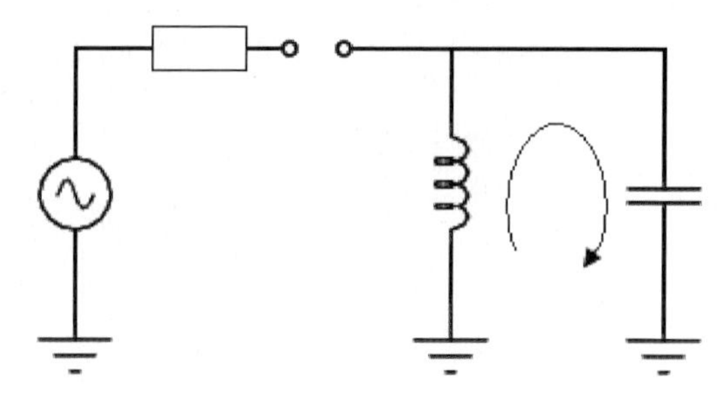

**Fig. 4.38** Schematic circuit of the reactor ring down phenomenon

ring down, the power swing and the normal fault condition. A proposed architecture of such a classifier is shown in Fig. 4.40.

As shown in Fig. 4.40, the classification algorithm employs two SVMs. The first SVM differentiates between the fault and the power swing condition. Composite voltage profiles during the fault, power swing and reactor ring down condition can be used as the feature space on which the first SVM operates. The classification algorithm for the SVM-1 operates on the basis of the detection of abrupt changes (Basseville and Nikoforov 2006). As evident from Fig. 4.37, if the voltage profile undergoes abrupt changes then it is indicative of fault condition. For the other case we can further segment the voltage profile and check the parameters of the segments. Here, we employ the second SVM operating on the composite feature space minus the fault signals. The classification algorithm of the SVM should be designed to check gradually decreasing voltage profile segments. That indicates reactor ring down phenomenon, otherwise it will be power swing.

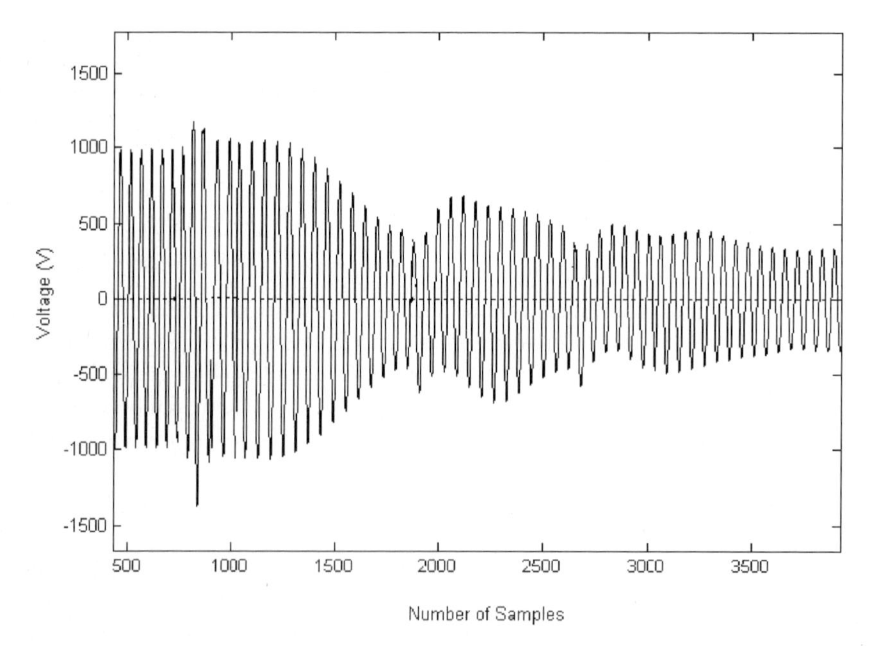


**Fig. 4.39** Voltage profile during reactor ring down phenomenon

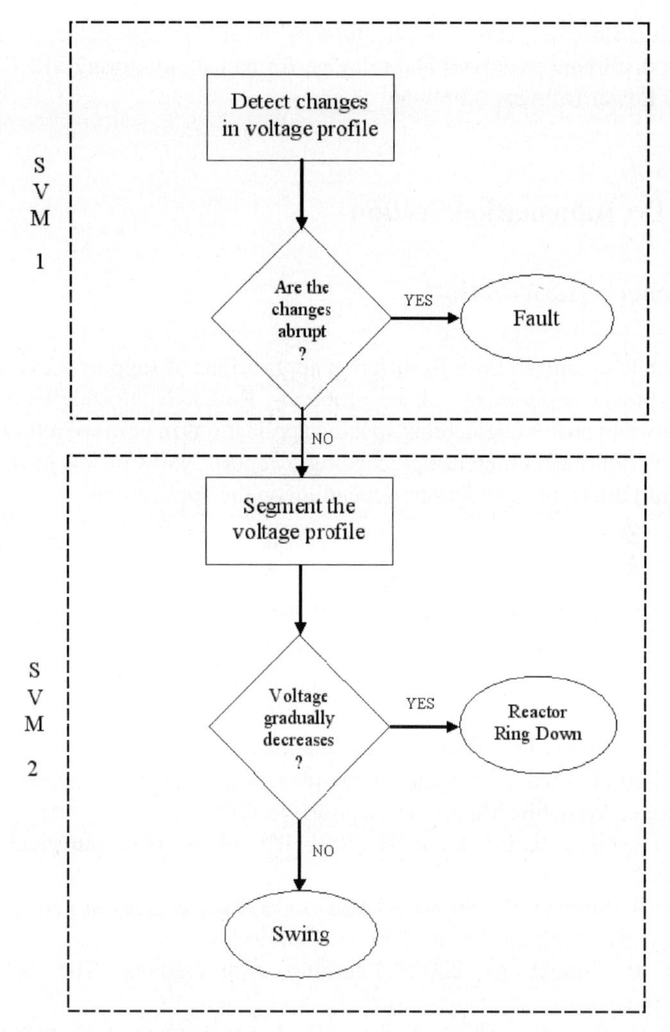


**Fig. 4.40** Architecture for SVM-based classifier for reactor ring down, power swing and fault condition

## References

Basseville M, Nikoforov IV (1993) Detection of abrupt changes – theory and applications. Prentice-Hall, Englewood Cliffs, NJ

Faiz J, Lotfi-Fard S (2006) A novel wavelet-based algorithm for discrimination of internal faults from magnetizing inrush currents in power transformers. IEEE Transactions on Power Delivery 21:1989–1996

Kundur P (1994) Power system stability and control. McGraw-Hill, New York

Taylor CW (1993) Power system voltage stability. McGraw-Hill, New York

Ukil A, Zivanovic R (2007) Application of abrupt change detection in power systems disturbance analysis and relay performance monitoring. IEEE Transactions on Power Delivery 22:59–66

# Section IV: Information Section

## 4.7 Research Information

This section is organized towards different applications of support vector machines in various power engineering related problems. Research information on general SVM theory and power engineering specific applications in terms of relevant books, publications (journal, conference proceedings), reports, software, etc have been categorized into different sub-sections depending on the applications.

### *4.7.1 General Support Vector Machine*

Burges CJC (1998) A tutorial on support vector machines for pattern recognition. Data Mining and Knowledge Discovery 2:121–167

Cristianini N, Taylor JS (2000) An introduction to support vector machines and other kernel-based learning methods. Cambridge University Press, Cambridge, Available: http://www.support-vector.net

Hastie T, Tibshirani R, Friedman JH (2001) The elements of statistical learning. Springer, New York

Schölkopf B, Burges CJC, Smola AJ (Eds.) (1998) Advances in kernel methods: support vector learning. The MIT Press, Cambridge

Schölkopf B, Smola A (2002) Learning with kernels. The MIT Press, Cambridge

Suykens J, Van Gestel T, De Brabanter J, De Moor B, Vandewalle J (2002) Least squares support vector machines. World Scientific Pub. Co., Singapore

Vapnik V (1995) The nature of statistical learning theory. Springer, New York

Vapnik V (1998) Statistical learning theory. Springer, New York

Vojislav K (2001) Learning and soft computing–support vector machines, neural networks and fuzzy logic models. The MIT Press, Cambridge

### *4.7.2 Support Vector Machine Software, Tool*

(The website addresses provided in this section are checked to be valid till the publishing date. However, those might be changed by their owners. Hence, readers are advised to check the validity.)

Chang CC, CJ Lin (2001) LIBSVM: a library for support vector machines. Software available from: http://www.csie.ntu.edu.tw/~cjlin/libsvm
http://www.csie.ntu.edu.tw/~cjlin/libsvmtools

Gunn SR (1997) Support vector machines for classification and regression. Technical report, Image Speech and Intelligent Systems Research Group, University of Southampton
Software available from: http://www.isis.ecs.soton.ac.uk/resources/svminfo

Kernel Machine: Various resources on kernel machines, support vector machines.
Available from: http://www.kernel-machines.org

Pelckmans K, Suykens J, Van Gestel T, De Brabanter J, Lukas L, Hamers B, De Moor B, Vandewalle J (2003) LS-SVMlab toolbox user's guide, version 1.5. Software available from: http://www.esat.kuleuven.ac.be/sista/lssvmlab

### *4.7.3 Load Forecasting*

Lin Z, Xian-Shan L, He-Jun Y (2004) Application of support vector machines based on time sequence in power system load forecasting. Power System Technology 28:38–41

Mohandes M (2002) Support vector machines for short-term electrical load forecasting. International Journal of Energy Research 26:335–345

Müller KR (1999) Predicting time series with support vector machines, Advances in kernel methods-support vector learning. MIT Press, MA

Pai PF, Hong WC (2005) Forecasting regional electricity load based on recurrent support vector machines with genetic algorithms. Electric Power Systems Research 74:417–425

Pai PF, Hong WC (2005) Support vector machines with simulated annealing algorithms in electricity load forecasting. Energy Conv. Manage. 46:2669–2688

Pan F, Cheng HZ, Yang JF, Zhang C, Pan ZD (2004) Power system short-term load forecasting based on support vector machines. Power System Technology 28: 39–42

Sun W, Lu JC, Meng M (2006) Application of time series based SVM model on next-day electricity price forecasting under deregulated power market. In Proc. Int. Conf. on Machine Learning and Cybernetics, pp. 2373–2378

Wu H, Chang X (2006) Power load forecasting with least squares support vector machines and chaos theory. In Proc. 6th World Congress on Intelligent Control and Automation, WCICA 2006, pp. 4369–4373

Xu H, Wang JH, Zheng SQ (2005) Online daily load forecasting based on support vector machines. Int. Conf. on Machine Learning and Cybernetics, ICMLC 2005, pp. 3985–3990

Yang J, Stenzel J (2006) Short-term load forecasting with increment regression tree. Electric Power Systems Research 76:880–888

Yang YX, Liu D (2005) Short-term load forecasting based on wavelet transform and least square support vector machines. Power System Tech. 29:60–64

Zhang MG (2005) Short-term load forecasting based on Support Vector Machines regression. Int. Conf. on Machine Learning and Cybernetics, ICMLC 2005, pp. 4310–4314

### 4.7.4 Disturbance & Fault Analysis

Dash PK, Samantaray SR, Panda G (2007) Fault classification and section identification of an advanced series-compensated transmission line using support vector machine. IEEE Transactions on Power Delivery 22:67–73

Jack LB, Nandi AK (2002) Fault detection using support vector machines and artificial neural networks: augmented by genetic algorithms. Mech. Syst. Signal Process. 16:373–390

Peng L, Da-Ping X, Yi-Bing L (2005) Study of fault diagnosis model based on multi-class wavelet support vector machines. In Proc. Int. Conf. Machine Learning and Cybernetics, 7:4319–4321

Pöyhönen S, Arkkio A, Jover P, Hyötyniemi H (2005) Coupling pairwise support vector machines for fault classification. Control Engg. Practice 13: 759–769

Salat R, Osowski S (2004) Accurate fault location in the power transmission line using support vector machine approach. IEEE Transactions on Power Systems 19:979–986

Thukaram D, Khincha HP, Vijaynarasimha HP (2005) Artificial neural network and support vector machine approach for locating faults in radial distribution systems. IEEE Transactions on Power Delivery 20:710–721

Yan WW, Shao HH (2002) Application of support vector machine non-linear classifier to fault diagnoses. In Proc. 4th World Congress Intelligent Control and Automation, Shanghai, China, pp. 2670–2697

Zhang J, He R (2004) A new algorithm of improving fault location based on SVM. In Proc. IEEE/PES Conf. Power Systems 2:609–612

### 4.7.5 Transient Analysis

Jayasekara B, Annakkkage UD (2007) Transient security assessment using multivariate polynomial approximation. Elec. Power Sys. Research 77:704–711

Li, D.H.; Cao, Y.J. (2005) SOFM based support vector regression model for prediction and its application in power system transient stability forecasting. In Proc. 7th Int. Conf. on Power Engg., IPEC 2005, 2:765–770

Moulin LS, Alves da Silva AP, El-Sharkawi MA, Marks II RJ (2001) Neural networks and support vector machines applied to power systems transient stability

analysis. International Journal of Engineering Intelligent Systems for Electrical Engineering and Communications 9:205–211
Moulin LS, Alves da Silva AP, El-Sharkawi MA, Marks II RJ (2002) Support Vector and multilayer perceptron neural networks applied to power systems transient stability analysis with input dimensionality reduction. In Proc. of the IEEE Power Engg. Society Transm. Distrib. Conf. 3 (SUMMER):1308–1313
Moulin LS, Alves Da Silva AP, El-Sharkawi MA, Marks II RJ (2004) Support vector machines for transient stability analysis of large-scale power systems. IEEE Transactions on Power Systems 19:818–825
Wang S, Wu S, Li Q, Wang X (2005) v-SVM for transient stability assessment in power systems. In Proc. Autonomous Decentralized Systems pp. 356–363
Zhonghong Y, Xiaoxin Z, Zhongxi W (2005) Fast transient stability assessment based on data mining for large-scale power system. In Proc. IEEE/PES Asia Pacific Conf. Transmission and Distributione, pp. 1–6

### 4.7.6 Harmonic Analysis

Li M, Kaipei L, Lanfang L (2005) Harmonic and inter-harmonic detecting based on support vector machine. In Proc. IEEE/PES Asia and Pacific Conf. Transm. Distrib., pp. 1–4
Lobos T, Kozina T, Koglin HJ (2001) Power systems harmonics estimation using linear least squares methods and SVD. IEE Proc. Generation, Transmission, Distribution 148:567–572
Zhan Y, Cheng H (2005) A robust support vector algorithm for harmonic and interharmonic analysis of electric power system. Electric Power Systems Research 73:393–400

### 4.7.7 Power Systems Equipments & Control

Hao L, Lewin PL, Tian Y, Dodd SJ (2005) Partial discharge identification using a support vector machine. Annual Report Conf. on Electrical Insulation and Dielectric Phenomena, pp. 414–417
Li X, Cao G, Zhu XJ (2006) Modeling and control of PEMFC based on least squares support vector machines. Energy Conversion Manage. 47:1032–1050
Schauder C, Mehta H (1991) Vector analysis and control of advanced static VAR compensators. IEE Conference Publication 345:266–272
Shutao Z, Baoshu L, Chengzong P, Jinsha Y (2006) Study on power instrument symbols identifying based on support vector machine. In Proc. Int. Conf. Power System Technology, PowerCon '06, pp. 1–4

### 4.7.8 Power Systems Operation

Chan WC, Chan CW, Cheung KC, Harris CJ (2001) On the modeling of nonlinear dynamic systems using support vector neural networks. Eng. Appl. Artif. Intell. 14:105–113

Chao Y, Neubauer C, Cataltepe Z, Brummel HG (2005) Support vector methods and use of hidden variables for power plant monitoring. In Proc. IEEE Int. Conf. Acoustics, Speech, Signal Processing, ICASSP '05, 5: v/693–v/696

Gottlieb C, Arzhanov V, Gudowski W, Garis N (2006) Feasibility study on transient identification in nuclear power plants using support vector machines. Nuclear Technology 155:67–77

Liu W, Han ZX (2005) Distribution system reconfiguration based on the support vector machine. Dianli Xitong Zidonghua/Automation of Electric Power Systems 29:48–52

Onoda, T.; Ito, N.; Yamasaki, H. (2006) Unusual condition mining for risk management of hydroelectric power plants. In Proc. 6th IEEE Int. Conf. Data Mining Workshops, ICDM '06, pp. 694–698

Sun W, Shen HY, Yang CG (2006) Comprehensive evaluation of power plants' competition ability with SVM Method. In Proc. Int. Conf. Machine Learning and Cybernetics, pp. 3568–3572

Wang R, Lasseter RH (2000) Re-dispatching generation to increase power system security margin and support low voltage bus. IEEE Transactions on Power Systems 15:496–501

Wang Y, Liu JH, Liu XJ (2006) Modeling of superheated steam temperature using sparse least squares support vector networks. In Proc. Int. Conf. Machine Learning and Cybernetics, pp. 3615–3620

Wang Y, Liu JH, Liu XJ, Tan W (2006) Research on power plant superheated steam temperature based on least squares support vector machines. In Proc. 6th World Cong. Intelligent Control and Automation, WCICA '06 1: 4752–4756

### 4.7.9 Power Quality

Hu GS, Xie J, Zhu FF (2005) Classification of power quality disturbances using wavelet and fuzzy support vector machines. Int. Conf. Mchine Learning and Cybernetics, ICMLC 2005, 7:3981–3984

Janik P, Lobos T, Schegner P (2004) Classification of power quality events using SVM networks. IEE Conference Publication 2:768–771

Janik P, Lobos T (2006) Automated classification of power-quality disturbances using SVM and RBF networks. IEEE Trans. Power Delivery 21:1663–1669

Lin WM, Wu CH, Lin CH, Cheng FS (2006) Classification of multiple power quality disturbances using support vector machine and one-versus-one approach. In Proc. Int. Conf. Power System Technology, PowerCon '06, 1:1–8

Peisheng G, Weilin W (2006) Power quality disturbances classification using wavelet and support vector machines. In Proc. 6th Int. Conf. Intelligent Systems Design and Applications, ISDA '06, 1:201–206

Vivek K, Gopa M, Panigrahi BK (2006) Knowledge discovery in power quality data using support vector machine and S-transform. In Proc. 3rd Conf. Information Technology: New Generations, ITNG 2006, pp. 507–512

### 4.7.10 Load Flow

Jin M, Renmu H, Hill DJ (2006) Load modeling by finding support vectors of load data from field measurements. IEEE Transactions Power Systems 21:726–735

### 4.7.11 Power Systems Oscillation

Chen J, Lie TT, Vilathgamuwa DM (2004) Damping of power system oscillations using SSSC in real-time implementation. International Journal of Electrical Power & Energy Systems 26:357–364

Kakimoto N, Sugumi M, Makino T, Tomiyama K (2006) Monitoring of interarea oscillation mode by synchronized phasor measurement. IEEE Transactions on Power Systems 21:260–268

Lee KC, Poon KP (1990) Analysis of power system dynamic oscillations with beat phenomenon by Fourier transformation. IEEE Transactions on Power Systems 5:148–153

### 4.7.12 Power Systems Security

Andersson, C.; Solem, J.E. (2006) Improvements in classification of power system security. In Proc. IEEE PES General Meeting 6:18–22

Gavoyiannis AE, Vogiatzis DG, Georgiadis DP, Hatziargyriou ND (2001) Combined support vector classifiers using fuzzy clustering for dynamic security assessment. In Proc. of the IEEE Power Engineering Society Transmission and Distribution Conference 2:1281–1286

Kim H, Singh C (2005) Power system probabilistic security assessment using Bayes classifier. Electric Power Systems Research 74:157–165

Marei MI, El-Saadany EF, Salama MMA (2004) Estimation techniques for voltage flicker envelope tracking. Electric Power Systems Research 70:30–37

Sidhu TS, Cui L (2000) Contingency screening for steady-state security analysis by using FFT and artificial neural networks. IEEE Transactions on Power Systems 15:421–426

Wehenkel L (1997) Machine learning approaches to power-system security assessment. Expert IEEE 12:60–72

### 4.7.13 Power Systems Stability

Andersson C, Solem JE, Eliasson B (2005) Classification of power system stability using support vector machines. In Proc. 2005 IEEE Power Engineering Society General Meeting 1:650–655

Boonprasert U, Theera-Umpon N, Rakpenthai C (2003) Support vector regression based adaptive power system stabilizer. In Proc. – IEEE Int. Symposium on Circuits and Systems 3:III371–III374

Niu L, Zhao J, Du Z, Jin X (2005) Application of time series forecasting algorithm via support vector machines to power system wide-area stability prediction. In Proc. IEEE/PES Transm. Distrib. Conf.: Asia Pacific, Iss. 2005:1–6

Quoc TT, Sabonnadiere JC, Fandino J (1993) Method for improving voltage stability based on critical voltage. In Proc. of the IEEE International Conference on Systems, Man and Cybernetics 3:746–750

### 4.7.14 Energy Management

Dong B, Cao C, Lee SE (2005) Applying support vector machines to predict building energy consumption in tropical region. Energy and Buildings 37:545–553

### 4.7.15 Energy Market

Dianmin Z, Feng G, Xiaohong G (2004) Application of accurate online support vector regression in energy price forecast. In Proc. 5th World Cong. Intelligent Control and Automation, WCICA 2004, 2:1838–1842

Zheng H, Zhang L, Xie L, Li X, Shen J (2004) SVM model of system marginal price based on developed independent component analysis. In Proc. Power System Technology Conf., PowerCon '04, 2:1437–1440

### 4.7.16 Renewable Energy

Hansen T, Wang CJ (2005) Support vector based battery state of charge estimator. Journal of Power Sources 141:351–358

Junping W, Quanshi C, Binggang C (2006) Support vector machine based battery model for electric vehicles. Energy Conversion and Management 47:858–864

Mohandes MA, Halawani TO, Rehman S, Hussain AA (2004) Support vector machines for wind speed prediction. Renewable Energy 29:939–947

### 4.7.17 Transformers

Ganyun Lv, Cheng H, Zhai H, Dong L (2005) Fault diagnosis of power transformer based on multi-layer SVM classifier. Elec. Power Sys. Res. 75:9–15
Koley C, Purkait P, Chakravorti S (2006) Wavelet-aided SVM tool for impulse fault identification in transformers. IEEE Trans. Power Deliv. 21:1283–1290
Lee FT, Cho MY, Shieh CS, Fang FU (2006) Particle swarm optimization-based SVM application: power transformers incipient fault syndrome diagnosis. In Proc. Int. Conf. on Hybrid Information Technology, ICHIT '06, pp. 468–472
Lv G, Cheng H, Zhai H, Dong L (2005) Fault diagnosis of power transformer based on multi-layer SVM classifier. Electric Power Systems Research 74:1–7

### 4.7.18 Rotating Machines

Poyhonen S, Negrea M, Arkkio A, Hyotyniemi H, Koivo H (2002) Support vector classification for fault diagnostics of an electrical machine. In Proc. 6th Int. Conf. Signal Processing 2:1719–1722
Poyhonen S, Negrea M, Arkkio A, Hyotyniemi H, Koivo H (2002) Fault diagnostics of an electrical machine with multiple support vector classifiers. In Proc. IEEE Int. Symp. Intelligent Control, pp. 373–378
Thomson WT, Fenger M (2001) Current signature analysis to detect induction motor faults. IEEE Transactions on Industry Applications 7:26–34
Widodo A, Yang BS, Han T (2007) Combination of independent component analysis and support vector machines for intelligent faults diagnosis of induction motors. Expert Systems with Applications 32:299–312
Widodo A, Yang BS (2007) Application of nonlinear feature extraction and support vector machines for fault diagnosis of induction motors. Expert Systems with Applications 33:241–250
Yang BS, Han T, Hwang WW (2005) Application of multi-class support vector machines for fault diagnosis of rotating machinery. Journal of Mechanical Science and Technology 19:845–858

### 4.7.19 Power Electronics

Al-Khalifa S, Maldonado-Bascon S, Gardner JW (2003) Identification of CO and NO/sub 2/ using a thermally resistive microsensor and support vector machine. Science, Measurement and Technology, IEE Proceedings 150:11–14
Boni, A.; Pianegiani, F.; Petri, D. (2007) Low-power and low-cost implementation of SVMs for smart sensors. IEEE Transactions on Instrumentation and Measurement 56:39–44
Genov R, Cauwenberghs G (2003) Kerneltron: support vector "machine" in silicon. IEEE Transactions on Neural Networks 14:1426–1434

Ling LP, Azli NA (2004) SVM based hysteresis current controller for a three phase active power filter. In Proc. Power Energy Conf., PECon '04, 132–136
Peng D, Lee FC, Boroyevich D (2002) A novel SVM algorithm for multilevel three-phase converters. In Proc. Power Electronics Special. Conf. 2:509–513
Rohwer JA, Abdallah CT, Christodoulou CG (2003) Least squares support vector machines for fixed-step and fixed-set CDMA power control. In Proc. IEEE Conference Decision and Control 5:5097–5102
Ruijian C, Zheng X, Yixin N (2004) A new current control method based on SVM in /spl alpha//spl beta/0 coordinate system. In Proc. IEEE Power Engineering Society General Meeting, pp. 353–357
Sanchis E, Maset E, Carrasco JA, Ejea JB, Ferreres A, Dede E, Esteve V, Jordan J, Garcia-Gil R (2005) Zero-current-switched three-phase SVM-controlled buck rectifier. IEEE Transactions on Industrial Electronics 52:679–688

# Chapter 5
# Signal Processing

## Section I: Theory

## 5.1 Introduction

The word signal, originated from the Latin word *signum*, indicates message or information. In our real world we deal with numerous such signals. Some of these are sounds, lights, electricity, electromagnetic signals etc. They have different forms and utilizations. However, these signals are like ingredients. We have to process or manipulate them to our needs. That's how we get into the signal processing domain.

Most real world physical signals we encounter come in the analog form. For example, acoustic signals are utilized for hearing, light signals for visualization, radio signals for transmitting information etc. To process them we need to apply mathematical operations on them. One common method is to convert the different physical signals into the electrical domain and then get hold of the enriched mathematical arsenal of the electrical engineering domain. This conversion can be done using a sensor or transducer. A transducer can be of many types depending on which domain it is operating on to convert the physical signal into the electrical signals. For example, microphones are used to convert acoustic signals into electrical signals, similarly photosensor works with light, antenna with electromagnetic signals, piezoelectric sensors with vibration signals and so on. Thus, from the raw physical domain we get into our well-known electrical domain. But still this is analog.

Before the digital revolution, all these analog signals used to be processed (by processing we mean signal conditioning, amplification, filtering etc) in an analog way using op-amps, resistors, inductors, capacitors etc. But now we are into the digital world where we can represent any analog quantity numerically. There are many formats, most common of them being the binary form, i.e., by using 0 and 1. What is the benefit? The answer is more precise and faster computation. Instead of our old bulky resistors, inductors, we can utilize digital computers. Op-amps can be replaced by the tiny microchips. In a nutshell, along with the mathematical arsenal, we also get immense computational power which is also mathematically enriched! So, from the analog signal processing (normally termed as SP), we get into the digital signal processing (DSP). These days signal processing inherently implies DSP, however, the name of this chapter is still signal processing? This is because to

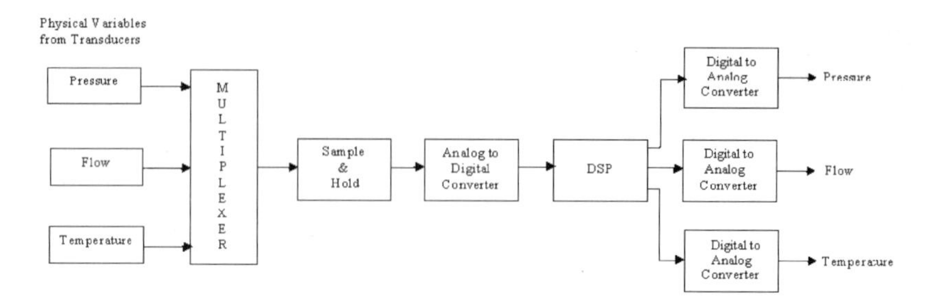


**Fig. 5.1** Digital signal processing

understand the DSP oftentimes we need to refer to its analog brother, so it's better to use the generic term. But to be honest, we would be bit biased towards DSP!

So, from the real world physical signal using sensors we get the analog signals. The analog signals are converted into the digital signals using the analog-to-digital converter (ADC). Then, we process the digital signals the way we like using different mathematical operations. Following this, most often we need to get back the analog signal, so we apply the digital-to-analog converter (DAC). This is how we solve numerous real world problems using digital signal processing. Figure 5.1 depicts it.

### *5.1.1 History and Background*

In the 1980s, introductions of the microprocessors like the Intel® 8086, Rockwell® 6502 etc resulted in a microprocessor oriented data processing revolution. This made the computation equipments cheaper and widely available. Home and personal computers (PCs) came into the market. In 1980, IBM® launched IBM-PC and in 1984 Macintosh® launched its PC version.

With this faster data processing (DP), the field of processing physical signals also flourished. The first special purpose digital signal processor (also termed as DSP) was introduced in the early 1980s. So, the boom period of DSP was also in the 1980s. DSP concentrated on the arithmetic processing of the analog quantities mostly in real time, while DP concentrated on non real time fast processing of stored numerical information.

From the 1980s to 1990s, the processing power of the DSP processors has increased by an order of magnitude, and price decreased. Companies like Analog Devices®, Texas Instrument® concentrated on the special purpose DSP processors.

As a result, the IBM-PC, a key outcome of the microprocessor revolution, is currently evolving into a multimedia PC enabled by the DSP algorithms and techniques.

## 5.1.2 Applications

Signal processing/DSP is a very interdisciplinary and rapidly growing subject. Owing to the fact that most disciplines encounter signals, DSP has a rather broad spectrum of applications as can be found in the references (Rabiner and Gold 1975; Lynn and Fuerst 1999; Ifeachor and Jervis 2001). Some of these are mentioned below.

- Data/Telecommunications
  - Data modems
  - Mobile communications (GSM, CDMA)
  - Signaling
  - Digital telephony
  - Echo cancellation/reduction
- Speech Processing
  - Speech recognition
  - Speech coding
  - Speech synthesis
- Image Processing
  - Digital imaging
  - Image compression (JPEG)
  - Image recognition
- Digital Audio
  - Compact disc (CD)
  - Compressed digital music (MP3)
  - Surround sound
  - MiniDisc
  - iPod™
  - Dolby® Prologic
- Digital Video
  - Digital movie (MPEG)
  - High Definition TV (HDTV)
  - Movie special effects
- Multimedia
  - PC Graphics/Fax, Modem
  - Internet/Email
  - Data compression
  - Video/Teleconferencing
  - Voice over IP (VOIP)

- Industrial
  - Digital process controller
  - Disc drives
  - Motor/drive control
  - Filtering
  - Noise cancellation
  - Computer aided design (CAD) and design tools
- Scientific
  - Data acquisition
  - Spectral analysis
  - Simulation and modeling
- Military
  - Radar
  - Sonar
  - Missile guidance system
  - Secure communication
- Biomedical
  - Electrocardiograph (ECG)
  - Electroencephalograph (EEG)
  - Computer Tomography (CT) scan
  - Magnetic Resonance Imaging (MRI)
  - Medical imaging, storage and retrieval
  - Digital X-ray
  - Digital hearing aids
- Automotive
  - Global Positioning System (GPS) navigation
  - Digital engine/gearbox management.

## 5.2 DSP Overview

The main components of a DSP system are depicted in Fig. 5.1. We shall discuss about them in details in this section to provide an overview of a general digital signal processing system.

### 5.2.1 Digital to Analog Converter (DAC)

A DAC converts a digital (voltage) value into the corresponding analog value. The following parameters are required to specify a DAC.

- **Accuracy**. This is a measure of the DAC as to how close the actual converted analog value is to the real value. For example, if the real value is 10 Volts, and the DAC converted value is 9.9 Volts, the error is 0.1 Volts or 1%. So, the accuracy of the DAC is 99%.
- **Resolution**. It is defined as an output quantity that refers to the difference between two output voltage levels generated by the application of adjacent digital input.
  $Resolution = \frac{V_{ref}}{2^n}$. $V_{ref}$ is the reference voltage and $n$ is the number of bits. For example, for a 3-bit DAC with $V_{ref} = 8$ Volts, we have resolution of 1 Volt, i.e., the smallest voltage change that the DAC can encounter is 1 Volt. To sense a smaller change, e.g., 0.5 Volt we must increase the number of bits to 4.
- **Monotonicity**. A monotonic DAC has output steps either increasing or remaining same for the corresponding increasing digital input steps. If such a converter skips or misses an output by decreasing an output step for a corresponding increasing input step, the converter is said to be non-monotonic. In Fig. 5.2, plot (a) and (b) show monotonic and non-monotonic DAC respectively.

There are usually two types of DAC. These are

- Binary weighted resistors
- R-2R Ladder network.

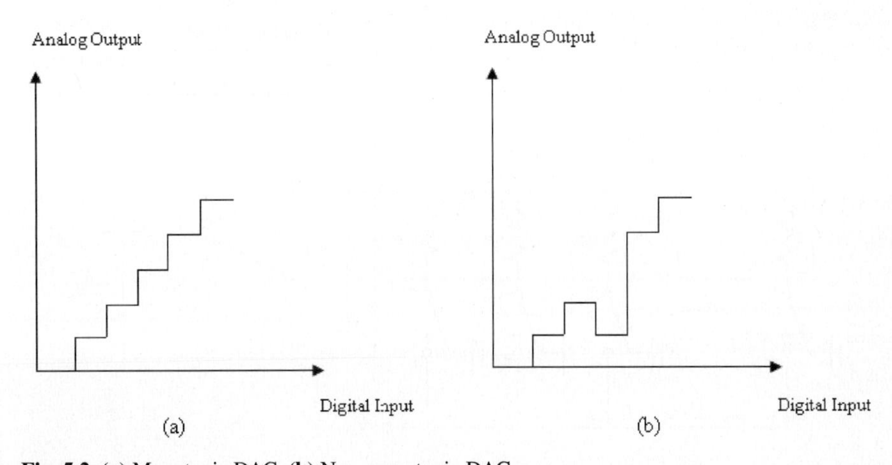


**Fig. 5.2** (**a**) Monotonic DAC, (**b**) Non-monotonic DAC

#### 5.2.1.1 Binary Weighted Resistors type DAC

Figure 5.3 shows a schematic diagram of the binary weighted resistor type DAC. The DAC has $n$ electronic switches $S_n$, where $n$ is the number of bits of the DAC. Corresponding digital inputs are $a_1, a_2, \ldots, a_n$. Using the reference voltage $V_{ref}$, the output analog value $V_{out}$ is given as.

$$V_{out} = V_{ref}\left(a_1.2^{-1} + a_2.2^{-2} + \ldots + a_n.2^{-n}\right) = V_{ref}.\sum_{i=1}^{n} a_i 2^{-i}. \quad (5.1)$$

For a 3-bit DAC ($n = 3$) with $V_{ref} = 8$ Volts, and for a digital value of 110 we get $V_{out} = V_{ref}.\left(1.2^{-1} + 1.2^{-2} + 0.2^{-3}\right) = 0.75\ V_{ref} = 6$ Volts.

#### 5.2.1.2 R-2R Ladder Network DAC

R-2R ladder network type DAC is shown in Fig. 5.4. The operating principle of R-2R ladder network type DAC is same as that of binary weighted resistor type (see 5.1). The only difference is that here only $R$ and $2R$ resistors are used along with the $3R$ resistor at the summing amplifier, while in binary weighted resistor type DAC different resistors values $(2^1R, 2^2R, \ldots, 2^nR)$ are used with the $R$ resistor at the summing amplifier. Hence, ladder network is more accurate than binary weighted resistors in which all the different valued resistors must be of correct precision.

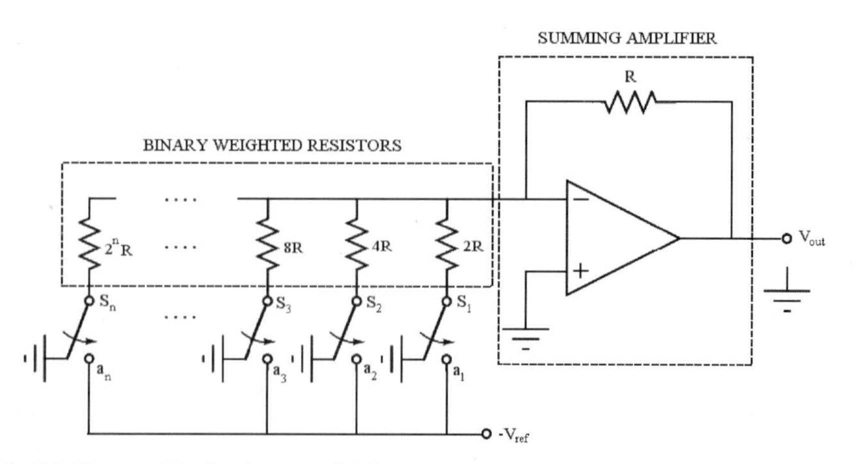


**Fig. 5.3** Binary weighted resistor type DAC

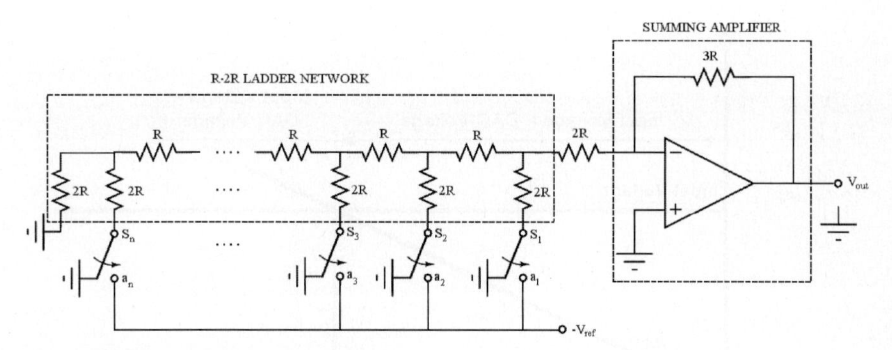


**Fig. 5.4** R-2R ladder network type DAC

## 5.2.2 Analog to Digital Converter (ADC)

An ADC converts an analog voltage value, obtained from the transducers, into a digital value. There are usually two types of ADC.

- Dual slope integration or Ramp type
- Successive Approximation type.

### 5.2.2.1 Ramp type ADC

This is the simplest ADC. The input voltage (analog) is compared against an external digital to analog converter (DAC) which has a ramp type waveform. A ramp type waveform is an increasing straight line passing through the origin. The ramp type external DAC signal is plotted for increasing digital value (corresponding to analog increasing time). The intersection point of the ramp and the input voltage gives the corresponding digital value for the input analog value. The operation is shown in Fig. 5.5.

### 5.2.2.2 Successive Approximation type ADC

Successive approximation type ADC uses an external DAC and an internal comparator. For the $n$-bits, each one is checked successively starting with the most significant bit (MSB) to the least significant bit (LSB). The schematic diagram of a successive approximation type ADC is shown in Fig. 5.6.

For the given input analog value $V_{in}$, for the $n$-bit ADC, first the MSB is set to 1 and the rest of the bits to 0. An internal $n$-bit DAC is used to convert the value to the comparative analog value, $V_{dac}$. $V_{dac}$ is then compared against the $V_{in}$ using the internal comparator. If $V_{in}$ is greater than or equal to $V_{dac}$, the bit is kept to 1, otherwise set to 0. Keeping this, then the next bit is set to 1 and the operation is repeated. This way all the bits from MSB to LSB are checked, resulting in the final analog output value. An example operation is shown in Fig. 5.7, with the tabular explanation in Table 5.1.

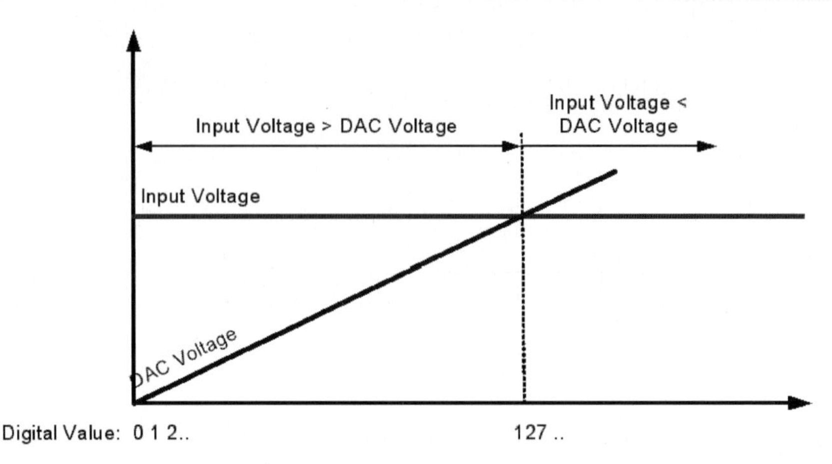


Fig. 5.5 Ramp type ADC operation

## 5.2.3 Quantization

The ADC quantizes an analog signal into a discrete amplitude and thereafter code it into an $n$-bit code.

An 8-bit ADC has $2^8 = 256$ different amplitude levels. Hence, the LSB represents $1/256 = 0.391\%$ of the amplitude range of the ADC and represents a source of noise in the digital representation of analog signals.

For a $\pm 5.12$ Volt amplitude range in the 8-bit ADC, the LSB repressents $10.24/256 = 40\,\text{mV}$.

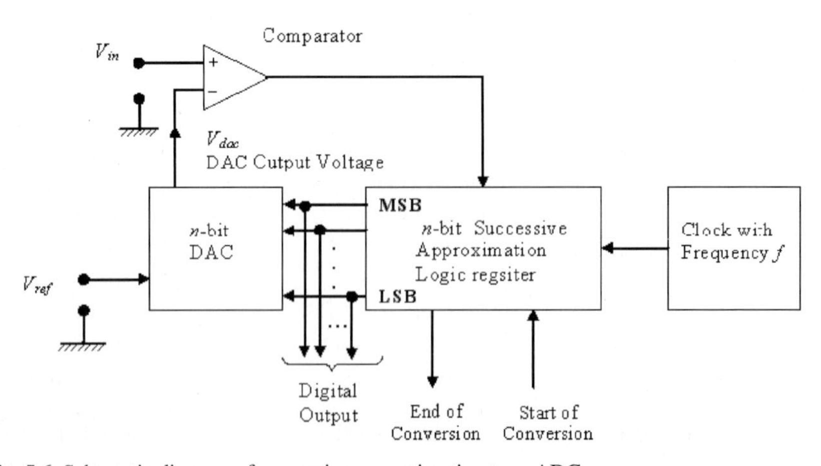


Fig. 5.6 Schematic diagram of successive approximation type ADC

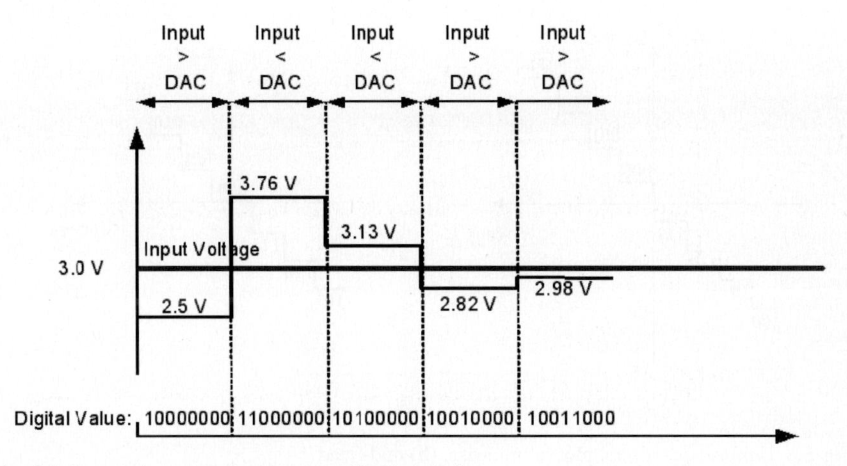


**Fig. 5.7** Successive approximation type ADC operation

For this ADC, the noise power represented by the LSB-error is

$$\sigma_n^2 = \frac{\Delta^2}{8}, \tag{5.2}$$

where, $\Delta$ is the LSB value in Volt.

#### 5.2.3.1 Uniform Quantizers

Uniform quantizers quantize uniformly with a unit value of $\Delta$. There are two types of uniform quantizers, mid-tread and mid-riser. These are shown in Fig. 5.8, plot (a) and (b) respectively. The mid-riser does not have a zero output level, whereas the mid-tread does have a zero output level but there are more levels above the zero level than there are below.

**Table 5.1** Explanation of the successive approximation type ADC operation Example, 8-bit ADC, $V_{in} = 3.0$ Volts

| Bit Value | $V_{dac}$ | $\mathbf{V}_{dac} <= \mathbf{V}_{in}$? | Keep Bit ? |
|---|---|---|---|
| 10000000 | 2.50 | YES | YES |
| 11000000 | 3.76 | NO | NO |
| 10100000 | 3.13 | NO | NO |
| 10010000 | 2.82 | YES | YES |
| 10011000 | 2.98 | YES | YES |
| 10011100 | 3.06 | NO | NO |
| 10011010 | 3.02 | NO | NO |
| 10011001 | 3.00 | YES | YES |

Final digital output = 10011001

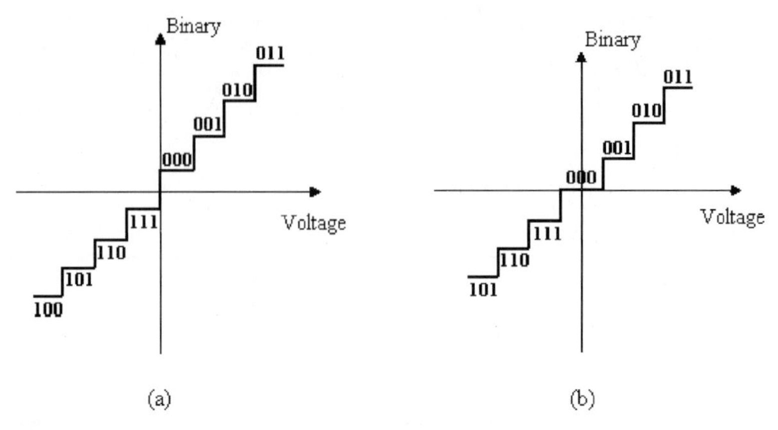


**Fig. 5.8** Uniform quantizers, plot (**a**) mid-rise, (**b**) mid-tread

#### 5.2.3.2 Quantization Signal to Noise Ratio

For a uniform quantizer and an $n$-bit ADC, the following relationship holds for the signal to noise ratio (SNR)[14].

$$SNR(dB) = 6n + 4.77 - 20\log_{10}\left(\frac{X_{max}}{\sigma_X}\right), \tag{5.3}$$

where,

$n$ is the number of bit,
$2X_{max}$ is the peak to peak range of the ADC in Volt,
$\sigma_X$ is the RMS (root mean square) value of the input signal in Volt.
If $X_{max} = 4\sigma_X$ and an $n$-bit ADC is used, then from (5.3) we get,

$$SNR(dB) = 6n - 7.2. \tag{5.4}$$

Therefore, the quantization SNR increases by 6 dB for every additional bit employed in the ADC.

## 5.3 Signals and Systems

So far we have seen the analog to digital conversion and vice versa. In this section we will see more into the details of signals and systems (Oppemheim et al. 1996; Lathi 2001).

[14] SNR, measured using decibels (dB), is typically used to judge the quality of a signal.

### 5.3.1 Discrete-Time Signals

In DSP, most often we deal with the discrete-time signals. What we exactly mean by discrete-time signal is that the signal is not continuous in time, rather composed of discrete points usually at equal intervals. We show a typical example with a sinusoid signal. Sinusoid signals are very important in DSP because we usually get a lot of important mathematical characteristics using them. A continuous sinusoid can be defined as

$$x(t) = A\sin(\omega t + \theta), \tag{5.5}$$

where, A is the amplitude of the sinusoid, $\omega$ is the frequency and $\theta$ is the phase angle. Figure 5.9 shows the continuous sinusoid.

The discrete-time version[15] of the above continuous sinusoid is

$$x[n] = A\sin[\omega n + \theta], \quad n = 0, 1, 2, \ldots \tag{5.6}$$

as shown in Fig. 5.10.

General form of the sinusoidal sequence is

$$x[n] = A\cos\left[\omega_0 n + \theta\right], \quad \forall n. \tag{5.7}$$

That is,

$$\begin{aligned} x[n] &= |A|e^{j(\omega_0 n + \theta)} \\ &= |A|\cos\left[\omega_0 n + \theta\right] + j|A|\sin\left[\omega_0 n + \theta\right]. \end{aligned} \tag{5.8}$$

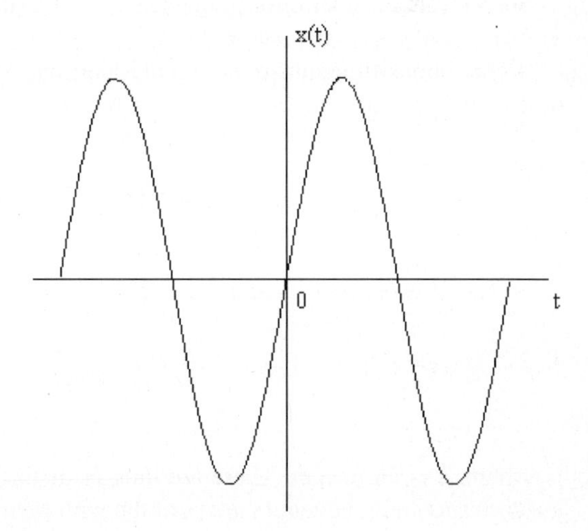


**Fig. 5.9** Continuous sinusoid signal

[15] Note that, continuous-time signals are generally represented using ( ), while discrete-time signals using [ ].

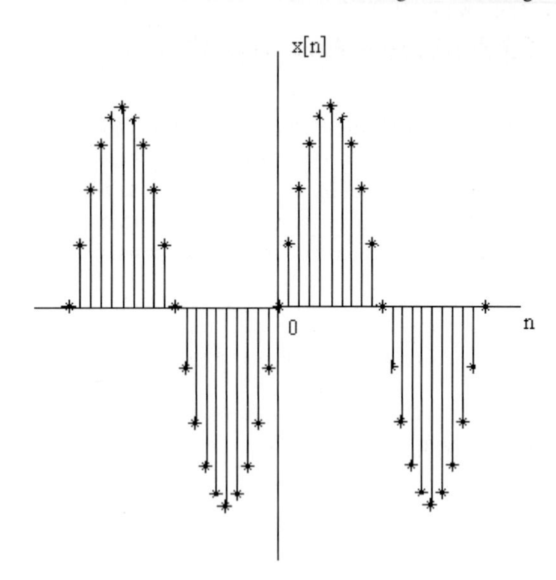


**Fig. 5.10** Discrete-time sinusoid signal

Here, $n$ is a dimensionless integer which indicates the discrete points (in time). So, $\omega_0$ has the dimension of radians. For a frequency of $(\omega_0 + 2\pi)$

$$x[n] = Ae^{j(\omega_0+2\pi)n} = Ae^{j\omega_0 n}e^{j2\pi n} = Ae^{j\omega_0 n}. \tag{5.9}$$

So, we can consider only frequencies in a frequency interval of length $2\pi$ such as, $-\pi < \omega_0 < \pi$ or $0 \leq \omega_0 < 2\pi$.

A continuous-time sinusoidal signal is periodic with period $2\pi/\omega_0$. In the discrete-time case a periodic sequence is one for which

$$x[n] = x[n + N], \quad \forall n, \tag{5.10}$$

where, $N$ is necessarily an integer. So,

$$\omega_0 N = 2\pi k, \tag{5.11}$$

where, $k$ is an integer. There are thus $N$ distinguishable frequencies for which the corresponding sequences are periodic with period $N$. One set of frequencies is

$$\omega_k = 2\pi k/N, \quad k = 0, 1, \ldots, N-1. \tag{5.12}$$

## 5.3.2 Important Discrete-time Signals

### 5.3.2.1 Unit Impulse

This is defined as

$$\begin{aligned} \delta[n] &= 0, \qquad n \neq 0, \\ &= 1, \qquad n = 0. \end{aligned} \tag{5.13}$$

A unit impulse is shown in Fig. 5.11.

### 5.3.2.2 Unit Step

Unit step function is defined as

$$\begin{aligned} u[n] &= 1, \qquad n \geq 0, \\ &= 0, \qquad n < 0. \end{aligned} \tag{5.14}$$

This is shown in Fig. 5.12.

The relationship between the unit step and the unit impulse signal (discrete-time signals can also be referred to as sequences) is as follows.

$$u[n] = \sum_{k=-\infty}^{n} \delta[k]. \tag{5.15}$$

$$\delta[n] = u[n] - u[n-1]. \tag{5.16}$$

### 5.3.2.3 Delayed Sequence

A delayed sequence is defined as

$$x[n] = x\left[n - n_0\right]. \tag{5.17}$$

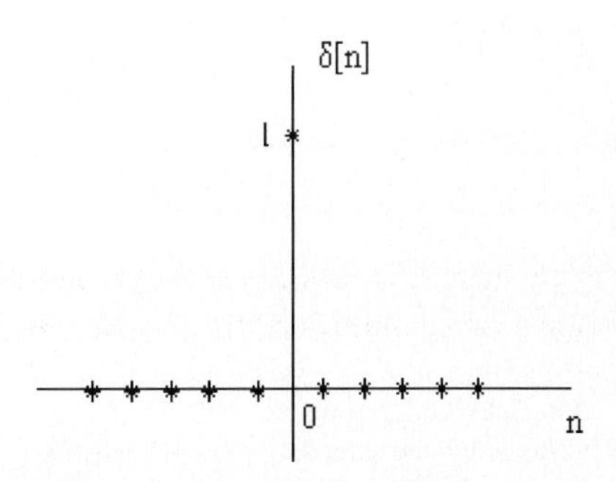


**Fig. 5.11** Unit impulse

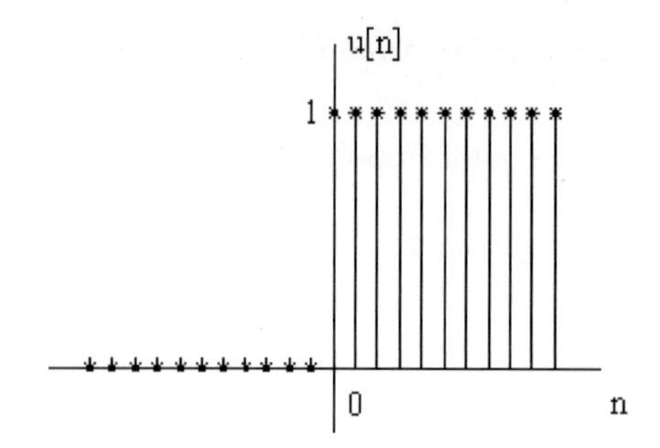


**Fig. 5.12** Unit step

A unit impulse delayed by a sample can be represented as $\delta\left[n-1\right]$. This is shown in Fig. 5.13.

#### 5.3.2.4 Representing Any Sequence

Using the above basic sequences, any arbitrary sequence can be expressed as a sum of scaled and delayed unit impulses. One example is shown below for an example sequence, shown in Fig. 5.14.

$$x\left[n\right]=A_{-4}\delta\left[n+4\right]+A_0\delta\left[n\right]+A_2\delta\left[n+2\right]-A_5\delta\left[n+5\right].$$

More generally, an arbitrary sequence can be expressed as

$$x[n]=\sum_{k=-\infty}^{\infty}x[k]\delta\left[n-k\right]. \tag{5.18}$$

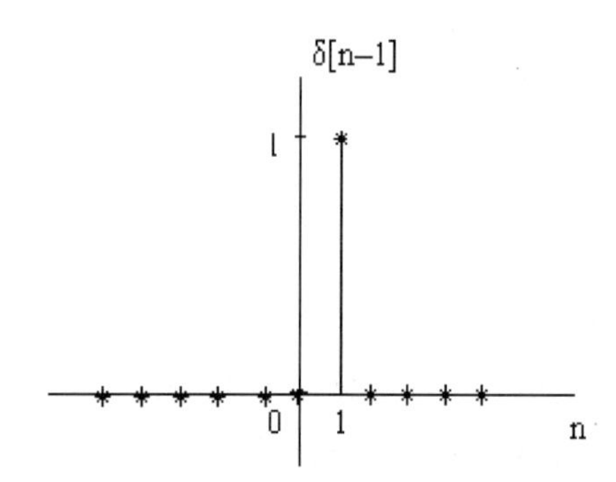


**Fig. 5.13** Unit impulse delayed by one sample

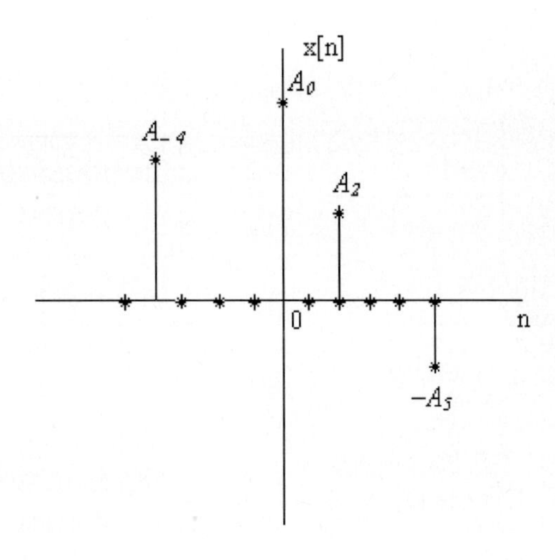


**Fig. 5.14** Representing an arbitrary sequence

### *5.3.3 Linear Shift-Invariant (LSI) System*

A system is a physical or conceptual entity possibly having more than one component with describable behavior (Lathi 2001). Aspects of the behavior of a system are called observable output. A system must have a boundary and at least one output. However, a system may have one or more inputs. Also, a system may have more than one interconnected components. A large part of system theory concerns how behavior of the composite system may be described from the knowledge of behavior of the elements. So, a system can be defined as

$$y[n] = T\left[x[n]\right]. \tag{5.19}$$

A system structure is shown in Fig. 5.15, where if the inside of the system is known it is termed as white box, otherwise black box.

#### 5.3.3.1 Linear Systems

A system is linear if it obeys the homogeneity and the additivity properties. A system has the homogeneity property if an amplitude change in the input results in an

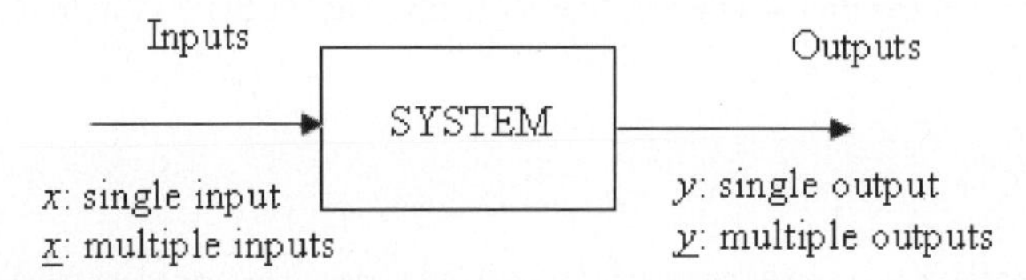


**Fig. 5.15** System structure

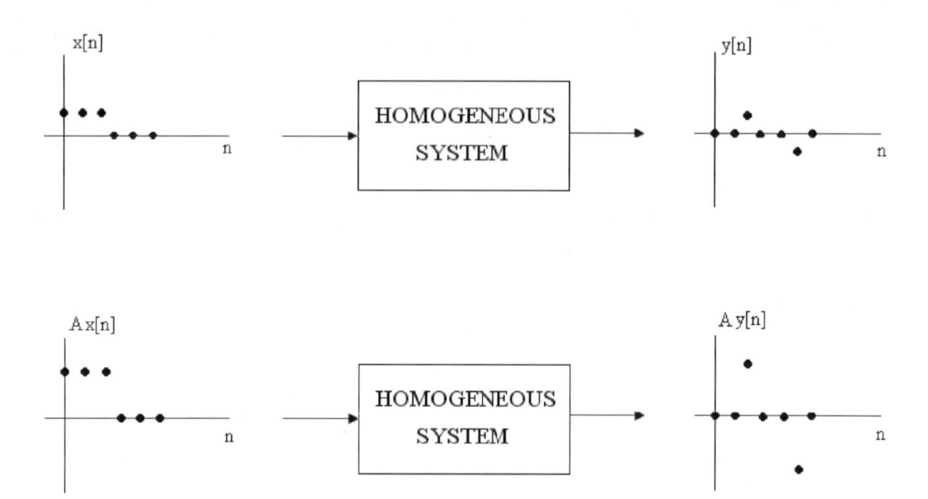


**Fig. 5.16** Homogeneous system

identical and corresponding amplitude change in the output. That is, if $x[n]$ results in $y[n]$, then $Ax[n]$ should result in $Ay[n]$, $A$ being a constant. The homogeneity property is shown in Fig. 5.16.

A system is said to be additive if multiple added inputs produce the corresponding added output. That is, if $x_1[n]$ results in $y_1[n]$, and $x_2[n]$ results in $y_2[n]$, then $x_1[n] + x_2[n]$ result in $y_1[n] + y_2[n]$. Example of an additive system is depicted in Fig. 5.17.

So, combining the above two properties and using (5.19), a linear system can be mathematically defined as

$$T[ax_1[n] + bx_2[n]] = aT[x_1[n]] + bT[x_2[n]] = ay_1[n] + by_2[n]. \quad (5.20)$$

#### 5.3.3.2 Shift Invariance

A system is said to be shift-invariant if a shift in the input produces an identical and corresponding shift in the output. That is, if $x[n]$ results in $y[n]$, then $x[n+k]$ should result in $y[n+k]$, $k$ being a constant. Shift-invariance is depicted in Fig. 5.18.

In DSP, only a linear system is not enough, we also need the shift invariance property. Then altogether we have linear shift-invariant (LSI)[16] system (Oppenheim et al. 1996).

[16] It is to be noted that in system theory, for continuous system, the equivalent thing is the linear time-invariant (LTI) system.

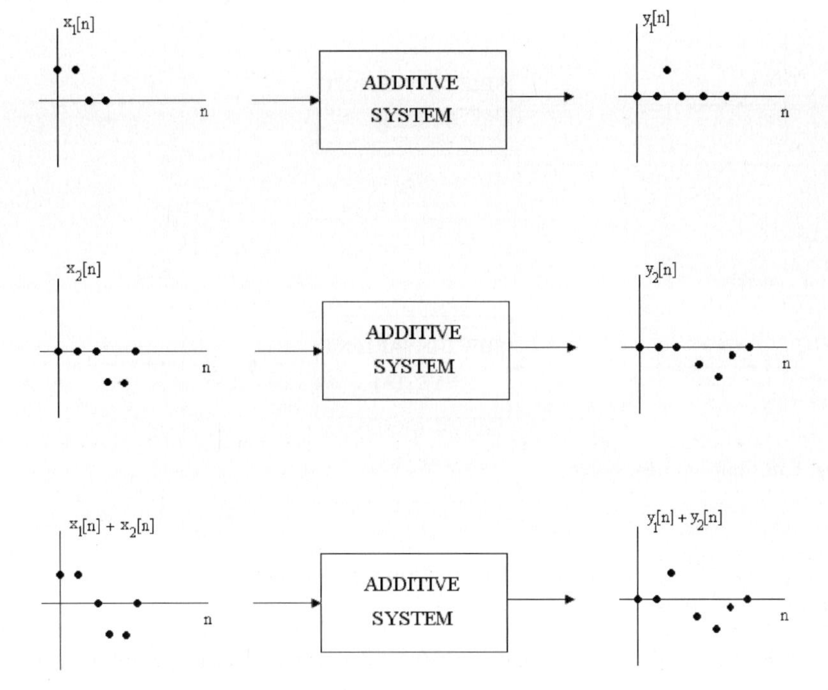


**Fig. 5.17** Additive system

#### 5.3.3.3 Impulse Response

The unit impulse response completely characterizes a linear system.

$$
\begin{aligned}
y[n] &= T\left[\sum_{k=-\infty}^{\infty} x[k]\delta[n-k]\right] \\
&= \sum_{k=-\infty}^{\infty} x[k]T\left[\delta[n-k]\right] \\
&= \sum_{k=-\infty}^{\infty} x[k]h_k[n].
\end{aligned}
\tag{5.21}
$$

If only linearity is imposed, $h_k\,[n]$ will depend on both $h$ and $k$. But for the LSI systems we also have the shift-invariance condition. So, if $h\,[n]$ is the response to $\delta\,[n]$, then the response to $\delta\,[n-k]$ is $h\,[n-k]$. Therefore,

$$y[n] = \sum_{k=-\infty}^{\infty} x[k]h\,[n-k]\,. \tag{5.22}$$

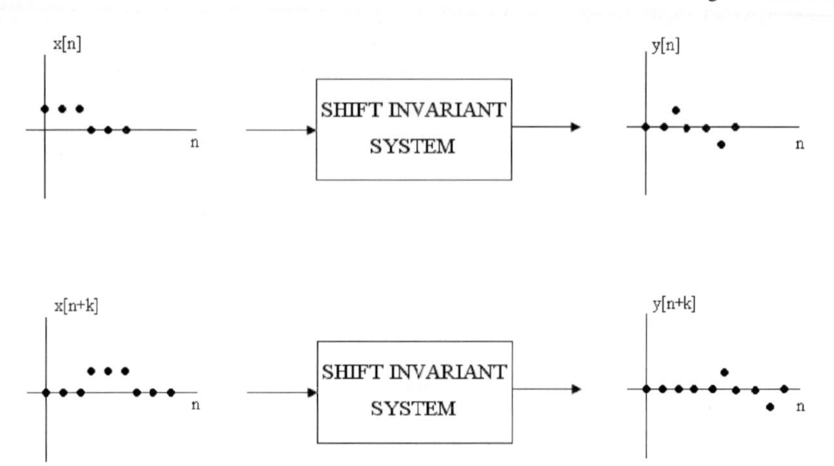


**Fig. 5.18** Shift-invariant system

Equation (5.22) is called the convolution sum $y[n] = x[n]^* h[n]$. We will discuss more about it later.

#### 5.3.3.4 Stability and Causality

A stable system is one for which every bounded input produces a bounded output. An LSI system is stable if and only if

$$\sum_{k=-\infty}^{\infty} |h[k]| < \infty. \quad (5.23)$$

A causal system is one which has a zero output until an input is applied. So, for a causal system

$$y[n] = 0, \quad \forall n < 0. \quad (5.24)$$

#### 5.3.3.5 Combination of LSI Systems

Two LSI systems in cascade form an LSI system with unit impulse response equal to the convolution of the two unit impulses of the cascading systems. The order in which the two sequences are convoluted is unimportant, as shown in Fig. 5.19.

Two LSI systems in parallel are equivalent to an LSI system with unit impulse response equal to the sum of the two unit impulse responses. This is depicted in Fig. 5.20.

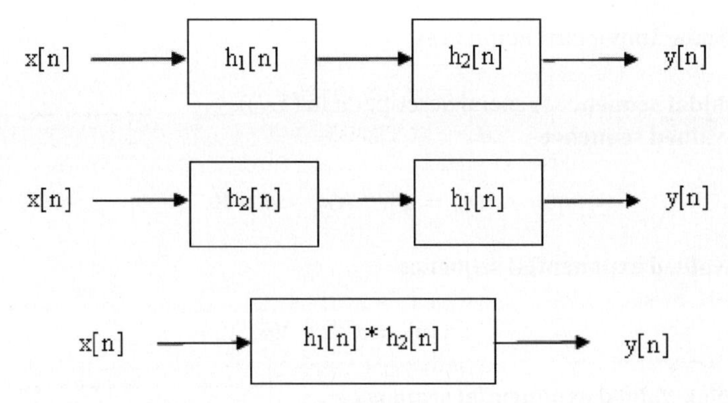


**Fig. 5.19** LSI systems in cascade

## *5.3.4 System Theory Basics*

Now we will see some basics of the system theory (Oppenheim et al. 1996; Lathi 2001; Hsu 1995). For this we will encounter both the discrete and continuous-time signals. In system theory, broadly we deal with two kinds of signals deterministic and random. Deterministic signal is one whose instantaneous value can be predicted in advance without any uncertainty. For example, the output value of the sinusoid signal, described in (5.5), can be computed for any known $A$, $\omega$, $\theta$ and given $t$. In contrast, if the value of the signal cannot be predicted at any moment, the signal is said to be random in nature. There is no explicit mathematical description available for the random signals which are rather characterized by their statistical properties. So, the system theory study is generally based on the deterministic signals, however, in certain cases analysis of the random signals are also necessary.

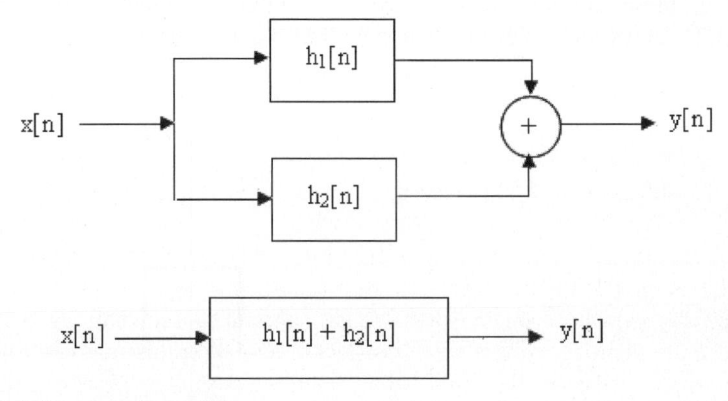


**Fig. 5.20** LSI systems in parallel

#### 5.3.4.1 Some Important Sequences

- **Sinusoidal sequence** (general description in (5.7)).
- **Real-valued sequence**

$$x(n) = a^n, \quad \forall n, \quad a \in \mathbf{R}. \tag{5.25}$$

- **Real-valued exponential sequence**

$$x(n) = e^{\sigma n}, \quad \forall n. \tag{5.26}$$

- **Complex-valued exponential sequence**

$$x(n) = e^{(\sigma + j\omega)n}, \quad \forall n. \tag{5.27}$$

- **Geometric sequence**

$$\sum_{n=0}^{\infty} x^n = \frac{1}{1-x}, \quad |x| < 1, \tag{5.28}$$

$$\sum_{n=0}^{N-1} x^n = \frac{1-x^N}{1-x}, \quad \forall x. \tag{5.29}$$

#### 5.3.4.2 Even and Odd Sequence

An even function or sequence is defined as

$$x_e(n) = x_e(-n). \tag{5.30}$$

Even functions or sequences generally have symmetry against the zero axis. Therefore, these are also called zero axis symmetry function or sequence. Figure 5.21 shows such an even function with zero axis symmetry.

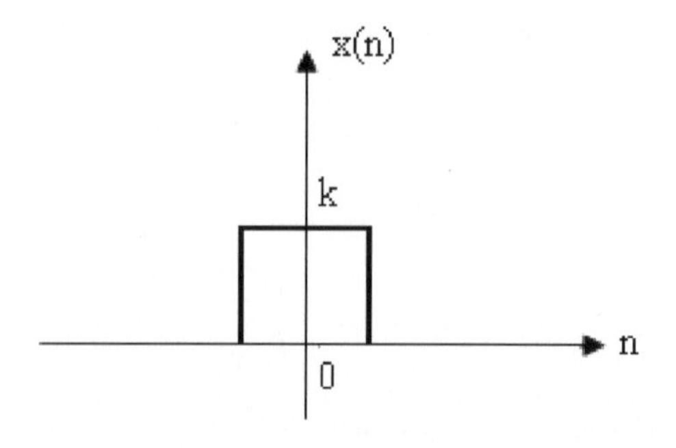


**Fig. 5.21** Even function

An odd function or sequence is defined as

$$x_o(n) = -x_o(-n). \tag{5.31}$$

Odd functions or sequences generally have symmetry against the zero point. Therefore, these are also called zero point symmetry function or sequence. Figure 5.22 shows such an odd function with zero point symmetry.

Even and odd sequences are defined for both the periodic and the aperiodic sequences, i.e., when the sequence repeats itself after certain finite time-interval or does not repeat itself respectively. However, any deterministic function or sequence can be resolved into even and odd sequences.

$$x(n) = x_e(n) + x_o(n). \tag{5.32}$$

Also,[17]

$$\begin{aligned} x(-n) &= x_e(-n) + x_o(-n) \\ &= x_e(n) - x_o(n). \end{aligned} \tag{5.33}$$

So, using (5.32–5.33) we get

$$x_e(n) = \frac{1}{2}\left[x(n) + x(-n)\right]. \tag{5.34}$$

$$x_o(n) = \frac{1}{2}\left[x(n) - x(-n)\right]. \tag{5.35}$$

Figure 5.23 shows a graphical demonstration of the even-odd decomposition.

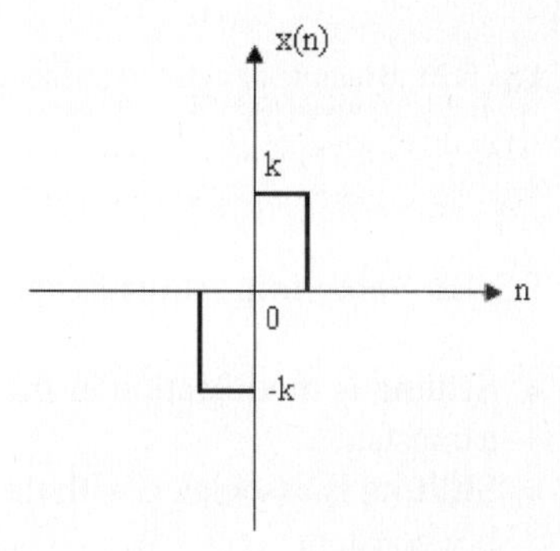


**Fig. 5.22** Odd function

[17] Negation of a function or sequence like $x(-n)$ is also known as folding of $x(n)$.

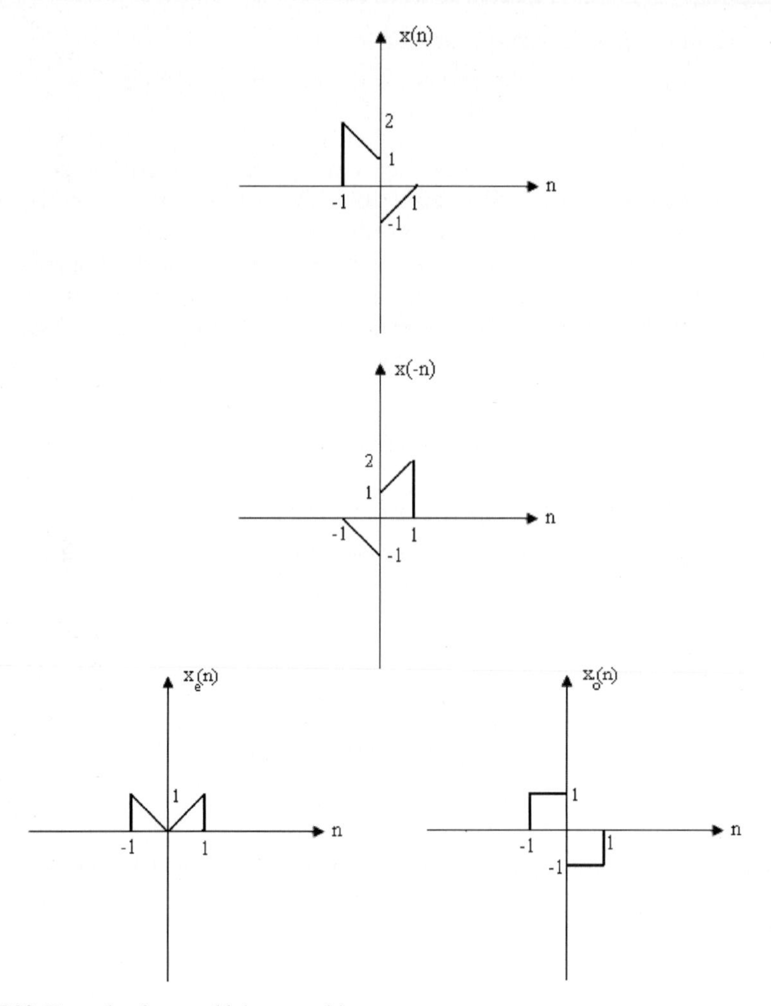


**Fig. 5.23** Example of even-odd decomposition

#### 5.3.4.3 Some Important Properties

- **Scaling** is modification of the amplitude of the function, like, $ax(n)$, where $a$ is a constant.
- **Shifting** is associated with the delaying of the sequence, like, $x(n-k)$, where $k$ is a constant.
- **Energy** of a signal is defined as

$$E = \sum_{n=-\infty}^{\infty} |x(n)|^2. \tag{5.36}$$

- **Power** of a signal (over finite interval) is defined as

$$P = \frac{1}{N}\sum_{n=0}^{N-1} |x(n)|^2. \tag{5.37}$$

- **Cross-correlation**, a measure of signal similarity, of two signals is defined as

$$r_{x,y}(l) = \sum_{n=-\infty}^{\infty} x(n)y(n-l). \tag{5.38}$$

- **Auto-correlation**, measure of self-similarity, is defined as

$$r_{x,x}(l) = \sum_{n=-\infty}^{\infty} x(n)x(n-l). \tag{5.39}$$

### *5.3.5 Convolution*

Convolution (Oppenheim et al. 2001) is a mathematical operation to combine two signals to form a final signal. This is like other mathematical operations like addition, multiplication only difference is that it is a little bit complicated. Convolution is applicable for both the continuous and discrete-time signals. However, in this section we will mostly concentrate on the discrete-time signals.

Convolution[18] is generally associated with the impulse response (see Sect. 5.3.3.3, (5.22)). When we talk about impulse response for a linear system, we can get the output as

$$y[n] = x[n]^{*}h[n], \tag{5.40}$$

where, $h\,[n]$ is the transfer function of the linear system. So, in simple words if we're given an LSI system and an input signal, we apply convolution operation on them as per (5.40) and get the corresponding output. In mathematical notation the convolution sum is expressed as in (5.22). Please note that the order of multiplication is unimportant in convolution. Let's clear the confusions on convolution by an example.

[18] A star in computer program generally indicates multiplication, while in equation it indicates a convolution.

### 5.3.5.1 Convolution Example

We are given an input signal $x[n] = \begin{bmatrix} 3, 11, 7, 0, -1, 4, 2 \\ \uparrow \end{bmatrix}$, $-3 \leq n \leq 3$, for an LSI system $h[n] = \begin{bmatrix} 2, 3, 0, -5, 2, 1 \\ \uparrow \end{bmatrix}$, $-1 \leq n \leq 4$. We need to find out the output $y[n]$.

Here, the arrows for the discrete-time signals indicate the zero-point. So, using this information we can represent the signals in Fig. 5.24 (a) and (b).

So, for $x[n]$ and $h[n]$, the range of $n$ is $-3 \leq n \leq 3$ and $-1 \leq n \leq 4$.

$$\begin{aligned} n &= [-3\ -2\ -1\ \ 0\ \ 1\ \ 2\ \ 3] \\ x[n] &= [\ 3\ \ 11\ \ 7\ \ 0\ -1\ \ 4\ \ 2]' \\ n &= [-1\ \ 0\ \ 1\ \ 2\ \ 3\ \ 4] \\ h[n] &= [\ 2\ \ 3\ \ 0\ -5\ \ 2\ \ 1]. \end{aligned}$$

So, $x[n]$ and $h[n]$ are respectively 7 and 6 points signals. So, the output $y[n]$ would be a $(7 + 6 - 1 =)$12-point signal. Range of $n$ for $y[n]$ would be from $[-3 + (-1) =] - 4$ to $[3 + 4 =]7$, i.e., $-4 \leq n \leq 7$. From this, it is evident that $y[n]$ is a 12-point signal. So, as a rule we can say

$$Length\ of\ y[n] = Length\ of\ x[n] + Length\ of\ h[n] - 1, \tag{5.41}$$

$$\begin{aligned} Range\ of\ y[n] = &-\{Negative\ minimum\ of\ x[n] + h[n]\}\ to \\ &+\{Positive\ maximum\ of\ x[n] + h[n]\} \end{aligned}. \tag{5.42}$$

The method to calculate the 12-point $y[n]$, $-4 \leq n \leq 7$ is as follows.

1. Iterate for axis point of $x[n]$ for the range $-3 \leq n \leq 3$.
2. At each iteration, multiply this value of $x[n]$ with the values of $h[n]$. This will give values for $y[n]$ for the range: axis point of $x[n]$+range of $h[n]$. For example, the most left value of $x[n]$ is 3 for the axis point $-3$. We multiply this value with the $h[n]$ values to get the output vector for this input point

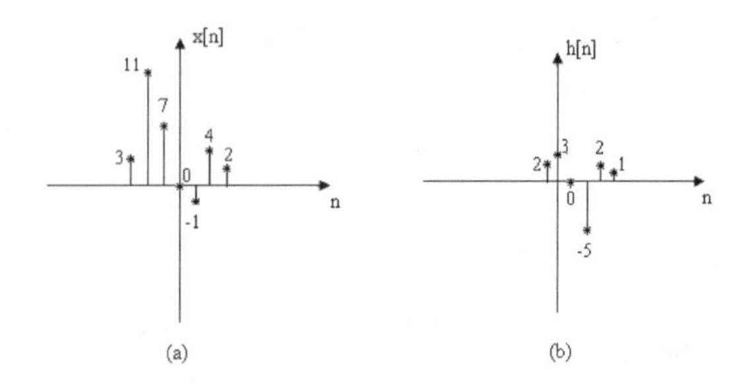


**Fig. 5.24** (**a**) $x[n]$ and (**b**) $h[n]$

as [6 9 0 − 15 6 3], and this covers the range of $-3 + [-1\ 0\ 1\ 2\ 3\ 4] = [-4\ -3\ -2\ -1\ 0\ 1]$ for $y[n]$. However, $y[n]$ ranges for $-4 \leq n \leq 7$. So, the rest six points (from 2 to 7) for $y[n]$ we fill up with zeros. So, for this first input point the output vector of $y[n]$ looks like [6 9 0 − 15 6 3 0 0 0 0 0 0].

3. This way we iterate for the whole range of the input $x[n]$ (i.e., we have 7 iterations for the 7 points of $x[n]$). And at the end we sum up the output vectors of $y[n]$, corresponding to the input axis point, to get the final $y[n]$. This is depicted in Table 5.2.

It is to be noted here that, we iterated over the range of $x[n]$ and at each iteration multiplied with the $h[n]$ vector. We could also do the reverse way. That is, we could iterate over the range of $h[n]$, taking each data point of $h[n]$ and multiply that with the $x[n]$ vector. In this case, we would have an iteration cycle of 6 corresponding to the 6 data points of $h[n]$. As a practice assignment the reader can compute the convolution in this reverse way as shown in Table 5.2 to get the final output $y[n]$ which should be the same. This would demonstrate the property of the invariance of the multiplication order in convolution.

In DSP, convolution is one of the most important operations. This is because, most often we will come up with a system description (the $h[n]$ vector). And we'll design a digital filter (the $x[n]$ vector). The output is nothing but the convolution as described above. Of course, as it seems, the convolution as computed above is computation intensive especially involving long sequences. We'll see that this time-domain convolution could be avoided with the equivalent multiplication operation in the frequency domain which will be described later. But if we are talking about time domain, then this is the convolution we rely upon. For example, when we want to compute the cross-correlation (see (5.38)) between two sequences, we actually fold one signal (interchange the negative and the positive axis point values around the 0-point value) and compute the convolution with the other signal. In case of autocorrelation (see (5.39)), we fold the signal and compute the convolution with the original signal.

**Table 5.2** Convolution example

| Sl. No. | $x[n]$ Value | Axis: Value | −4 | −3 | −2 | −1 | 0 | 1 | 2 | 3 | 4 | 5 | 6 | 7 |
|---|---|---|---|---|---|---|---|---|---|---|---|---|---|---|
| 1 | 3 | | 6 | 9 | 0 | −15 | 6 | 3 | 0 | 0 | 0 | 0 | 0 | 0 |
| 2 | 11 | | 0 | 22 | 33 | 0 | −55 | 22 | 11 | 0 | 0 | 0 | 0 | 0 |
| 3 | 7 | | 0 | 0 | 14 | 21 | 0 | −35 | 14 | 7 | 0 | 0 | 0 | 0 |
| 4 | 0 | | 0 | 0 | 0 | 0 | 0 | 0 | 0 | 0 | 0 | 0 | 0 | 0 |
| 5 | −1 | | 0 | 0 | 0 | 0 | −2 | −3 | 0 | 5 | −2 | −1 | 0 | 0 |
| 6 | 4 | | 0 | 0 | 0 | 0 | 0 | 8 | 12 | 0 | −20 | 8 | 4 | 0 |
| 7 | 2 | | 0 | 0 | 0 | 0 | 0 | 0 | 4 | 6 | 0 | −10 | 4 | 2 |
| Sum $y[n]$ | | | 6 | 31 | 47 | 6 | −51 | −5 | 41 | 18 | −22 | −3 | 8 | 2 |

## 5.4 Laplace, Fourier, Z-Transform

Signal processing is full of *transforms*. For example, we will often encounter Laplace transform, Fourier transform, Z transform, Hilbert transform, Discrete Cosine transform and so on. The nearest similarity of a transform is a function. A function procedurally changes one or more variables into one variable. For example, $x = 5t + 2$, in this function, $x$ is a function of $t$ which is described by the relation. In the case of a transform, both the input and the output have multiple values. In other word, a transform takes a set of value(s) and alters them to values of different forms. However, the different forms are equivalent and transform (and inverse transform) is the bridge in between. Transforms are not only used in signal processing but in many fields. We will revise some of the major transforms in this section.

### *5.4.1 Laplace Transform*

Laplace transform for a causal signal (see Sect. 5.3.3.4) $x(t)$ is defined as

$$L(s) = \int_0^\infty x(t)e^{-st}dt. \tag{5.43}$$

Here, $x(t) = 0$ for $t < 0$. The result of the finite integral shown in (5.43) is known as the Laplace transform and usually represented in terms of $s$. Hence, Laplace transform values are sometimes also referred as the $s$-domain values corresponding to their time ($t$) domain values. The term $e^{-st}$ in (5.43) should be dimensionless. Hence, the unit of $s$ is time-inverse (e.g., second$^{-1}$).

#### 5.4.1.1 Properties of Laplace Transform

1. **Linearity**

$$L[af_1(t) + bf_2(t)] = aF_1(s) + bF_2(s). \tag{5.44}$$

2. **Scaling**

$$Lf(at) = \frac{1}{s}F\left(\frac{s}{a}\right). \tag{5.45}$$

3. **Time Shift**

$$Lf(t-\tau) = e^{-s\tau}F(s). \tag{5.46}$$

4. **Frequency Differentiation**

$$L[(-t)^n f(t)] = \frac{d^n}{ds^n} F(s). \tag{5.47}$$

5. **Time Differentiation**

$$L\left[\frac{d}{dt} f(t)\right] = s\ F(s) - f(0), \tag{5.48}$$

where, $f(0)$ represents the initial value.

6. **Frequency Shift**

$$L[e^{-at} f(t)] = F(s+a). \tag{5.49}$$

#### 5.4.1.2 Laplace Transforms of Common Signals

1. Unit Impluse
$f(t) = \delta(t)$ (see Sect. 5.3.2.1).

$$F(s) = \int_0^\infty \delta(t) e^{-st} dt = e^{-st}|_{t=0}^{\infty} = 1. \tag{5.50}$$

2. Unit Step
$f(t) = u(t)$ (see Sect. 5.3.2.2).

$$F(s) = \int_0^\infty u(t) e^{-st} dt = -\frac{1}{s} e^{-st}|_{t=0}^{\infty} = \frac{1}{s}\ (\mathrm{Re}(s) > 0). \tag{5.51}$$

3. Exponential signal
$f(t) = e^{-at}$.

$$F(s) = \int_0^\infty e^{-at} e^{-st} dt = -\frac{1}{s+a} e^{-(s+a)t}|_{t=0}^{\infty} = \frac{1}{s+a} (s > -a). \tag{5.52}$$

4. Delayed signal
$f(t) = x(t-a)$.

$$\begin{aligned} F(s) &= \int_0^\infty x(t-a) e^{-st} dt = \int_a^\infty x(t-a) e^{-st} dt = \int_0^\infty x(\tau) e^{-s(\tau+a)} d\tau \\ &= e^{-sa} \int_0^\infty x(\tau) e^{-s\tau} d\tau = e^{-sa} X(s) \end{aligned} \tag{5.53}$$

**Table 5.3** Common Laplace transforms

| Time-domain Function | Laplace Transform | |
|---|---|---|
| $1$ | $\frac{1}{s},$ | $(s > 0)$ |
| $t^n$ | $\frac{n!}{s^{n+1}},$ | $(n = 0, 1, 2, 3, ..)$ |
| $e^{at}$ | $\frac{1}{s-a},$ | $(s > a)$ |
| $\sin(at)$ | $\frac{a}{s^2+a^2},$ | $(s > 0)$ |
| $\cos(at)$ | $\frac{s}{s^2+a^2},$ | $(s > 0)$ |
| $\sinh(at)$ | $\frac{a}{s^2-a^2},$ | $(s > \|a\|)$ |
| $\cosh(at)$ | $\frac{s}{s^2-a^2},$ | $(s > \|a\|)$ |
| $e^{at}t^n$ | $\frac{n!}{(s-a)^{n+1}},$ | $(s > a)$ |
| $e^{at}\sin(bt)$ | $\frac{b}{(s-a)^2+b^2},$ | $(s > a)$ |
| $e^{at}\cos(bt)$ | $\frac{s-a}{(s-a)^2+b^2},$ | $(s > a)$ |
| $t\sin(at)$ | $\frac{2as}{(s^2+a^2)^2},$ | $(s > 0)$ |
| $t\cos(at)$ | $\frac{s^2-a^2}{(s^2+a^2)^2},$ | $(s > 0)$ |

Table 5.3 enlists the common Laplace transforms. And Table 5.4 shows the common inverse Laplace transforms.

#### 5.4.1.3 Use of Laplace Transform

Laplace transform is commonly used in system theory for the analysis of the LTI (continuous)/LSI (discrete) systems. Figure 5.25 shows a single input single output (SISO) system (Oppenheim et al. 1996; Lathi 2001).

Laplace transform of the transfer function (system description) of the SISO system $H(s)$ is usually calculated as

$$H(s) = \frac{Y(s)}{X(s)}. \tag{5.54}$$

Using (5.54) we can also calculate the output of a system response. For example, let's assume we know the system transfer function $H(s)$.

**Table 5.4** Common Inverse Laplace transforms

| Laplace Transform | Inverse Laplace Transform |
|---|---|
| $\frac{1}{s}$ | $1$ |
| $\frac{1}{s^n}$ | $\frac{t^{n-1}}{(n-1)!} \quad (n = 1, 2, 3, \ldots)$ |
| $\frac{1}{s-a}$ | $e^{at}$ |
| $\frac{1}{(s-a)^n}$ | $\frac{e^{at}t^{n-1}}{(n-1)!}$ |
| $\frac{1}{s^2+a^2}$ | $\frac{1}{a}\sin(at)$ |
| $\frac{s}{s^2+a^2}$ | $\cos(at)$ |
| $\frac{1}{s^2-a^2}$ | $\frac{1}{a}\sinh(at)$ |
| $\frac{s}{s^2-a^2}$ | $\cosh(at)$ |
| $\frac{1}{(s-a)^2+b^2}$ | $\frac{1}{b}e^{at}\sin(bt)$ |
| $\frac{s-a}{(s-a)^2+b^2}$ | $e^{at}\cos(bt)$ |
| $\frac{s}{(s^2+a^2)^2}$ | $\frac{1}{2a}t\sin(at)$ |
| $\frac{1}{(s^2+a^2)^2}$ | $\frac{1}{2a^3}(\sin(at) - at\cos(at))$ |

1. **Impulse response**
   $x(t) = \delta(t)$. So, $X(s) = 1$.
   Impluse response (output) $= Y(s) = X(s)H(s) = H(s)$.
   So, impulse response in time domain is $y(t) = L^{-1}H(s) = h(t)$.
2. **Step response**
   $x(t) = u(t)$. So, $X(s) = \frac{1}{s}$.
   Step response (output) $= Y(s) = X(s)H(s) = \frac{1}{s}H(s)$.
   So, step response of the system in time domain is $y(t) = L^{-1}\left[\frac{1}{s}H(s)\right]$.

For example, let's consider transfer function of a system is $h(t) = e^{-2t}$. We want to compute the impulse and the step response for this system. The impulse response from above discussion is the same as the system transfer function, $y(t) = e^{-2t}$. This is shown in Fig. 5.26.

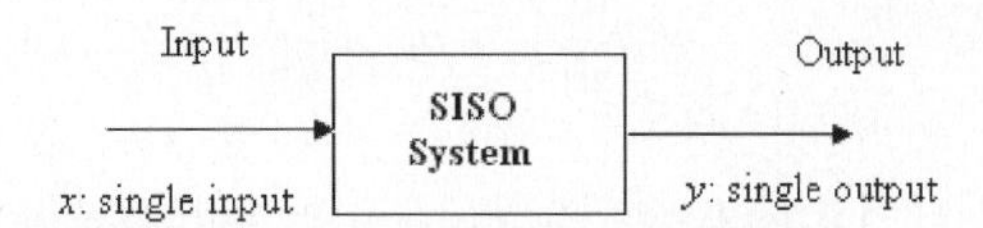


**Fig. 5.25** SISO system

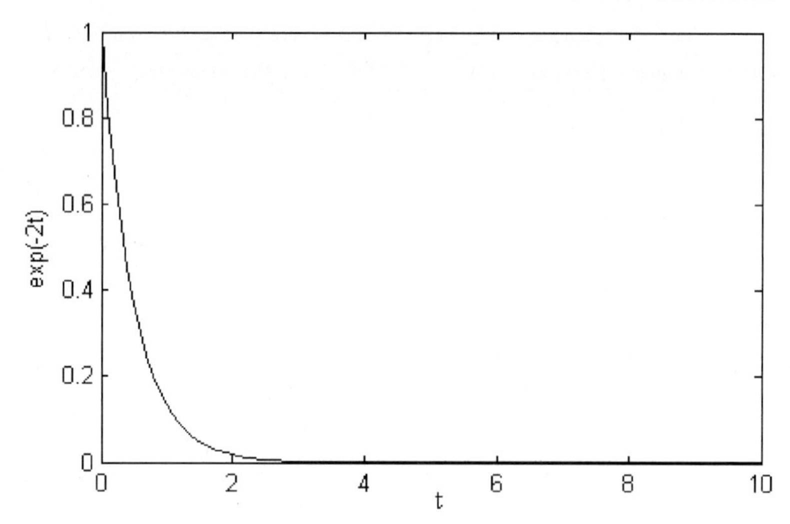


**Fig. 5.26** Impulse response of the system with transfer function $h(t) = e^{-2t}$

This is why if impulse response is known for a system, we say that we know everything about the system. However, in practice it's often very hard to measure the pure impulse response of the system. A more practical measure is the step response.

Step response of the same system would be

$$Y(s) = \frac{1}{s}H(s) = \frac{1}{s(s+2)} = \frac{1}{2}\left[\frac{(s+2)-s}{s(s+2)}\right] = \frac{1}{2}\left[\frac{1}{s} - \frac{1}{(s+2)}\right].$$

$$\therefore y(t) = L^{-1}Y(s) = \frac{1}{2}\left[u(t) - e^{-2t}\right].$$

The above step response is shown in Fig. 5.27.

Another interesting application of the Laplace transform is to transform and analyze systems described by differential equations. This is very useful because often in system theory and system modeling differential equations are used to describe the system behavior. To accomplish this, generally the time differentiation property of the Laplace transform (see (5.48)) is used. For example, consider a system is described as follows.

$$\begin{aligned} a_n\frac{d^n y}{dt^n} + a_{n-1}\frac{d^{n-1}y}{dt^{n-1}} + \ldots + a_2\frac{d^2y}{dt^2} + a_1\frac{dy}{dt} + a_0 y = \\ b_m\frac{d^m x}{dt^m} + b_{m-1}\frac{d^{m-1}x}{dt^{m-1}} + \ldots + b_2\frac{d^2x}{dt^2} + b_1\frac{dx}{dt} + b_0 x \end{aligned} \quad (5.55)$$

Taking Laplace transform on both sides of (5.55) we get,

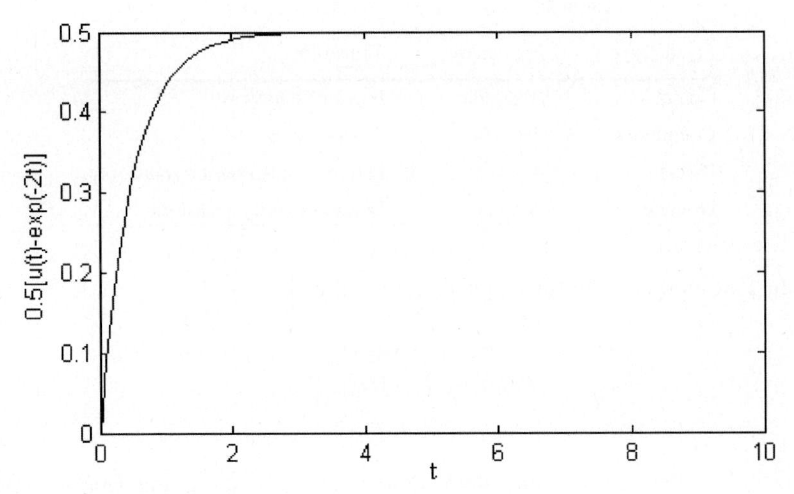


**Fig. 5.27** Step response of the system with transfer function $h(t) = e^{-2t}$

$$\begin{gathered}\left(a_n s^n + a_{n-1} s^{n-1} + \ldots + a_2 s^2 + a_1 s + a_0\right) Y(s) = \\ \left(b_m s^m + b_{m-1} s^{m-1} + \ldots + b_2 s^2 + b_1 s + b_0\right) X(s)\end{gathered}. \tag{5.56}$$

So, the transfer function of the above system is

$$H(s) = \frac{Y(s)}{X(s)} = \frac{b_m s^m + b_{m-1} s^{m-1} + \ldots + b_2 s^2 + b_1 s + b_0}{a_n s^n + a_{n-1} s^{n-1} + \ldots + a_2 s^2 + a_1 s + a_0} = \frac{\sum_{i=0}^{m} b_i s^i}{\sum_{j=0}^{n} a_j s^j}. \tag{5.57}$$

## 5.4.2 Fourier Transform

Fourier transform (Oppenheim et al. 1996, 2001) has a key role to play in signal processing. Originally, the Fourier transform was postulated to represent any continuous periodic signal as a sum of (in general infinite number) of sinusoids. However, in signal processing we are concerned not just about the continuous periodic signals only. We can actually categorize the signals that we encounter in signal processing (both analog and digital) as *continuous* and *discrete*, which can be *periodic* or *aperiodic*. Accordingly, the general term Fourier transform can be classified into four categories as described in Table 5.5. However, the basis of the Fourier transform are the sinusoids. That is why, we said before that sinusoids are one of the most significant signals in the signal processing domain.

**Table 5.5** Fourier transform classification

| Signal Type | Periodicity | Transform |
|---|---|---|
| Continuous | Aperiodic | Fourier Transform |
| Continuous | Periodic | Fourier Series |
| Discrete | Aperiodic | Discrete Time Fourier Transform |
| Discrete | Periodic | Discrete Fourier Transform |

The Fourier transform for a signal $f(t)$ is defined as

$$F(\omega) = \int_{-\infty}^{\infty} f(t)e^{-j\omega t}dt. \tag{5.58}$$

Here, $j$ is the complex operator[19] defined as $j = \sqrt{-1}$. For causal signals, Fourier transform is a subset of the Laplace transform.

$$F(\omega) = F(s)|_{s=j\omega}, \quad if\ f(t) = 0\ for\ t < 0. \tag{5.59}$$

The explanation for this is as follows. See Sect. 5.4.1 that unit of the Laplace operator $s$ used in the Laplace transform is the inverse of time. Hence, $s$ can be thought of as a complex frequency, $s = \sigma + j\omega$, where $\sigma$ and $\omega$ are real variables. Hence, the term in the Laplace transform, $e^{-st} = e^{-(\sigma+j\omega)t} = e^{-\sigma t}.e^{-j\omega t}$. That is, the Laplace transform (with the term $e^{-st}$) incorporates both the exponentials (owing to the real term $e^{-\sigma t}$) and the sinusoids (coming from the complex term $e^{-j\omega t}$)[20], two fundamental and most important signals. In comparison, the Fourier transform (see (5.58)) incorporates only the sinusoids.

The inverse Fourier transform is defined as

$$f(t) = \frac{1}{2\pi}\int_{-\infty}^{\infty} F(\omega)e^{j\omega t}d\omega. \tag{5.60}$$

A sequence can be represented using the Fourier transform as

$$x(n) = \frac{1}{2\pi}\int_{-\infty}^{\infty} X(e^{j\omega})e^{j\omega n}d\omega, \tag{5.61}$$

where, $X(e^{j\omega})$ is given by

[19] This is similar to complex number notation $i$ in mathematics. In electrical engineering as $i$ is generally associated with the current flow, the complex operator is represented using the symbol $j$.
[20] As per Euler's formulae, $\cos\omega t = 1/2\left(e^{j\omega t} + e^{-j\omega t}\right)$, $\sin\omega t = 1/2j\left(e^{j\omega t} - e^{-j\omega t}\right)$.

$$X(e^{j\omega}) = \sum_{n=-\infty}^{\infty} x(n)e^{-j\omega n}. \tag{5.62}$$

It is clear from (5.62) that $X(e^{j\omega})$ is a continuous function of $\omega$ and also a periodic function with period $2\pi$. In general, the Fourier transform is a complex-valued function of $\omega$. Accordingly, the complex-valued $X(e^{j\omega})$ can be represented in the rectangular form as,

$$X(e^{j\omega}) = X_{real}(e^{j\omega}) + jX_{imag}(e^{j\omega}). \tag{5.63}$$

The polar format is

$$X(e^{j\omega}) = \left|X(e^{j\omega})\right| e^{j\angle X(e^{j\omega})}. \tag{5.64}$$

$\left|X(e^{j\omega})\right|$ is the magnitude of the Fourier transform, and $\angle X(e^{j\omega})$ is the phase of the Fourier transform. The magnitude and the phase are also referred to as the amplitude and the angle of the Fourier transform. This is how the Fourier transform converts the *time-domain* information into the *frequency-domain.*

The *frequency response*, $H(e^{j\omega})$, of an LTI-system is the Fourier transform of the impulse response $h(n)$.

$$H(e^{j\omega}) = \sum_{n=-\infty}^{\infty} h(n)e^{-j\omega n}, \tag{5.65}$$

and

$$h(n) = \frac{1}{2\pi} \int_{-\infty}^{\infty} H(e^{j\omega})e^{j\omega n} d\omega. \tag{5.66}$$

However, for $H(e^{j\omega})$ to exist, $x(n)$ must be absolutely summable. That is,

$$\sum_{n=-\infty}^{\infty} |x(n)| < \infty. \tag{5.67}$$

The group delay is defined as the first derivative of the transfer function.

$$\tau(\omega) = \text{grd}\left[H\left(e^{j\omega}\right)\right] = -\frac{d}{d\omega}\left\{\arg\left[H\left(e^{j\omega}\right)\right]\right\}. \tag{5.68}$$

The convolution sum can be represented as

$$y(n) = \sum_{k=-\infty}^{\infty} h(n-k)x(k), \tag{5.69}$$

$$Y\left(e^{j\omega}\right)=H\left(e^{j\omega}\right)X\left(e^{j\omega}\right). \tag{5.70}$$

#### 5.4.2.1 Fourier Transform Properties

We consider a sequence $x(n)$ and the corresponding Fourier transform as $X\left(e^{j\omega}\right)$. Then, the following properties could be observed.

1. For sequence $x^*(n)$, the Fourier transform would be $X^*(e^{-j\omega})$.
2. For sequence $x^*(-n)$, the Fourier transform would be $X^*\left(e^{j\omega}\right)$.
3. For sequence $\mathrm{Re}\{x(n)\}$, the Fourier transform would be $X_e(e^{j\omega})$ (conjugate-symmetric part of $X(e^{j\omega})$).
4. For sequence $j\mathrm{Im}\{x(n)\}$, the Fourier transform would be $X_o(e^{j\omega})$ (conjugate-antisymmetric part of $X(e^{j\omega})$).
5. For sequence $x_e(n)$ (conjugate-symmetric part of $x(n)$), the Fourier transform would be $X_{real}\left(e^{j\omega}\right)$.
6. For sequence $x_o(n)$ (conjugate-antisymmetric part of $x(n)$), the Fourier transform would be $jX_{imag}\left(e^{j\omega}\right)$.
   The following properties apply only when $x(n)$ is real.
7. For any real $x(n)$, the Fourier transform is conjugate-symmetric, i.e., $X\left(e^{j\omega}\right)=X^*\left(e^{j\omega}\right)$.
8. For any real $x(n)$, in the Fourier transform, the real part is even, i.e., $X_{real}(e^{j\omega})=X_{real}(e^{-j\omega})$.
9. For any real $x(n)$, in the Fourier transform, the imaginary part is odd, i.e., $X_{imag}(e^{j\omega})=-X_{imag}(e^{-j\omega})$.
10. For any real $x(n)$, the magnitude of the Fourier transform is even, i.e., $\left|X(e^{j\omega})\right|=\left|X(e^{-j\omega})\right|$.
11. For any real $x(n)$, the phase of the Fourier transform is odd, i.e., $\angle X(e^{j\omega})=-\angle X(e^{-j\omega})$.
12. For any real $x_e(n)$ (even part of $x(n)$), the Fourier transform would be $X_{real}(e^{j\omega})$.
13. For any real $x_o(n)$ (odd part of $x(n)$), the Fourier transform would be $jX_{imag}(e^{j\omega})$.

#### 5.4.2.2 Fourier Transform Theorems

Let us consider the sequences $x(n)$, $y(n)$ and their respective Fourier transforms, $X(e^{j\omega})$, $Y(e^{j\omega})$.

1. **Linearity:** for the sequence $a\,x(n)+b\,y(n)$, the Fourier transform would be $a\,X(e^{j\omega})+b\,Y(e^{j\omega})$.
2. **Time-delay:** for the sequence $x(n-n_0)$ ($n_0$ is an integer), the Fourier transform would be $e^{-j\omega n_0}X(e^{j\omega})$.
3. **Frequency-delay:** for the sequence $e^{j\omega_0 n}x(n)$, the Fourier transform would be $X\left(e^{j(\omega-\omega_0)}\right)$.
4. **Time-reversal:** for the sequence $x(-n)$, the Fourier transform would be $X(e^{-j\omega})$.

5. **Time-scaling:** for the sequence $x\left(\frac{n}{k}\right)$, the Fourier transform would be $X(e^{jk\omega})$.
6. **Frequency differentiation:** for the sequence $n\,x(n)$, the Fourier transform would be $j\frac{d}{d\omega}X(e^{j\omega})$.
7. **Convolution:** for the sequence $x(n)^{*}y(n)$, the Fourier transform would be $X(e^{j\omega})\,Y(e^{j\omega})$.
8. **Multiplication:** for the sequence $x(n)y(n)$, the Fourier transform would be $\frac{1}{2\pi}\int\limits_{-\pi}^{\pi} X\left(e^{j\theta}\right) Y\left(e^{j(\omega-\theta)}\right) d\theta$.
9. **Parseval's theorem**(Oppenheim et al. 2001):

$$\sum_{n=-\infty}^{\infty} |x(n)|^2 = \frac{1}{2\pi}\int\limits_{-\pi}^{\pi} \left|X(e^{j\omega})\right|^2 d\omega, \tag{5.71}$$

$$\sum_{n=-\infty}^{\infty} x(n)y^*(n) = \frac{1}{2\pi}\int\limits_{-\pi}^{\pi} X\left(e^{j\omega}\right) Y^*\left(e^{j\omega}\right) d\omega. \tag{5.72}$$

For real $x(n)$, the energy would be $= \frac{1}{\pi}\int\limits_{0}^{\pi} \left|X(e^{j\omega})\right|^2 d\omega$.

#### 5.4.2.3 Examples

1. Determine the frequency response of a moving average system:

$$h(n) = \begin{cases} \frac{1}{M}, & 0 \le n < M \\ 0, & otherwise \end{cases}.$$

Using (5.65) and the description of the moving average system, we get the frequency response as

$$H(e^{j\omega}) = \sum_{n=-\infty}^{\infty} h(n)e^{-j\omega n} \tag{5.73}$$

$$= \frac{1}{M}\sum_{n=0}^{M-1} e^{-j\omega n} = \frac{1}{M}\cdot\frac{1-e^{-j\omega M}}{1-e^{-j\omega}} = \frac{1}{M}\cdot\frac{e^{-j\omega M/2}\left(e^{j\omega M/2}-e^{-j\omega M/2}\right)}{e^{-j\omega/2}\left(e^{j\omega/2}-e^{-j\omega/2}\right)}$$

$$= \frac{1}{M}\cdot\frac{e^{-j\omega(M-1)/2}\left(e^{j\omega M/2}-e^{-j\omega M/2}\right)/2j}{\left(e^{j\omega/2}-e^{-j\omega/2}\right)/2j} = \frac{1}{M}e^{-j\omega(M-1)/2}\frac{\sin(\omega M/2)}{\sin(\omega/2)}$$

The moving average system, the magnitude and the phase responses are shown in Fig. 5.28 to 5.30.

2. The system function for an LTI-system is defined as $H(e^{j\omega}) = \frac{1-e^{-j2\omega}}{1+e^{-j4\omega}}$, $-\pi < \omega \le \pi$. Determine the output for an input $x(n) = \sin(\pi n/4)$.
We have $H(e^{j\omega}) = \frac{1-e^{-j2\omega}}{1+e^{-j4\omega}}$.

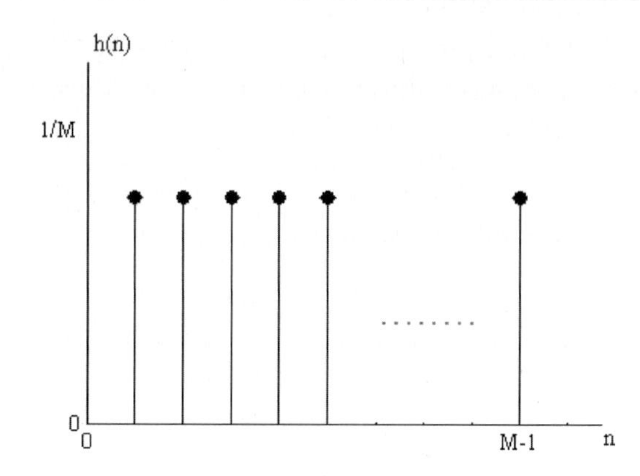


**Fig. 5.28** Moving average system

So,

$$H(e^{j\frac{\pi}{4}}) = \frac{1-e^{-j2\frac{\pi}{4}}}{1+e^{-j4\frac{\pi}{4}}} = \frac{1-(\cos(\pi/2) - j\sin(\pi/2))}{1+\frac{1}{2}(\cos(\pi) - j\sin(\pi))} = \frac{1+j}{1-\frac{1}{2}} \quad . \tag{5.74}$$
$$= 2\sqrt{2}\angle\frac{\pi}{4}$$

Hence, the output for the input $x(n) = \sin(\pi n/4)$ would be, $y(n) = 2\sqrt{2}\sin\left(\frac{\pi n}{4} - \frac{\pi}{4}\right) = 2\sqrt{2}\sin\left[\frac{\pi}{4}(n-1)\right]$. The delay of $\pi/4$ comes from (5.74).

3. Determine the impulse response for the rectangular pulse defined as (shown in Fig. 5.31)

$$X(e^{j\omega}) = \begin{cases} 1, & |\omega| \le \omega_c \\ 0, & \omega_c < |\omega| \le \pi \end{cases} .$$

Using the property of periodicity of $2\pi$ (see (5.60)), we have

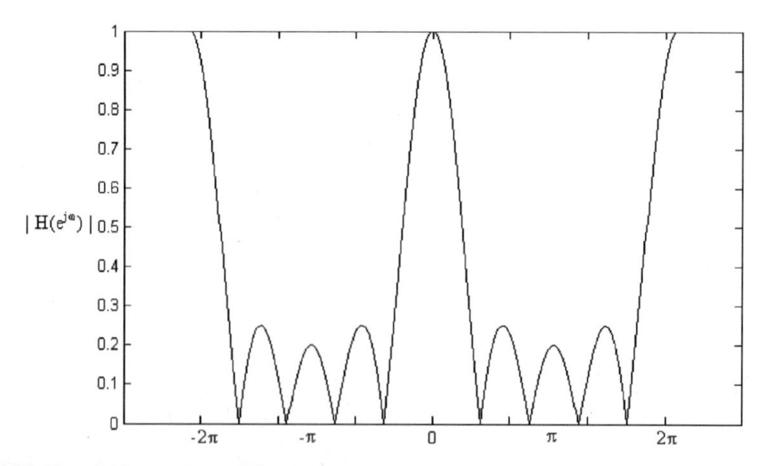


**Fig. 5.29** Magnitude response of the moving average system

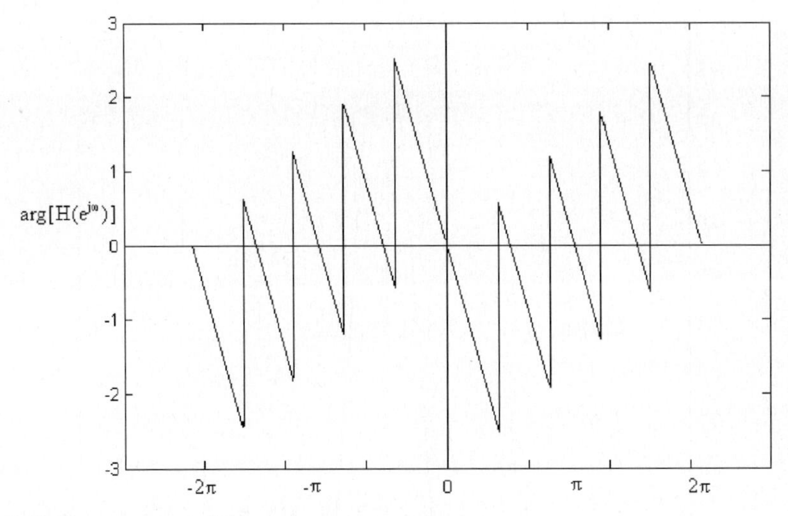


**Fig. 5.30** Phase response of the moving average system

$$x(n) = \frac{1}{2\pi}\int_{-\pi}^{\pi} X(\omega)e^{j\omega n}d\omega = \frac{1}{2\pi}\int_{-\omega_c}^{\omega_c} e^{j\omega n}d\omega = \frac{\sin(\omega_c n)}{\pi n}, -\infty < n < \infty \tag{5.75}$$

The function in (5.75) is called a *sinc* [21] function (Oppenheim et al. 1996, 2001). It is plotted in Fig. 5.32.

#### 5.4.2.4 Fourier Transform Pairs

The time and frequency domains are dual. That is, there is a frequency domain representation for every time-domain signal and vice versa. Example 3 in the previous

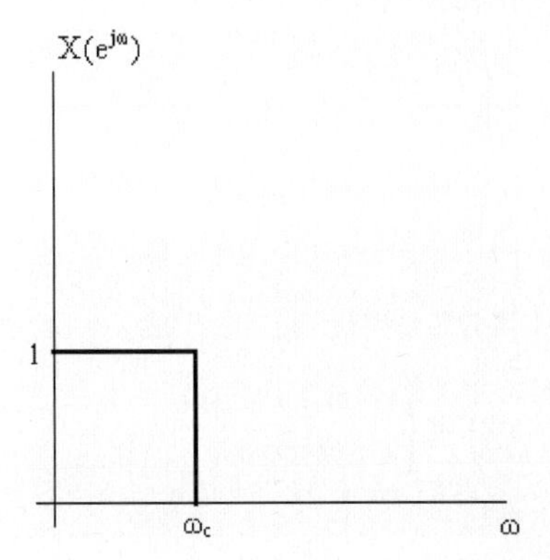


**Fig. 5.31** Impulse response of a rectangular pulse

[21] Pronounced as *sink*

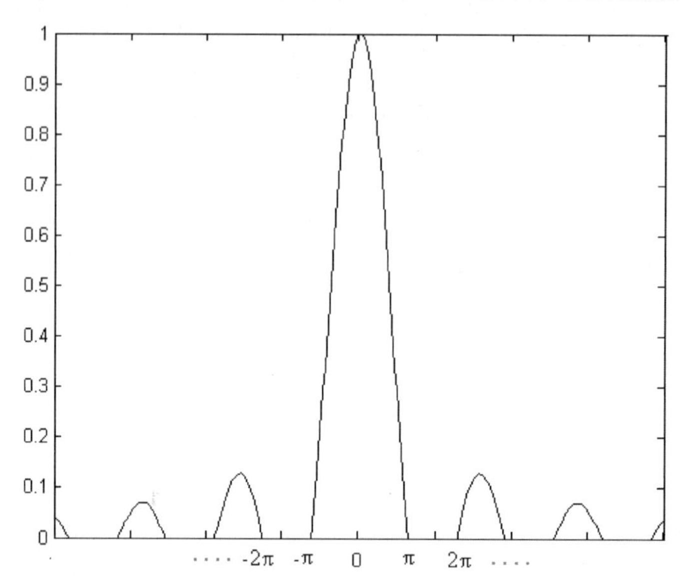


**Fig. 5.32** Sinc function

Sect. 5.4.2.3 shows that the frequency domain representation of a rectangular pulse is the sinc function in time domain. Also, the sinc function in frequency domain results in a time-domain rectangular pulse. Signal pairs conforming to this property are said to be Fourier transform pairs (Oppenheim et al. 2001). Common Fourier transform pairs are listed in Table 5.6.

**Table 5.6** Common Fourier Transform Pairs

| Sequence | Fourier Transform |
|---|---|
| $\delta(n)$ | $1$ |
| $\delta(n-n_0)$ | $e^{-j\omega n_0}$ |
| $1(-\infty < n < \infty)$ | $\sum_{k=-\infty}^{\infty} 2\pi\delta(\omega+2\pi k)$ |
| $a^n u(n) \quad (\lvert a\rvert < 1)$ | $\frac{1}{1-ae^{-j\omega}}$ |
| $u(n)$ | $\frac{1}{1-e^{-j\omega}} + \sum_{k=-\infty}^{\infty} \pi\delta(\omega+2\pi k)$ |
| $(n+1)a^n u(n) \quad (\lvert a\rvert < 1)$ | $\frac{1}{(1-ae^{-j\omega})^2}$ |
| $\frac{r^n \sin\omega_p(n+1)}{\sin\omega_p} u(n) \quad (\lvert r\rvert < 1)$ | $\frac{1}{1-2r\cos\omega_p e^{-j\omega}+r^2e^{-j2\omega}}$ |
| $\frac{\sin\omega_c n}{\pi n}$ | $X(e^{j\omega}) = \begin{cases} 1, & \lvert\omega\rvert < \omega_c, \\ 0, & \omega_c < \lvert\omega\rvert \le \pi \end{cases}$ |
| $x(n) = \begin{cases} 1, & 0 \le n \le M \\ 0, & otherwise \end{cases}$ | $\frac{\sin[\omega(M+1)/2]}{\sin(\omega/2)} e^{-j\omega M/2}$ |
| $e^{j\omega_0 n}$ | $\sum_{k=-\infty}^{\infty} 2\pi\delta(\omega-\omega_0+2\pi k)$ |
| $\cos(\omega_0 n+\phi)$ | $\pi \sum_{k=-\infty}^{\infty} \left[e^{j\phi}\delta(\omega-\omega_0+2\pi k)+e^{-j\phi}\delta(\omega+\omega_0+2\pi k)\right]$ |

### 5.4.3 Z-Transform

The Laplace transform is a generalization of the Fourier transform (FT) for continuous-time systems. The Z-transform (ZT) (Oppenheim et al. 1996, 2001) is a generalization of the Fourier transform for discrete-time signals. Z-transform for a discrete sequence $x[n]$ is defined as

$$X(z) = \sum_{n=-\infty}^{\infty} x[n]\, z^{-n}. \tag{5.76}$$

Here, $z$ is complex. The relationship between the ZT and the FT is as follows.

$$\text{Complex } z = r\, e^{j\omega}. \tag{5.77}$$

Therefore,

$$X(re^{j\omega}) = \sum_{n=-\infty}^{\infty} x[n]\, (re^{j\omega})^{-n} = \sum_{n=-\infty}^{\infty} x[n]\, r^{-n} re^{j\omega n}. \tag{5.78}$$

So, we have

$$\text{ZT} = \text{FT} \quad if \quad |z| = 1. \tag{5.79}$$

Figure 5.33 shows the description of $z$ in the complex $Z$-plane, specifically the unit circle corresponding to the (5.79). It is to be noted from Fig. 5.33 that the FT is defined only on the unit circle, while ZT on the whole $Z$-plane. That's why FT is a special case of ZT.

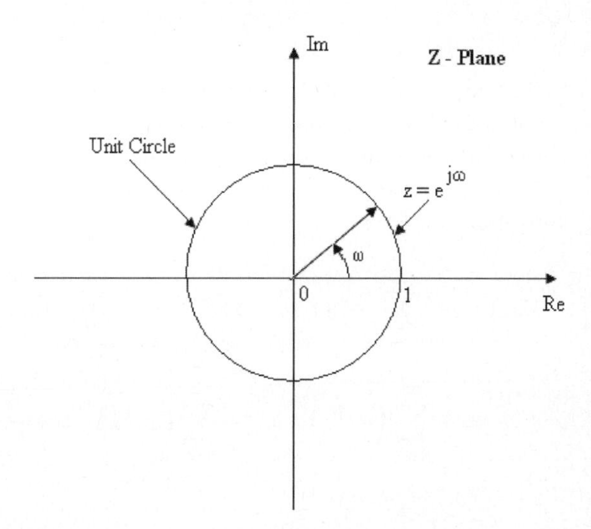


**Fig. 5.33** The unit circle in the complex $z$-plane

#### 5.4.3.1 Convergence of the Z-Transform

The set of values of $z$ for which the ZT converges, is called the *Region of Convergence* (ROC). Therefore, for convergence,

$$\sum_{n=-\infty}^{\infty} \left|x[n] r^{-n}\right| < \infty. \tag{5.80}$$

Therefore, ZT can converge although the FT does not converge. In general, the ZT converges in the following areas

$$R_{\infty-} < |z| < R_{\infty+}. \tag{5.81}$$

Let us see an example to find out the ROC for $x[n] = a^n u[n]$. Following the definition of the ZT,

$$X(z) = \sum_{n=-\infty}^{\infty} a^n u[n] z^{-n} = \sum_{n=0}^{\infty} \left(az^{-1}\right)^n = \frac{1}{1 - az^{-1}} = \frac{z}{z-a}. \tag{5.82}$$

For convergence we must satisfy the condition

$$\sum_{n=0}^{\infty} \left|az^{-1}\right|^n < \infty. \tag{5.83}$$

This implies that

$$\begin{aligned} &\left|az^{-1}\right| < 1 \\ &\therefore |z| > a \\ &\therefore R_{\infty-} = a, \quad R_{\infty+} = \infty \end{aligned} \tag{5.84}$$

This sequence is called a *right-sided sequence*. The ROC is indicated in Fig. 5.34 by the shaded area. The unit circle (i.e., circle with radius $|z| = 1$) is inside the ROC. Hence, the FT is existing for this particular example.

However, if we consider a sequence like $x[n] = -a^n u[-n-1]$, we get the corresponding ZT as

$$\begin{aligned} X(z) &= -\sum_{n=-\infty}^{\infty} a^n u[-n-1] z^{-n} = -\sum_{n=-\infty}^{-1} \left(az^{-1}\right)^n \\ &= -\sum_{n=1}^{\infty} \left(a^{-1}z\right)^n = 1 - \sum_{n=0}^{\infty} \left(a^{-1}z\right)^n = 1 - \frac{1}{1-a^{-1}z} = \frac{z}{z-a} \end{aligned} \tag{5.85}$$

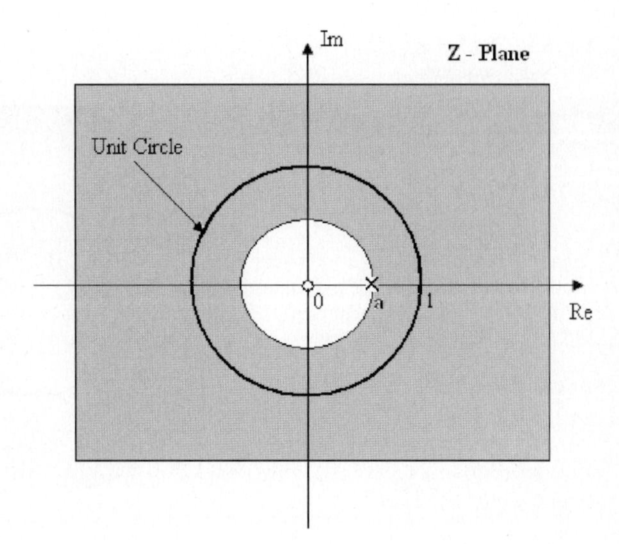


**Fig. 5.34** ROC of a right-sided sequence

The sum in (5.85) converges if $\left|a^{-1}z\right| < 1$, or $|z| < |a|$. So, this is an example of a *left-sided sequence*. The ROC is shown in Fig. 5.35 by the shaded area.

Note that, for $|a| < 1$, the sequence $-a^n u\,[-n-1]$ would grow exponenttially. For $|a| > 1$, the unit circle would be included as $n \to -\infty$, and thus, the FT would exist. It is to be noted that the value of the ZT is same for the right- and left-handed sequences except for the respective ROCs. However, the complete description of the ZT comprises of both the value and the ROC. Without any one of them the answer is incomplete (and wrong!).

We consider another interesting example for a sequence $x\,[n] = \left(-\frac{1}{3}\right)^n u\,[n] - 2^n u\,[-n-1]$. Using the arguments of the above two examples, the ZT for this sequence would be

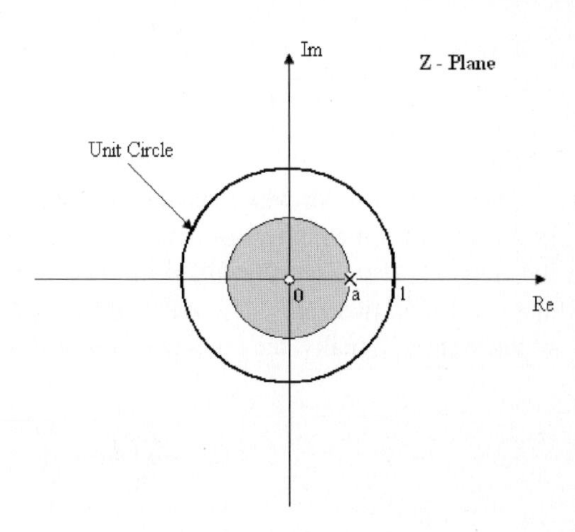


**Fig. 5.35** ROC of a left-sided sequence

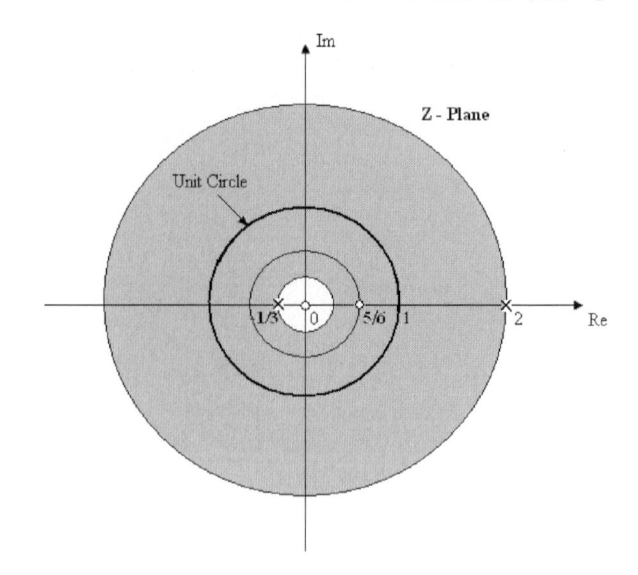


**Fig. 5.36** ROC of a both-sided sequence

$$X(z) = \frac{1}{1 + \frac{1}{3}z^{-1}} + \frac{1}{1 - 2z^{-1}}. \tag{5.86}$$

The corresponding ROC would be, $|z| > \frac{1}{3}$ (for the first term on the right-hand side of (5.86)), and $|z| < 2$ (for the second term on the right-hand side of (5.86)). The complete ROC would be $\frac{1}{3} < |z| < 2$, which is an annular region as shown in Fig. 5.36 by the shaded area. This is called *both-sided sequence*. As this ROC contains the unit circle, the FT exists and also the system is stable.

From (5.86) we get

$$X(z) = \frac{1}{1 + \frac{1}{3}z^{-1}} + \frac{1}{1 - 2z^{-1}} = \frac{2z\left(z - \frac{5}{6}\right)}{\left(z + \frac{1}{3}\right)(z - 2)}. \tag{5.87}$$

The values of $z$ for which the denominator of (5.87) is zero are called the *poles* of $X(z)$ and for the numerator being zero, they are called the *zeros*. The stability of the system is also determined by the poles. This sequence has two poles, at $z = -\frac{1}{3}$, and $z = 2$, and two zeros, at $z = 0$, and $z = \frac{5}{6}$. Figure 5.37 shows the pole-zero plot of the system. Usually, the pole-zero description is described as

$$X(z) = \frac{P(z)}{Q(z)}. \tag{5.88}$$

Equation (5.88) is often used for system analysis, we will see later.

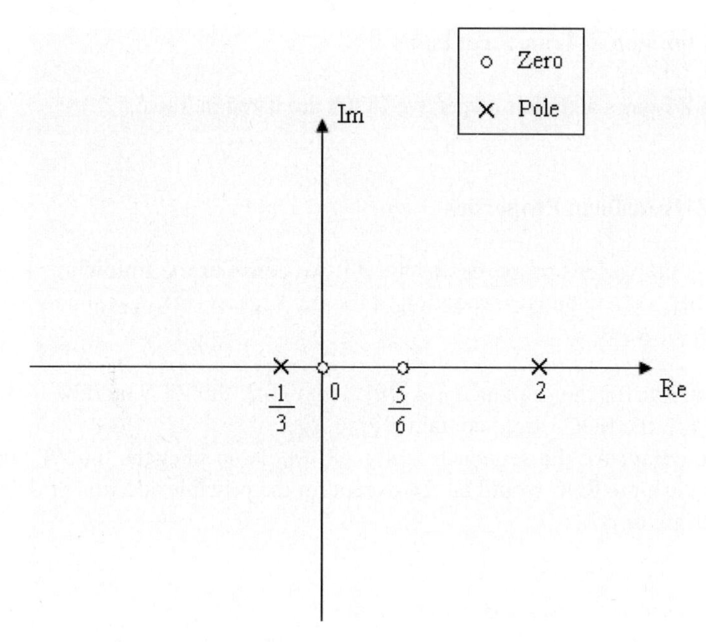


**Fig. 5.37** Pole-zero plot of the both-sided sequence

### 5.4.3.2 Properties of the ROC

1. If $x[n]$ is a right-sided sequence ($x[n] = 0$ for $n < N < \infty$), and $X(z)$ is convergent, then ROC has the form

$$|z| > p_{\max} \text{ or } \infty > |z| > p_{\max}, \tag{5.89}$$

where $p_{\max}$ is the largest magnitude of any of the poles of $X(z)$.

2. If $x[n]$ is a left-sided sequence ($x[n] = 0$ for $n > N > -\infty$), and $X(z)$ is convergent, then ROC has the form

$$|z| < p_{\min} \text{ or } 0 < |z| < p_{\min}, \tag{5.90}$$

where $p_{\min}$ is the smallest magnitude of any of the poles of $X(z)$.

3. If $x[n]$ is a both-sided sequence (of infinite duration), and $X(z)$ is convergent, then ROC has the form

$$p_1 < |z| < p_2, \tag{5.91}$$

where $p_1$ and $p_2$ are the magnitudes of two poles of $X(z)$.

4. If $x[n]$ is a finite sequence ($N_1 < n < N_2$), and $X(z)$ is convergent, then ROC is the entire $z$-plane except at $z = 0$ or $z = \infty$.
5. There exist no poles within the ROC.

#### 5.4.3.3 Common Z-Transform Pairs

Common ZT-pairs and their respective ROCs are listed in Table 5.7.

#### 5.4.3.4 Z-Transform Properties

To discuss the various properties of the ZT, we consider the following sequences, $x[n]$, $x_1[n]$, $x_2[n]$, the corresponding ZT's are $X(z)$, $X_1(z)$, $X_2(z)$ and the ROCs are $R_X$, $R_{X_1}$, $R_{X_2}$.

1. **Linearity:** for the sequence $a_1x_1[n] + a_2x_2[n]$, the ZT would be $a_1X_1(z) + a_2X_2(z)$, the ROC would contain $R_{X_1} \cap R_{X_2}$.
2. **Time-delay:** for the sequence $x[n - n_0]$ ($n_0$ is an integer), the ZT would be $z^{-n_0}X(z)$, the ROC would be $R_X$ except for the possible addition or deletion of the origin or $\infty$.

**Table 5.7** Common Z-transform pairs and their ROCs

| Sequence | Transform | ROC |
|---|---|---|
| $\delta[n]$ | $1$ | All $z$ |
| $\delta[n-m]$ | $z^{-m}$ | All $z$ except 0 (if m > 0) or $\infty$ (if m < 0) |
| $u[n]$ | $\frac{1}{1-z^{-1}}$ | $\|z\| > 1$ |
| $-u[-n-1]$ | $\frac{1}{1-z^{-1}}$ | $\|z\| < 1$ |
| $a^n u[n]$ | $\frac{1}{1-az^{-1}}$ | $\|z\| > a$ |
| $-a^n u[-n-1]$ | $\frac{1}{1-az^{-1}}$ | $\|z\| < a$ |
| $na^n u[n]$ | $\frac{az^{-1}}{(1-az^{-1})^2}$ | $\|z\| > a$ |
| $-na^n u[-n-1]$ | $\frac{az^{-1}}{(1-az^{-1})^2}$ | $\|z\| < a$ |
| $\cos[\omega_0 n]u[n]$ | $\frac{1-[\cos\omega_0]z^{-1}}{1-[2\cos\omega_0]z^{-1}+z^{-2}}$ | $\|z\| > 1$ |
| $\sin[\omega_0 n]u[n]$ | $\frac{[\sin\omega_0]z^{-1}}{1-[2\cos\omega_0]z^{-1}+z^{-2}}$ | $\|z\| > 1$ |
| $\left[r^n \cos[\omega_0 n]\right]u[n]$ | $\frac{1-[r\cos\omega_0]z^{-1}}{1-[2r\cos\omega_0]z^{-1}+z^{-2}}$ | $\|z\| > r$ |
| $\left[r^n \sin[\omega_0 n]\right]u[n]$ | $\frac{[r\sin\omega_0]z^{-1}}{1-[2r\cos\omega_0]z^{-1}+z^{-2}}$ | $\|z\| > r$ |
| $\begin{cases} a^n, & 0 \le n \le N-1 \\ 0, & otherwise \end{cases}$ | $\frac{1-a^N z^{-N}}{1-az^{-1}}$ | $\|z\| > 0$ |

3. **Sample-delay:** for the sequence $z_0^n x[n]$, the ZT would be $X(z/z_0)$, the ROC would be $|z_0|\, R_X$.
4. **Time-reversal:** for the sequence $x[-n]$, the ZT would be $X(1/z)$, the ROC would be $1/R_X$.
5. **Sample differentiation:** for the sequence $nx[n]$, the ZT would be $-z\frac{d}{dz}X(z)$, the ROC would be $R_X$ except for the possible addition or deletion of the origin or $\infty$.
6. **Convolution:** for the sequence $x_1[n]^* x_2[n]$, the ZT would be $X_1(z)X_2(z)$, the ROC would contain $R_{X_1} \cap R_{X_2}$.
7. **Conjugate:** for the complex conjugate sequence $x^*[n]$, the ZT would be $X^*(z^*)$, (i.e., the complex conjugate of the $z$). The ROC would be $R_X$.
8. **Real sequence:** for the real sequence $\mathrm{Re}\{x[n]\}$, the ZT would be $\frac{1}{2}[X(z) + X^*(z^*)]$, the ROC would contain $R_X$.
9. **Imaginary sequence:** for the imaginary sequence $\mathrm{Im}\{x[n]\}$, the ZT would be $\frac{1}{2j}\left[X(z) - X^*(z^*)\right]$, the ROC would contain $R_X$.
10. **Initial value theorem:**

$$\begin{aligned} &if\ x[n] = 0, n < 0, \\ &\lim_{z\to\infty} X(z) = X(0) \end{aligned} . \tag{5.92}$$

### 5.4.3.5 Transfer Function using the Z-Transform

The system or transfer function for an LSI-system can be expressed in terms of the ZT. Hence, we would be able to utilize all the ZT properties. If the system function for an LSI-system is $h[n]$, then for an input $x[n]$, the output $y[n] = x[n]^* h[n]$. Using the convolution property of the ZT,

$$\begin{aligned} Y(z) &= X(z)H(z) \\ \therefore H(z) &= \frac{Y(z)}{X(z)} \end{aligned} . \tag{5.93}$$

Now, we consider the $N$-th order difference equation:

$$\sum_{k=0}^{N} a_k y[n-k] = \sum_{k=0}^{M} b_k x[n-k]. \tag{5.94}$$

Applying the ZT,

$$H(z) = \frac{\sum\limits_{k=0}^{M} b_k z^{-k}}{\sum\limits_{k=0}^{N} a_k z^{-k}} = \left(\frac{b_0}{a_0}\right) \cdot \frac{\prod\limits_{k=1}^{M} (1 - c_k z^{-1})}{\prod\limits_{k=1}^{N} (1 - d_k z^{-1})}. \tag{5.95}$$

In (5.95), each factor in the numerator contributes a zero at $z = c_k$ and a pole at $z = 0$, while each factor in the denominator contributes a pole at $z = d_k$ and a zero at $z = 0$.

The frequency response of a system with transfer function $H(z)$ can be determined by substituting $z = e^{j\omega}$ in $H(z)$. This is equivalent to determining the response in the $z$-plane by traversing the $z$-plane on the unit-circle. Unlike $H(e^{j\omega})$, $H(z)$ exists for systems that may not be BIBO (bounded input bounded output) (Oppenheim et al. 1996, Lathi 2001) stable.

#### 5.4.3.6 Inverse Z-Transform

Inverse of the ZT means to get back the time-domain sequence $x[n]$ from the ZT value $X(z)$. Definition of the inverse ZT is as follows

$$x[n] = \frac{1}{2\pi j} \oint_C X(z) z^{n-1} dz \tag{5.96}$$

where, $C$ is a closed contour-integral in an anti-clockwise direction which includes the origin of the $z$-plane.

The inverse ZT can be obtained through

1. **Inspection method**
   In this method, we try to express $X(z)$ as a sum of finite number of elements like,

$$X(z) = X_1(z) + X_2(z) + \ldots + X_n(z). \tag{5.97}$$

   Then by inspection for each term of the right-hand side of (5.97) (and consulting the ZT table like Table 5.7), we try to find the inverse ZT as a sum like

$$x[n] = x_1[n] + x_2[n] + \ldots + x_n[n]. \tag{5.98}$$

2. **Partial fraction expansion**
   In the cases when $X(z)$ can be expressed as a ratio of polynomials involving $z^{-1}$ (e.g., see (5.93, 5.95)), to obtain the partial fraction expansion of $X(z)$ it is more convenient to rewrite the expression for $X(z)$ as

$$X(z) = \left(\frac{b_0}{a_0}\right) \frac{\prod_{k=1}^{M} (1 - c_k z^{-1})}{\prod_{k=1}^{N} (1 - d_k z^{-1})}. \tag{5.99}$$

   The $c_k$'s and $d_k$'s are the nonzero zeros and poles of $X(z)$. If $M < N$, and the poles are all first order, then $X(z)$ could be expresssed as

$$X(z) = \sum_{k=1}^{N} \frac{A_k}{1 - d_k z^{-1}}, \tag{5.100}$$

and,

$$A_k = (1 - d_k z^{-1})\, X(z)|_{z=d_k}\,. \tag{5.101}$$

If $M \geq N$ then,

$$X(z) = \sum_{r=0}^{M-N} B_r z^{-r} + \sum_{k=1}^{N} \frac{A_k}{1 - d_k z^{-1}}. \tag{5.102}$$

If $X(z)$ has a multiple-order pole of order $s$ at $z = d_i$ and $M \geq N$ then

$$X(z) = \sum_{r=0}^{M-N} B_r z^{-r} + \sum_{k=1, k\neq i}^{N} \frac{A_k}{1 - d_k z^{-1}} + \sum_{m=1}^{s} \frac{C_m}{(1 - d_l z^{-1})^m}, \tag{5.103}$$

with

$$C_m = \frac{1}{(s-m)!(-d_i)^{s-m}} \left\{ \frac{d^{s-m}}{dw^{s-m}} \left[ (1 - d_l w)^s X(w^{-1}) \right] \right\}_{w=d_l^{-1}}. \tag{5.104}$$

3. **Power series expansion**
We use long-division to determine a sequence for $x[n]$. The ROC determines if the result is a right-, left- or both-sided sequence. That is, we expand the expression of $X(z)$ as a power series like

$$\begin{aligned} X(z) &= \sum_{n=-\infty}^{\infty} x[n] z^{-n} \\ &= \ldots + x[-2]z^2 + x[-1]z^1 + x[0]z^0 + x[1]z^{-1} + x[2]z^{-2} + \ldots \end{aligned} \tag{5.105}$$

For the expansion in (5.105), by inspection, we can determine any particular value of sequence by finding the coefficient of the appropriate power of $z^{-1}$. This approach is very useful for a finite-length sequence.

#### 5.4.3.7 Examples

1. For a system defined as $y[n] = 0.81 y[n-2] + x[n] - x[n-2]$, find
   i. Transfer function
   ii. Impulse response

iii. Step response
iv. Frequency response.

i. Taking the ZT of the system we get,

$$Y(z) = 0.81z^{-2}Y(z) + X(z) - z^{-2}X(z). \quad (5.106)$$

So, the transfer function in $z$-domain is

$$H(z) = \frac{Y(z)}{X(z)} = \frac{1 - z^{-2}}{1 - 0.81z^{-2}} = \frac{(z+1)(z-1)}{(z+0.9)(z-0.9)}. \quad (5.107)$$

The ROC would be $|z| > 0.9$. So, this sequence is a right-sided sequence.

ii. The unit impulse response would be $Y(z) = X(z)H(z) = H(z)$.

$$\begin{aligned} Y(z) = \frac{1 - z^{-2}}{1 - 0.81z^{-2}} &= \frac{1 - z^{-2}}{\left(1 - 0.9z^{-1}\right)\left(1 + 0.9z^{-1}\right)} \\ &= \frac{A_1}{\left(1 - 0.9z^{-1}\right)} + \frac{A_2}{\left(1 + 0.9z^{-1}\right)} + C \end{aligned}. \quad (5.108)$$

Using the partial fraction rules as described in the previous section,

$$A_1 = \left.\frac{1 - z^{-2}}{1 + 0.9z^{-1}}\right|_{z^{-1}=1/0.9} = -0.1173,$$

$$A_2 = \left.\frac{1 - z^{-2}}{1 - 0.9z^{-1}}\right|_{z^{-1}=-1/0.9} = -0.1173.$$

Rewriting (5.108),

$$(1 - z^{-2}) = A_1(1 + 0.9z^{-1}) + A_2(1 - 0.9z^{-1}) + C(1 - 0.81z^{-2}). \quad (5.109)$$

From (5.109), equating the coefficients on both sides we get,

$$0.81C = 1, \text{ so } C = 1.2346. \quad (5.110)$$

So, (5.108) can be rewritten as

$$Y(z) = 1.2346 - \frac{0.1173}{(1 - 0.9z^{-1})} - \frac{0.1173}{(1 + 0.9z^{-1})}. \quad (5.111)$$

So, in time-domain, the impulse response would be

$$\begin{aligned} y[n] &= 1.2346\delta[n] - 0.1173(0.9)^n u[n] + 0.1173(-0.9)^n u[n] \\ &= 1.2346\delta[n] - 0.1173(0.9)^n\{1 + (-1)^n\}u[n] \end{aligned} \quad . \tag{5.112}$$

iii. For the step response, the input is $x[n] = u[n]$, i.e., the output would be $Y(z) = X(z)H(z) = \frac{1}{1-z^{-1}}H(z)$.

$$\begin{aligned} Y(z) &= \frac{1-z^{-2}}{1-0.81z^{-2}} \cdot \frac{1}{1-z^{-1}} = \frac{(1+z^{-1})(1-z^{-1})}{(1-0.9z^{-1})(1+0.9z^{-1})(1-z^{-1})} \\ &= \frac{(1+z^{-1})}{(1-0.9z^{-1})(1+0.9z^{-1})} = \frac{A_1}{(1-0.9z^{-1})} + \frac{A_2}{(1+0.9z^{-1})} \end{aligned} . \tag{5.113}$$

Using partial fraction expansion,

$$A_1 = \left.\frac{1+z^{-1}}{1-0.9z^{-1}}\right|_{z^{-1}=-1/0.9} = -0.0556,$$

$$A_2 = \left.\frac{1+z^{-1}}{1+0.9z^{-1}}\right|_{z^{-1}=1/0.9} = 1.0556.$$

So, rewriting (5.113),

$$Y(z) = -\frac{0.0556}{(1+0.9z^{-1})} + \frac{1.0556}{(1-0.9z^{-1})}. \tag{5.114}$$

So, in time-domain, the step response would be

$$\begin{aligned} y[n] &= -0.0556(-0.9)^n u[n] + 1.0556(0.9)^n u[n] \\ &= \left[1.0556(0.9)^n - 0.0556(-0.9)^n\right] u[n] \end{aligned} \quad . \tag{5.115}$$

iv. The transfer function in $z$-domain is $H(z) = \frac{1-z^{-2}}{1-0.81z^{-2}}$. The ROC is $|z| > 0.9$. Hence, the ROC encloses the unit circle as shown in Fig. 5.38. So, the FT exists, hence, replacing $z = e^{j\omega}$, we get the frequency response of the system as

$$H(e^{j\omega}) = \frac{1-e^{-j2\omega}}{1-0.81e^{-j2\omega}}. \tag{5.116}$$

**2.** A finite-length system is defined as $h[n] = \left[\begin{matrix} -1, 2, 1, 2, -3, 0, 1 \\ \uparrow \end{matrix}\right]$. Find

i. Impulse response
ii. Step response.

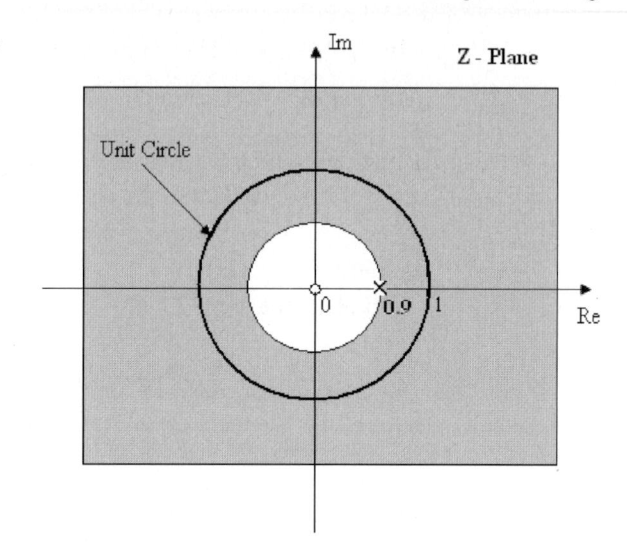


**Fig. 5.38** ROC of the transfer function

i. This finite-length sequence can be rewritten as

$$H(z) = \sum_{n=-2}^{4} h[n] z^{-n} = -z^2 + 2z + 1 + 2z^{-1} - 3z^{-2} + z^{-4}. \quad (5.117)$$

The impulse response is given as $Y(z) = X(z)H(z) = H(z)$. So, in time-domain the impulse response is given as

$$y[n] = \begin{bmatrix} -1, 2, \underset{\uparrow}{1}, 2, -3, 0, 1 \end{bmatrix}$$
$$= -\delta[n+2] + 2\delta[n+1] + \delta[n] + 2\delta[n-1] - 3\delta[n-2] + \delta[n+4] \quad (5.118)$$

ii. For the step response, the input is $x[n] = u[n]$, i.e., the output would be $Y(z) = X(z)H(z) = \frac{1}{1-z^{-1}} H(z)$.

$$Y(z) = \frac{-z^2 + 2z + 1 + 2z^{-1} - 3z^{-2} + z^{-4}}{1 - z^{-1}}$$
$$= \frac{-z^2}{1 - z^{-1}} + \frac{2z}{1 - z^{-1}} + \frac{1}{1 - z^{-1}} + \frac{2z^{-1}}{1 - z^{-1}} + \frac{-3z^{-2}}{1 - z^{-1}} + \frac{z^{-4}}{1 - z^{-1}} \quad (5.119)$$

By inspecting each terms of (5.119), we get the time-domain step response as

$$y[n] = -u[n+2] + 2u[n+1] + u[n] + 2u[n-1] - 3u[n-2] + u[n-4]. \quad (5.120)$$

**3.** A causal LSI system has system function $H(z) = \frac{1+z^{-1}}{(1-\frac{1}{2}z^{-1}) + (1+\frac{1}{4}z^{-1})}$.

i. What is the ROC?
ii. Is the system stable?
iii. Find the input which produces the output

$$y[n] = -\tfrac{1}{3}\left(-\tfrac{1}{4}\right)^n u[n] - \tfrac{4}{3}(2)^n u[-n-1].$$

i. Causal LSI system is right-sided sequence. So, the ROC has form $|z| > p_{\max}$, where $p$ indicates the poles of the system. The system has poles at $z = -\frac{1}{4}, z = \frac{1}{2}$. So, $p_{\max} = \frac{1}{2}$ (maximum between the poles). So, the ROC is $|z| > \frac{1}{2}$.
ii. As the ROC is $|z| > \frac{1}{2}$, it encloses the unit circle. So, the system is stable.
iii. The output is $y[n] = -\frac{1}{3}\left(-\frac{1}{4}\right)^n u[n] - \frac{4}{3}(2)^n u[-n-1]$. So,

$$Y(z) = -\tfrac{1}{3}.\frac{1}{1+\frac{1}{4}z^{-1}} + \tfrac{4}{3}.\frac{1}{1+2z^{-1}} = \frac{1+z^{-1}}{(1+\frac{1}{4}z^{-1})(1+2z^{-1})}. \quad (5.121)$$

Hence, the corresponding input in $z$-domain is

$$\begin{aligned} X(z) = \frac{Y(z)}{H(z)} &= \frac{(1+z^{-1})}{(1+\frac{1}{4}z^{-1})(1+2z^{-1})} \div \frac{(1+z^{-1})}{(1-\frac{1}{2}z^{-1}) + (1+\frac{1}{4}z^{-1})} \\ &= \frac{1-\frac{1}{2}z^{-1}}{1-2z^{-1}} = \frac{1}{1-2z^{-1}} - \frac{1}{2}\frac{z^{-1}}{1-2z^{-1}} \end{aligned} \quad (5.122)$$

So, in time-domain the input is

$$x[n] = 2^n u[n] - (\tfrac{1}{2})2^{n-1}u[n-1] = 2^n u[n] - 2^{n-2}u[n-1]. \quad (5.123)$$

## 5.5 DSP Fundamentals

Transforms, both the analog and the discrete one, are the building blocks of signal processing. Using them, now we look into the more details of the DSP, concentrating especially on the discrete-time signals.

### 5.5.1 Discrete Fourier Series

While discussing about the Fourier transform in Sect. 5.4.2, we mentioned that any continuous periodic signal can be represented as a sum of (in general infinite number) of sinusoids. This can be also extended to the discrete-time periodic signals which can be defined as

$$x_N[n] = x_N[n + rN] \quad \forall\, integer\ r. \tag{5.124}$$

Here, the subscript $N$ indicates how many samples are there in the period. This periodic discrete-time sequence cannot be represented using the $Z$-transform because no convergence can be attained in this case. In this case, we represent $x_N[n]$ by a series of complex exponentials

$$e_k(n) = e^{j\left(\frac{2\pi}{N}\right)nk} = e_k(n + rN), \tag{5.125}$$

which is periodic in $k$ with period $N$. The harmonic frequencies are represented as

$$\left\{\frac{2\pi k}{N}, k = 0, 1, 2, \ldots, N-1\right\}. \tag{5.126}$$

The Fourier-series representation of a periodic sequence (of period $N$) would comprise of $N$ harmonically related complex exponentials.

$$x_N[n] = \frac{1}{N}\sum_{k=0}^{N-1} x_N[k]e^{j\left(\frac{2\pi}{N}\right)nk}. \tag{5.127}$$

We consider the operator $W_N = e^{-j\left(\frac{2\pi}{N}\right)}$ which is the $N$-th root of unity. Using this, the synthesis equation of the discrete-Fourier series (DFS) (Oppenheim et al. 1996, 2001) can be represented as

$$x_N[n] = \frac{1}{N}\sum_{k=0}^{N-1} x_N[k]W_N^{-kn}, \quad n = 0, \pm 1, \pm 2, \ldots \tag{5.128}$$

And the analysis equation of the DFS can be expressed as

$$x_N[k] = \sum_{n=0}^{N-1} x_N[n]W_N^{kn}, \quad k = 0, \pm 1, \pm 2, \ldots. \tag{5.129}$$

The complex-valued sequence $\{x_N[k], k = 0, \pm 1, \pm 2, \ldots\}$ consists of the DFS coefficients. $x_N[k]$ can therefore be interpreted as evenly spread samples of one period of the FT of $x_N[n]$, taken on the unit circle. The $N$ equally spaced frequencies between $\omega = 0$ and $\omega = 2\pi$ has a frequency spacing of $\frac{2\pi}{N}$.

Let's see an example. We want to find the DFS of the periodic sequence $x[n] = \begin{bmatrix} \ldots, 0, 1, 2, 3, 0, 1, 2, 3, \ldots \\ \uparrow \end{bmatrix}$.

The fundamental period is 4, i.e., $N = 4$. We have

$$x[k] = \sum_{n=0}^{3} x[n] W_4^{kn}, \quad k = 0, \pm 1, \ldots \tag{5.130}$$

$$W_4 = e^{-j\frac{2\pi}{4}} = -j. \tag{5.131}$$

So, the DFS coefficients are computed as follows.

$$x[0] = \sum_{n=0}^{3} x[n] W_4^0 = \sum_{n=0}^{3} x[n] = 6. \tag{5.132}$$

$$x[1] = \sum_{n=0}^{3} x[n] W_4^n = \sum_{n=0}^{3} x[n](-j)^n = -2 + 2j. \tag{5.133}$$

$$x[2] = \sum_{n=0}^{3} x[n](-j)^{2n} = -2. \tag{5.134}$$

$$x[3] = \sum_{n=0}^{3} x[n](-j)^{3n} = -2 - 2j. \tag{5.135}$$

### *5.5.2 Discrete-Time Fourier Transform (DTFT)*

DFS can be applied on the discrete-time periodic signal. However, in DSP we can have both periodic and aperiodic signals (more precisely sequences). So, for all kinds of discrete-time sequences, the version of the Fourier transform is called the Discrete-Time Fourier Transform (DTFT) (Oppenheim et al. 1996, 2001) which also uses the complex exponential. DTFT for a sequence $x[n]$ is defined as

$$X[j\omega] = \sum_{n=-\infty}^{\infty} x[n] e^{-j\omega n}. \tag{5.136}$$

Comparing this to the definition of the ZT (see (5.76)), we can see that the DTFT is the case, $X\left[j\omega\right] = X[z]|_{z=e^{j\omega}}$. In general, from the ZT we can evaluate the FT if the ROC encompasses the unit circle.

#### 5.5.2.1 Properties of the DTFT

1. **Periodicity**
   The DTFT is periodic in $\omega$ with period $2\pi$, i.e.,

$$X[e^{j\omega}] = X[e^{j(\omega+2\pi)}]. \tag{5.137}$$

   The implication of the periodicity property is that we need only one period of $X\left[e^{j\omega}\right]$. In other words, for the Fourier analysis we only require $\omega \in [0, 2\pi]$ or $\omega \in [-\pi, \pi]$, not the whole domain $[-\infty, \infty]$.
2. **Symmetry**
   For a real-valued sequence $x[n]$, the corresponding DTFT $X\left[e^{j\omega}\right]$ is conjugate symmetric. That is,

$$X[e^{-j\omega}] = X^*[e^{j\omega}]. \tag{5.138}$$

$$\begin{aligned} \mathrm{Re}\{X[e^{-j\omega}]\} &= \mathrm{Re}\{X[e^{j\omega}]\}, \\ \mathrm{Im}\{X[e^{-j\omega}]\} &= -\mathrm{Im}\{X[e^{j\omega}]\}. \end{aligned} \tag{5.139}$$

$$\begin{aligned} |X[e^{-j\omega}]| &= |X[e^{j\omega}]|, \\ \angle X[e^{-j\omega}] &= -\angle X[e^{j\omega}]. \end{aligned} \tag{5.140}$$

   The implication of the symmetry property is that to plot the DTFT $X\left[e^{j\omega}\right]$, we need only the half-period, i.e., $\omega \in [0, \pi]$.

#### 5.5.2.2 DTFT Example

We want to evaluate the DTFT of the sequence $x[n] = 0.2^n u[n]$.
As per (5.136), the DTFT is given as

$$X\left[j\omega\right] = \sum_{n=-\infty}^{\infty} x[n] e^{-j\omega n} = \frac{e^{j\omega}}{e^{j\omega} - 0.2}. \tag{5.141}$$

Using (5.141), we plot the magnitude, phase responses, real and imaginary parts of the DTFT in Fig. 5.39–5.42 respectively.

### *5.5.3 Discrete Fourier Transform*

The ZT and the DTFT have the following two features in common.

1. These transforms are defined for infinite length sequences.
2. Most importantly, they are functions of continuous variables ($z$ and $\omega$).

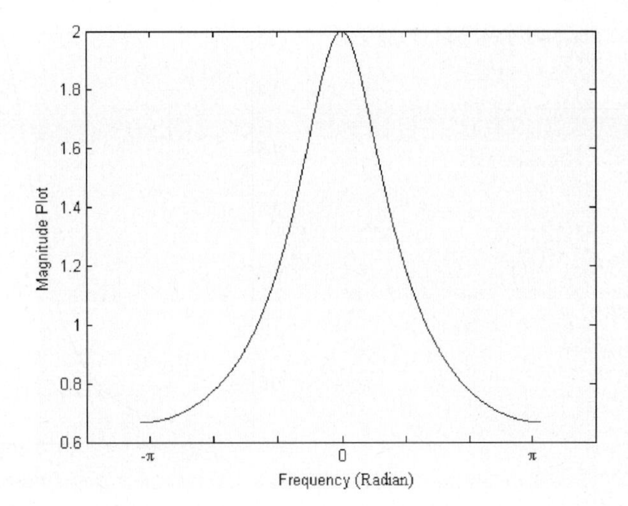


**Fig. 5.39** Magnitude plot of the DTFT

From the numerical computation point of view, these two features are especially troublesome because then one has to evaluate infinite sums of uncountably infinite frequencies. Therefore, the ZT and the DTFT are not numerically computable transforms.

From the Fourier analysis we know that a periodic function (or sequence) can always be represented by a linear combination of harmonically related complex exponentials (which is a form of sampling), giving us the DFS. We can then extend the DFS to the finite-duration sequences which leads us to a new transform, the Discrete Fourier Transform (DFT) (Oppenheim et al. 1996, 2001). DFT can actually be obtained by sampling the DTFT in the frequency domain or the ZT on the unit-circle.

DFT of a sequence $x[n]$ is defined as

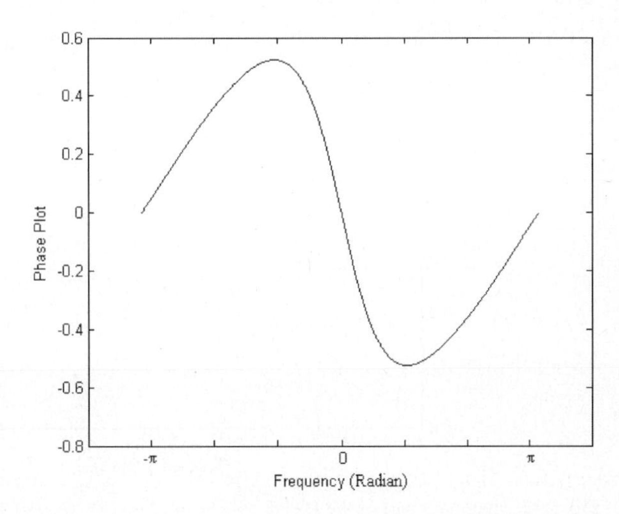


**Fig. 5.40** Phase plot of the DTFT

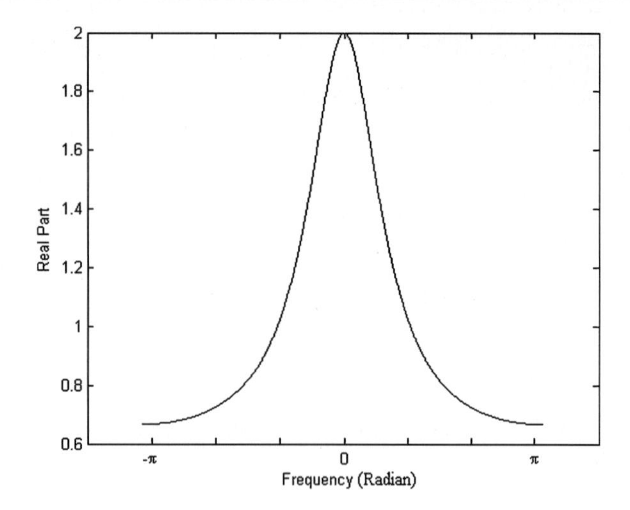


**Fig. 5.41** Real part of the DTFT

$$X[k] = \sum_{n=0}^{N-1} x[n] e^{-j\frac{2\pi}{N}kn}, \qquad n = 0, 1, 2, \ldots. \tag{5.142}$$

Comparing this with the DTFT (see (5.136)), we see the relationship between the DFT and the DTFT as

$$DFT = DTFT|_{\omega=\frac{2\pi k}{N}} . \tag{5.143}$$

$2\pi/N$ is like the fundamental frequency, and there are a finite number of harmonics, the frequencies being $\frac{2\pi k}{N}$, $k = 0, 1, 2, \ldots, N-1$. So, with these finite number

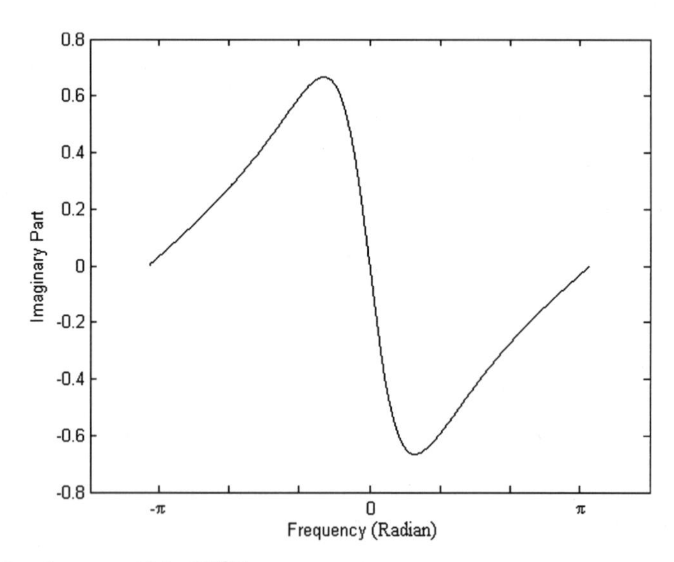


**Fig. 5.42** Imaginary part of the DTFT

of frequencies $\left(\frac{2\pi k}{N}\right)$ the first problem stated in the aforesaid paragraph is solved. And we solve the second problem by discretizing $\omega$ into $\frac{2\pi k}{N}$, $k = 0, 1, 2, \ldots, N-1$.

The Inverse Discrete Fourier Transform (IDFT) is defined as

$$x[n] = \frac{1}{N}\sum_{n=0}^{N-1} X[k]\, e^{j\frac{2\pi}{N}kn}, \quad n = 0, 1, 2, \ldots. \tag{5.144}$$

Thus, the DFT avoids the two major problems of the ZT and the DTFT and is numerically computable transform. However, the numerical computation of the DFT for long sequences is prohibitively time-consuming. Therefore, several algorithms have been developed to efficiently (and fast) calculate the DFT. One of the best is the Fast Fourier Transform, popularly known as the FFT algorithm (Cooley and Tukey 1965).

#### 5.5.3.1 Steps to Compute the DFT and the IDFT

Let's consider that we have a $n$-point discrete-time sequence defined as $x[n] = \{x_0, x_1, x_2, \ldots, x_{n-1}\}$. The steps to calculate the DFT are as follows.

1. We consider the following important relationships.

$$e^{-j\frac{2\pi}{N}kn} = e^{\left(-j\frac{2\pi}{N}\right)kn} = W_N^{kn}, \tag{5.145}$$

where, $W_N$ is the $N$-th root of unity, i.e.,

$$W_N = e^{-j\frac{2\pi}{N}} = \cos\left(\frac{2\pi}{N}\right) - j\sin\left(\frac{2\pi}{N}\right). \tag{5.146}$$

For our given sequence, $n = N$. So,

$$W_N = e^{-j\frac{2\pi}{n}} = \cos\left(\frac{2\pi}{n}\right) - j\sin\left(\frac{2\pi}{n}\right). \tag{5.147}$$

2. With these information we calculate a $n \times n$ matrix (for the given $n$-point sequence) of the $W_N$, which we call the DFT matrix. The DFT matrix $W$ can be defined as

$$W = \begin{bmatrix} 1 & 1 & 1 & 1 & \cdots & 1 \\ 1 & W_N^1 & W_N^2 & W_N^3 & \cdots & W_N^{N-1} \\ 1 & W_N^2 & W_N^4 & W_N^6 & \cdots & W_N^{2(N-1)} \\ \cdot & \cdot & \cdot & \cdot & \cdots & \cdot \\ \cdot & \cdot & \cdot & \cdot & \cdots & \cdot \\ \cdot & \cdot & \cdot & \cdot & \cdots & \cdot \\ 1 & W_N^{N-1} & W_N^{2(N-1)} & W_N^{3(N-1)} & \cdots & W_N^{(N-1)(N-1)} \end{bmatrix}. \tag{5.148}$$

The DFT matrix $W$ is reciprocal, i.e., rows $\equiv$ columns.

3. From the DFT matrix, we calculate the DFT $X[k]$ as

$$X[k] = W.x^T = \begin{bmatrix} 1 & 1 & 1 & 1 & \dots & 1 \\ 1 & W_N^1 & W_N^2 & W_N^3 & \dots & W_N^{N-1} \\ 1 & W_N^2 & W_N^4 & W_N^6 & \dots & W_N^{2(N-1)} \\ \cdot & \cdot & \cdot & \cdot & \dots & \cdot \\ \cdot & \cdot & \cdot & \cdot & \dots & \cdot \\ \cdot & \cdot & \cdot & \cdot & \dots & \cdot \\ 1 & W_N^{N-1} & W_N^{2(N-1)} & W_N^{3(N-1)} & \dots & W_N^{(N-1)(N-1)} \end{bmatrix} \begin{bmatrix} x_0 \\ x_1 \\ x_2 \\ \cdot \\ \cdot \\ \cdot \\ x_{n-1} \end{bmatrix} \quad (5.149)$$

Here, $T$ indicates the transpose[22] of the matrix.
For calculating the IDFT, we use the following steps.

1. We define the IDFT matrix $W^*$, i.e., the complex conjugate of the DFT matrix $W$. In principle, in the DFT matrix (defined by (5.148)), we replace $j$ with $-j$ and vice versa (i.e., change the signs of $j$). It is to be noted that for real case, the terms of $W$ like $W_N^1$ would be complex (represented in general as $a + jb$).
2. Using the IDFT matrix $W^*$, we calculate the IDFT as

$$x[n] = \frac{1}{N}.W^*.X[k]^T . \quad (5.150)$$

Here, $T$ indicates the transpose of the matrix.

#### 5.5.3.2 Example

**1.** We define a discrete-time sequence as $x[n] = \{1, 2, 3\}$, and would like to compute the DFT.
In this case, $n = N = 3$.
So, $W_N^1 = \cos\left(\frac{2\pi}{3}\right) - j\sin\left(\frac{2\pi}{3}\right) = -0.5 - j0.866 = 1\angle - 120°$.

$$W_N^2 = 1\angle - 120° \times 1\angle - 120° = 1\angle - 240° = -0.5 + j0.866.$$
$$W_N^4 = 1\angle - 480° = -0.5 - j0.866.$$

So, the DFT matrix

[22] Transpose of a matrix (indicated by $T$) can be computed by interchanging the rows and the columns. For example, if $A = \begin{bmatrix} 1 & 2 & 3 \\ 4 & 5 & 6 \end{bmatrix}$, then $A^T = \begin{bmatrix} 1 & 4 \\ 2 & 5 \\ 3 & 6 \end{bmatrix}$.

$$W = \begin{bmatrix} 1 & 1 & 1 \\ 1 & W_N^1 & W_N^2 \\ 1 & W_N^2 & W_N^4 \end{bmatrix} = \begin{bmatrix} 1 & 1 & 1 \\ 1 & (-0.5 - j0.866) & (-0.5 + j0.866) \\ 1 & (-0.5 + j0.866) & (-0.5 - j0.866) \end{bmatrix}.$$

$$\therefore X[k] = \begin{bmatrix} 1 & 1 & 1 \\ 1 & (-0.5 - j0.866) & (-0.5 + j0.866) \\ 1 & (-0.5 + j0.866) & (-0.5 - j0.866) \end{bmatrix} \begin{bmatrix} 1 \\ 2 \\ 3 \end{bmatrix}$$

$$= \begin{bmatrix} 6 \\ -1.5 + j0.866 \\ -1.5 - j0.866 \end{bmatrix}.$$

From this DFT $X[k]$, we would like to calculate the IDFT to cross-check. So, the IDFT matrix is

$$W^* = \begin{bmatrix} 1 & 1 & 1 \\ 1 & (-0.5 + j0.866) & (-0.5 - j0.866) \\ 1 & (-0.5 - j0.866) & (-0.5 + j0.866) \end{bmatrix}.$$

$$\therefore x[n] = \frac{1}{3} \begin{bmatrix} 1 & 1 & 1 \\ 1 & (-0.5 + j0.866) & (-0.5 - j0.866) \\ 1 & (-0.5 - j0.866) & (-0.5 + j0.866) \end{bmatrix} \begin{bmatrix} 6 \\ -1.5 + j0.866 \\ -1.5 - j0.866 \end{bmatrix}$$

$$= \begin{bmatrix} 1 \\ 2 \\ 3 \end{bmatrix}.$$

**2.** A system is defined by the finite-duration system function $h[n] = \{1, 2, 3, 4\}$. Consider an input to the signal $x[n] = \{1, 2, 3\}$. Compute the output $y[n]$.

The system function $h[n] = \{1, 2, 3, 4\}$.
In this case, $n = N = 4$.
So, $W_N^1 = \cos\left(\frac{2\pi}{4}\right) - j\sin\left(\frac{2\pi}{4}\right) = -j$.
So, the DFT matrix

$$W = \begin{bmatrix} 1 & 1 & 1 & 1 \\ 1 & -j & (-j)^2 & (-j)^3 \\ 1 & (-j)^2 & (-j)^4 & (-j)^6 \\ 1 & (-j)^3 & (-j)^6 & (-j)^9 \end{bmatrix} = \begin{bmatrix} 1 & 1 & 1 & 1 \\ 1 & -j & -1 & j \\ 1 & -1 & 1 & -1 \\ 1 & j & -1 & -j \end{bmatrix}.$$

$$\therefore H[k] = \begin{bmatrix} 1 & 1 & 1 & 1 \\ 1 & -j & -1 & j \\ 1 & -1 & 1 & -1 \\ 1 & j & -1 & -j \end{bmatrix} \begin{bmatrix} 1 \\ 2 \\ 3 \\ 4 \end{bmatrix} = \begin{bmatrix} 10 \\ -2 + j2 \\ -2 \\ -2 - 2j \end{bmatrix}.$$

The input is $x[n] = \{1, 2, 3\}$. But here, cannot use the 3-point DFT (and use the computation of the exmaple 1). Why? Because, we have to first make this input signal compatible to the system function. That is, we have to convert this 3-point input signal into a 4-point signal (equivalent to the system function) by *zero-padding* (Oppenheim et al. 2001) it. Zero-padding means we add additional points with zero coefficients. In that way, we can increase the data point of the sequence without distorting the information content because data point with zero-coefficient does not contribute any unwanted information. So, the new zero-padded input is, $x[n] = \{1, 2, 3, 0\}$. So,

$$X[k] = \begin{bmatrix} 1 & 1 & 1 & 1 \\ 1 & -j & -1 & j \\ 1 & -1 & 1 & -1 \\ 1 & j & -1 & -j \end{bmatrix} \begin{bmatrix} 1 \\ 2 \\ 3 \\ 0 \end{bmatrix} = \begin{bmatrix} 6 \\ -2 - j2 \\ 2 \\ -2 + 2j \end{bmatrix}.$$

In the frequency domain, the output $Y[k] = H[k] X[k]$. So, using inner (scaler) product

$$Y[k] = \begin{bmatrix} 10 \\ -2 + j2 \\ -2 \\ -2 - 2j \end{bmatrix} \cdot \begin{bmatrix} 6 \\ -2 - j2 \\ 2 \\ -2 + 2j \end{bmatrix} = \begin{bmatrix} 60 \\ 8 \\ -4 \\ 8 \end{bmatrix}.$$

In time-domain, the output is obtained using the IDFT operation, as

$$y[n] = \frac{1}{4} \begin{bmatrix} 1 & 1 & 1 & 1 \\ 1 & -j & -1 & j \\ 1 & -1 & 1 & -1 \\ 1 & j & -1 & -j \end{bmatrix} \begin{bmatrix} 60 \\ 8 \\ -4 \\ 8 \end{bmatrix} = \begin{bmatrix} 18 \\ 16 \\ 10 \\ 16 \end{bmatrix}.$$

#### 5.5.3.3 Fast Fourier Transform (FFT)

Fast Fourier transform, mostly known as FFT, is a fast way to calculate the DFT. It was first introduced by Cooley and Tukey in the 1960s (Cooley and Tukey 1965). The end results of the DFT and the FFT are same. FFT algorithms are based on the fundamental principle of decomposing the computation of the discrete Fourier transform of a sequence of length $N$ into successively smaller discrete Fourier transforms.

An $N$-point DFT requires $N^2$ complex multiplications and $N(N-1)$ complex additions compared to an $N$-point FFT where the computation is proportional to $N \log N$. Hence, it is much faster.

The mathematical treatment of the FFT algorithm is skipped in the scope and aim of this text. However, detailed discussions about the FFT can be found in the references (Ramirez 1985; Duhamel and Vetterli 1990).

### 5.5.4 Circular Convolution

From the example 2 of the previous section, one question arises. From our discussion of the convolution (Sect. 5.3.5), we can imagine that we could also calculate the output using a general convolution operation on the sequences $h[n]$ and $x[n]$. In that case, it should produce an output signal of length $(4+3-1=)6$ points. But we only get a 4-point output signal!

The reason is that the DFT operation is based on a periodic sequence. The DFT operation on an $N$-point time-domain sequence deals the sequence as an infinitely long sequence with $N$ periods. Mathematically, when a periodic sequence is shifted left or right, a certain number of samples, a wrap-around occurs in each period of the sequence. This is called a *circular shift* or rotation of the sequence in the interval $0 \leq n \leq N-1$.

For the circular shift, we convert a sequence $x[n]$ into its periodic extension $\tilde{x}[n] (= x_N(n))$. And then, shift it by $m$ samples,

$$\tilde{x}[n-m] = x((n-m))_N \,. \tag{5.151}$$

The periodic shift is then converted into an $N$-point sequence called the circular shift of $x[n]$. That is why, we zero-padded the input signal in the example 2 in Sect. 5.5.3.2 to make both the system function and the input sequence 4-point ones. Then, we could apply the 4-point DFT operation to get the 4-point output.

We consider two finite-duration sequences $x_1[n]$ and $x_2[n]$, and convolute to form a third sequence $x_3[n]$. All the three sequences have their DFTs given by $X_1[k]$, $X_2[k]$ and $X_3[k]$. Therefore, in the frequency domain

$$X_3[k] = X_1[k]X_2[k]. \tag{5.152}$$

And

$$x_3[n] = \sum_{m=0}^{N-1} x_{1_N}(m)x_{2_N}(n-m), \quad 0 \leq n \leq N-1. \tag{5.153}$$

Equivalently,

$$x_3[n] = \sum_{m=0}^{N-1} x_1\left[(m)_N\right]x_2\left[(n-m)_N\right], \quad 0 \leq n \leq N-1. \tag{5.154}$$

But we have

$$((m))_N = m \quad for \quad 0 \leq m \leq N-1. \tag{5.155}$$

Therefore,

$$x_3[n] = \sum_{m=0}^{N-1} x_1(m)x_2\left[(n-m)_N\right], \quad 0 \leq n \leq N-1. \tag{5.156}$$

The latter part of (5.156) is called an $N$-point *circular convolution*. Normally, for two sequences of length $N$ each, the result of convolution is a third signal of length $2N-1$, while a circular convolution results in an $N$ point signal.

#### 5.5.4.1 Example

Redo example 2 of Sect. 5.5.3.2 using the circular convolution.

We have $h[n]$ and $x[n]$. We want to calculate the output in time-domain using the circular convolution as

$$y[n] = h[n] \otimes x[n] = \sum_{m=0}^{3} h[m]x\left[|n-m|_3\right], \; 0 \leq n \leq 3. \tag{5.157}$$

So, we represent the sequence $x[n]$ as a periodic one, zero-padded to make $x[n]$ and $h[n]$ of same-length.

For $n=0$:

$$\begin{aligned} n &= 0\,1\,2\,3\,\overset{\downarrow}{0}\,1\,2\,3 \\ x[n] &= 1\,\underbrace{2\,3\,0\,1}_{\leftarrow}\,2\,3\,0 \end{aligned}$$

$$y[0] = \sum_{m=0}^{3} h[m]x\left[|0-m|_3\right] = \{1,2,3,4\}\{1,0,3,2\} = 1+0+9+8 = 18.$$

For $n=1$:

$$\begin{aligned} n &= 0\,1\,2\,3\,0\,\overset{\downarrow}{1}\,2\,3 \\ x[n] &= 1\,2\,\underbrace{3\,0\,1\,2}_{\leftarrow}\,3\,0 \end{aligned}$$

$$y[1] = \sum_{m=0}^{3} h[m]x\left[|1-m|_3\right] = \{1,2,3,4\}\{2,1,0,3\} = 2+2+0+12 = 16.$$

For $n=2$:

$$\begin{aligned} n &= 0\,1\,2\,3\,0\,1\,\overset{\downarrow}{2}\,3 \\ x[n] &= 1\,2\,3\,\underbrace{0\,1\,2\,3}_{\leftarrow}\,0 \end{aligned}$$

$$y[2] = \sum_{m=0}^{3} h[m]x[|2-m|_3] = \{1,2,3,4\}\{3,2,1,0\} = 3+4+3+0 = 10.$$

For $n = 3$:

$$n = 0\,1\,2\,3\,0\,1\,2\,\overset{\downarrow}{3}$$
$$x[n] = 1\,2\,3\,0\,\underbrace{1\,2\,3\,0}_{\leftarrow}$$

$$y[3] = \sum_{m=0}^{3} h[m]x[|3-m|_3] = \{1,2,3,4\}\{0,3,2,1\} = 0+6+6+4 = 16.$$
$$y[n] = \{y[0], y[1], y[2], y[3]\} = \{18, 16, 10, 16\}.$$

So, we see that $y[n]$ calculated using the circular convolution is exactly same as that calculated using the DFT-IDFT technique in the example 2 in Sect. 5.5.3.2. The circular convolution represents the time-domain operation while the DFT-IDFT technique the frequency domain one, anyway they are equivalent.

#### 5.5.4.2 Normal Convolution vs. Circular Convolution

Of course normal convolution and the circular convolution have relationship. Let's redo again the above example. First, we compute the normal convolution. For that, we have

$$n = 0\,1\,2\,3$$
$$h[n] = 1\,2\,3\,4\text{'}$$
$$n = 0\,1\,2$$
$$x[n] = 1\,2\,3^{\cdot}$$

So, using the normal convolution operation we will have a $(4+3-1=)$ 6-point output sequence, ranging from $(0+0=)0 \le n \le 5(=3+2)$. Table 5.8 shows the convolution operation.

So, from the normal convolution we get a 6-point output as $y[n] = [1, 4, 10, 16, 17, 12]$. Now, we would like to compute the 4-point circular convolution using this 6-point sequence. The philosophy, as stated earlier, is that now we treat the sequence

**Table 5.8** Convolution operation for $h[n]$ and $x[n]$

| Sl. No. | $x[n]$ Value | Axis: Value | 0 | 1 | 2 | 3 | 4 | 5 |
|---|---|---|---|---|---|---|---|---|
| 1 | 1 | | 1 | 2 | 3 | 4 | 0 | 0 |
| 2 | 2 | | 0 | 2 | 4 | 6 | 8 | 0 |
| 3 | 3 | | 0 | 0 | 3 | 6 | 9 | 12 |
| $Sum\,y[n]$ | | | 1 | 4 | 10 | 16 | 17 | 12 |

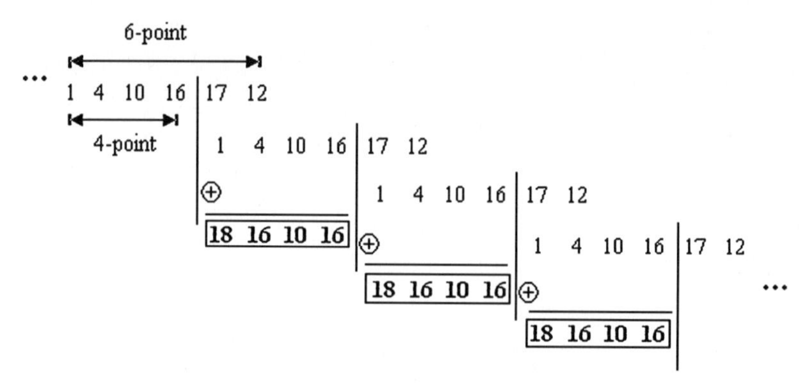


**Fig. 5.43** Calculation of the circular convolution from normal convolution

as a periodic one. For each period, we consider only four points. But we cannot discard any information. So, what about the last 2 points from the original 6-point output? In the circular convolution they don't get a space in the 4-point period. So, we accommodate them at the next period. This way, we keep the information content intact periodically. This is the underlying philosophy of the circular convolution (time-domain) and the DFT (frequency domain). This is demonstrated in Fig. 5.43.

So, we arrive at the same result as before by considering the sequence as a periodic one.

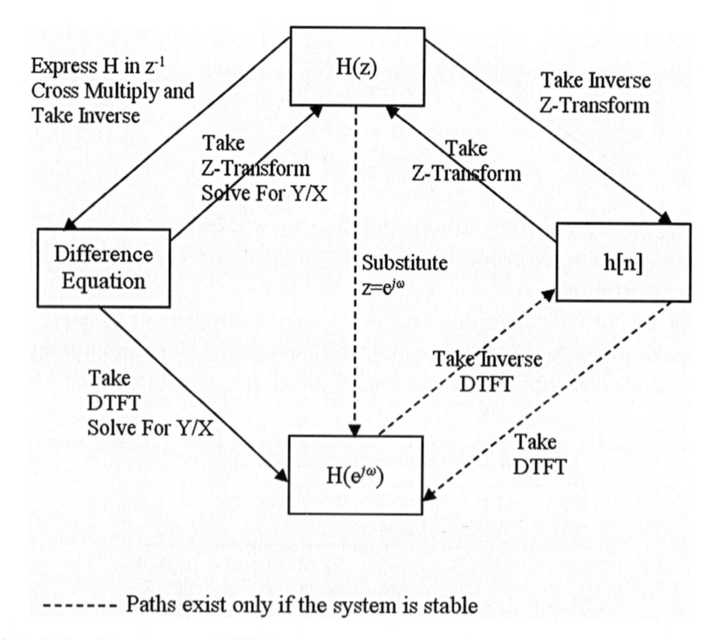


**Fig. 5.44** Relationships between different system representations

### 5.5.5 *Synopsis*

In this section, so far we've seen the fundamental operations in DSP like the DTFT, DFT, etc. We've also seen important transforms like the Laplace transform, FT, ZT for representing the signals. In Fig. 5.44, we can visualize the relationships between the different system representations and the bridging operations.

It is to be noted in Fig. 5.44 that the dashed paths exist only if the system is stable. Stability of a system can be defined in many ways, but they are all equivalent. Some are mentioned below.

- Mathematically, a system with system function $h[n]$ is said to be stable if $\sum_{-\infty}^{\infty} |h[n]| < \infty$.
- From the ZT point of view, a system is said to be stable if the unit circle is contained in the ROC of the ZT of the system, $H[z]$.
- For a causal LTI (or LSI) system (i.e., when $h[n] = 0$ for $n = 0$), it is said to be stable if all the system poles are inside the unit circle.

## 5.6 Sampling

### 5.6.1 *Introduction*

Let's begin with an example. We consider the system shown in Fig. 5.45 represented using the Laplace transform. It's actually an integrator system due to the term $1/s$.

So, the transfer function for this system in the $s$-domain would be given by

$$H(s) = \frac{Y(s)}{X(s)} = \frac{K(s)}{1 + \alpha K(s)} = \frac{\frac{1}{s}}{1 + \frac{\alpha}{s}} = \frac{1}{s + \alpha}. \tag{5.158}$$

So, the impulse response would be $Y(s) = H(s).1 = H(s)$. In time-domain,

$$y(t) = L^{-1}[H(s)] = e^{-\alpha\, t} u(t), \quad t > 0. \tag{5.159}$$

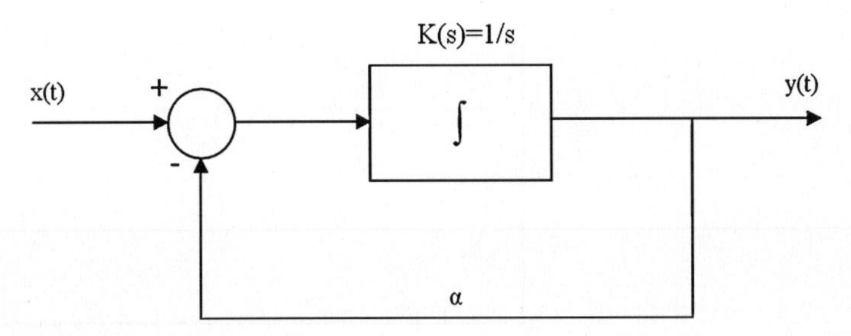


**Fig. 5.45** Integrator system

The time-domain impulse response is shown in Fig. 5.46.

Now, let's consider a *sampled* system with one delay element, as shown in Fig. 5.47.

The system shown in Fig. 5.47 produces signals only at times equal to multiples of the sampling period. Let $x\,[nT]$ be a sampled unit impulse, so that

$$x[nT] = \delta[nT] = \begin{cases} 1, & n = 0 \\ 0, & n \neq 0 \end{cases} \tag{5.160}$$

So, by inspection we get the output as

$$\begin{aligned} y[0] &= 1 \\ y[T] &= \beta \\ y[2T] &= \beta^2. \\ \vdots \quad & \quad \vdots \\ y[nT] &= \beta^n \end{aligned} \tag{5.161}$$

The plot for the (5.161) is shown in Fig. 5.48.

The response plotted in Fig. 5.48 is actually the sampled impulse response. Now, let's put say $\beta = e^{-\alpha\,t}$, so that $\beta^n = e^{-\alpha\,n\,t}$. If we now compare the discrete-time output with the continuous-time output, we find that they coincide at time $nT$. That is,

$$y[nT] = y(t)|_{t=nT}\,. \tag{5.162}$$

So, the two systems are equivalent. This is depicted by Fig. 5.49.

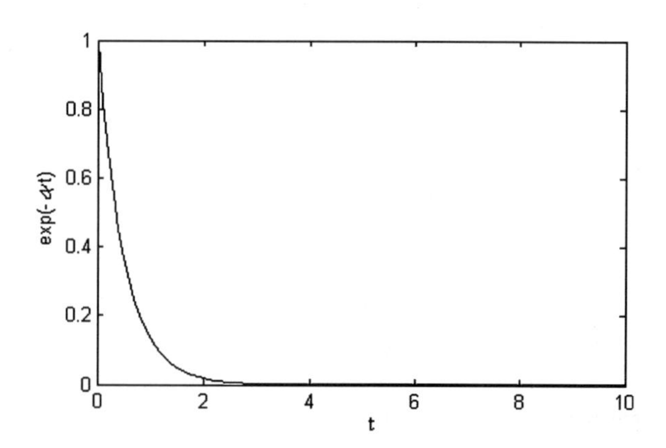


**Fig. 5.46** Time-domain impulse response of the integrator system

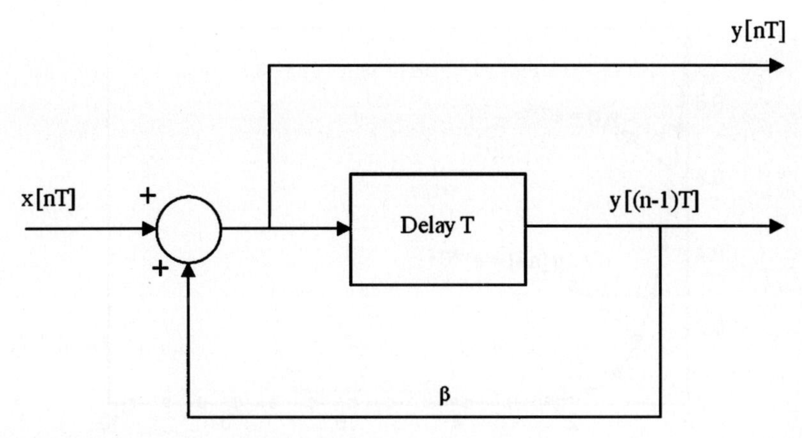


**Fig. 5.47** Sampled system with delay element

## 5.6.2 The Sampling Theorem

We have seen that the sampled system can generate the desired signal at times *nT*. However, the concern is could the continuous-time signal be recovered from these sample values? The answer is conditionally yes. The continuous-time signal can be restored if we weigh an infinite series of unit impulses with the sample values and pass that through an ideal lowpass filter of appropriate bandwidth. The sampled signal can be described as

$$f_T\,[t] = \delta_T\,[t]\,f(t) = \sum \delta\,[t - nT]f(nT). \tag{5.163}$$

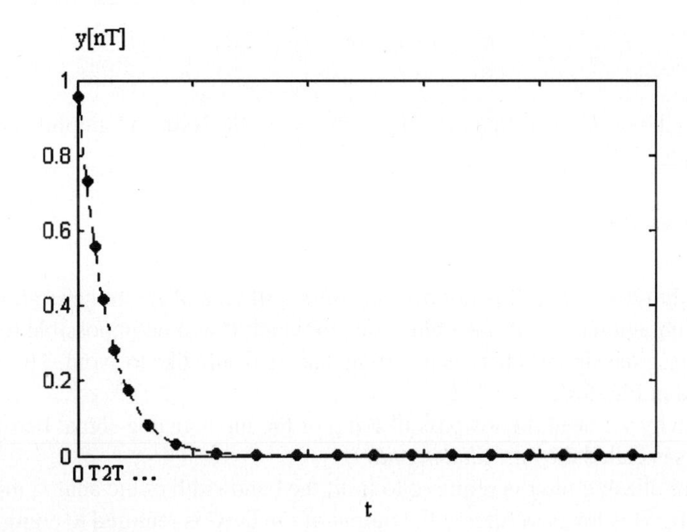


**Fig. 5.48** Sampled signal

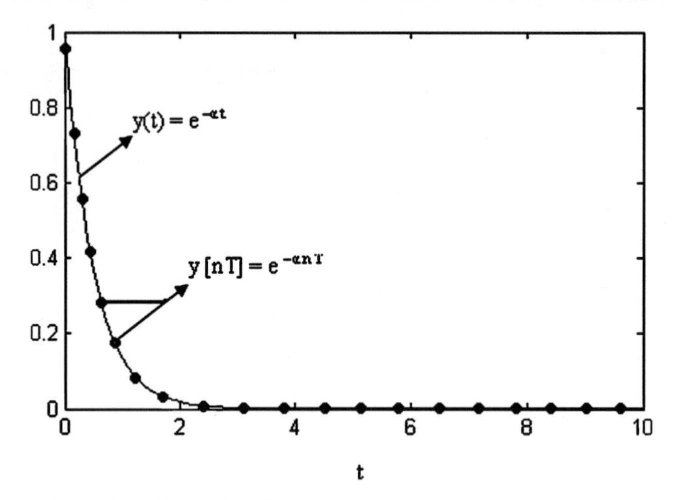


**Fig. 5.49** Sampled and continuous impulse response

This is shown in Fig. 5.50.

The original signal, with spectrum $F(\omega)$ and bandwidth $B$ Hz, can be recovered from the sampled signal spectrum $F_T(\omega)$ by passing the latter through an ideal lowpass filter of bandwidth $BW$, where

$$B < BW < f_S/2, \tag{5.164}$$

provided

$$B < f_S/2 \quad \text{or} \quad f_S > 2B. \tag{5.165}$$

Here, $f_S$ is the sampling frequency. This is known as the Nyquist Sampling criterion.

### 5.6.3 Aliasing

If the condition $B < f_S/2$ is not met, *aliasing* will take place. In general, aliasing means a contamination will take place due to which it will be impossible to reconstruct the original signal. This is something that we would like to avoid. The aliasing is depicted in Fig. 5.51.

That's why we need the lowpass filtering of the input analog signal before sampling. This is called the *anti-aliasing filter*.

The anti-aliasing filter is required to limit the bandwidth of the analog input signal to $B$ Hz. This lowpass filter at the output of the DAC is required to continuously interpolate the sampled output signal to provide an analog output.

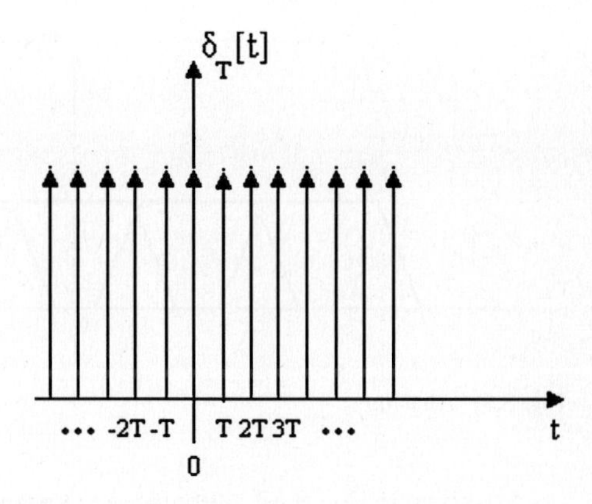


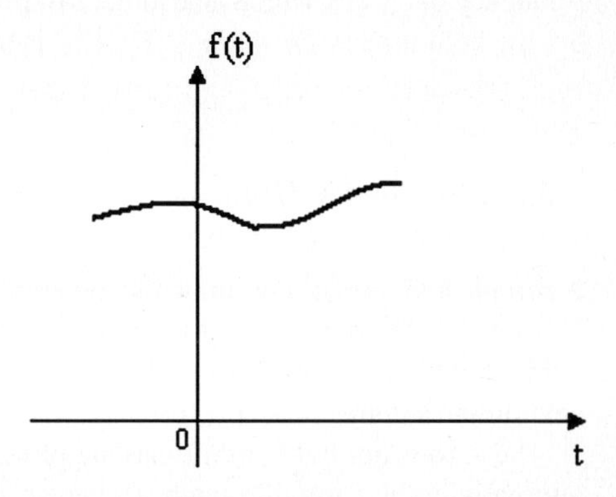


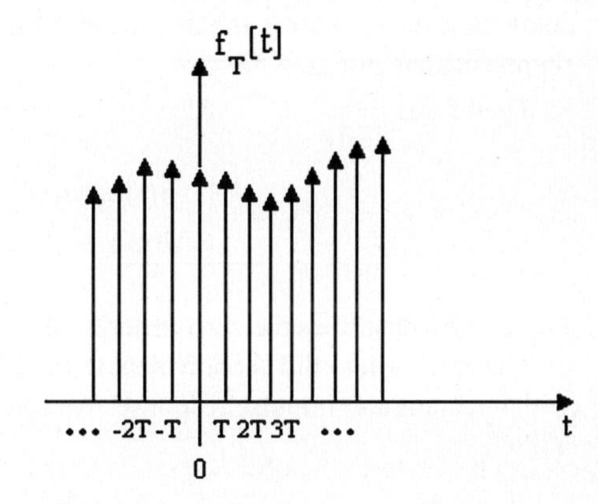


**Fig. 5.50** Sampled signal representation

## *5.6.4 Sample and Hold*

A sample and hold circuit is employed at the input before the ADC to hold the input value steady while the conversion is taking place. The simplest sample and hold circuit is a switch ($S$) and a holding capacitor ($C_H$). This is shown in Fig. 5.52.

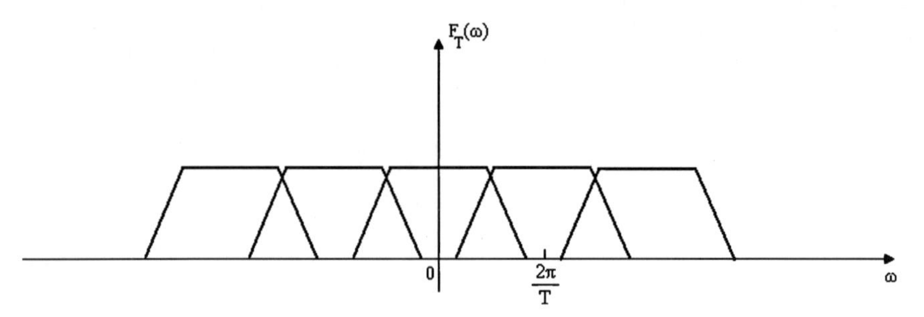


**Fig. 5.51** Aliasing

As shown in Fig. 5.52, we sample a continuous-time signal $f(t)$, with a sampling frequency of $f_S$ Hz, which is also the sampling rate $1/T_S$ ($T_S$ is the sampling time). So, the hold time is $T_H = T_S - T_C$, ($T_C$ is the conversion time).

### *5.6.5 Zero-order Hold*

A sample and hold circuit may also be employed at the output, after the DAC. This is equivalent to some form of lowpass filtering. In this case, it is called a zero-order hold circuit because it keeps (holds) the output value of the DAC constant until the next output sample.

The zero-order hold circuit may be physically external to the DAC or it may be internally in the form of a latch. However, the zero-order hold circuit causes some distortion of the spectrum which must be compensated for. The higher the sampling frequency, the smaller is the distortion. Mathematically, the zero-order hold can be represented as in Fig. 5.53.

Therefore,

$$\begin{aligned} b(t) &= a(t)^* h(t), \\ B(\omega) &= A(\omega).H(\omega). \end{aligned} \tag{5.166}$$

Figure 5.54 describes the zero-order hold in a graphical way.

The zero-order hold is equivalent to the filtering of the output impulses by a filter with a rectangular impulse response. We can determine the transfer function of such a filter.

$$\begin{aligned} H(\omega) &= \int_{-\infty}^{\infty} h(t) e^{-j\omega t} dt = \int_0^T e^{-j\omega t} dt = \frac{1}{j\omega}\left[1 - e^{-j\omega T}\right] \\ &= \frac{2}{\omega}.\frac{1}{2j}\left[e^{j\omega T/2} - e^{-j\omega T/2}\right]\left(e^{-j\omega T/2}\right) \\ &= \frac{2}{\omega}\sin\left(\frac{\omega T}{2}\right)\left(e^{-j\omega T/2}\right) \end{aligned} \tag{5.167}$$

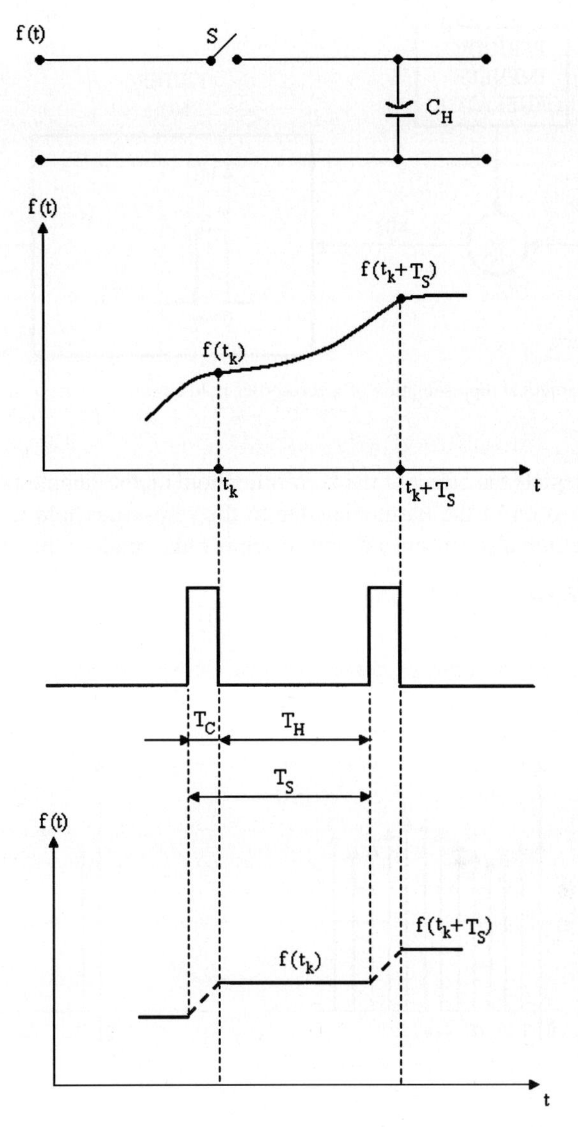


**Fig. 5.52** Sample and hold circuit and operation

$$= T\frac{\sin\left(\frac{\omega T}{2}\right)}{\left(\frac{\omega T}{2}\right)}\left(e^{-j\omega T/2}\right).$$

So, the magnitude response is

$$|H(\omega)| = T\frac{\sin\left(\frac{\omega T}{2}\right)}{\left(\frac{\omega T}{2}\right)}. \tag{5.168}$$

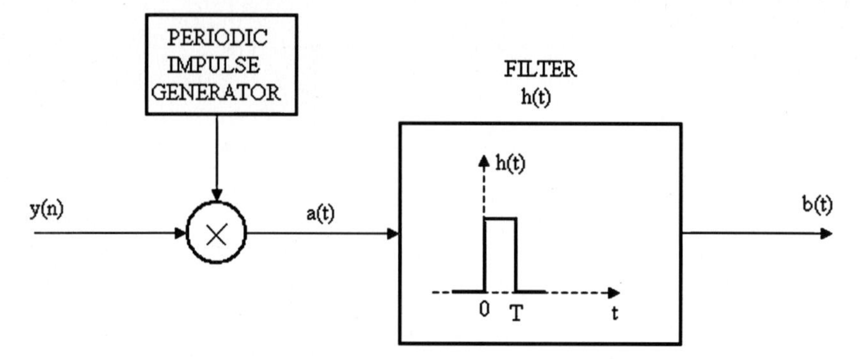


**Fig. 5.53** Mathematical representation of a zero-order hold circuit

Figure 5.55 depicts the effect of the zero-order hold on the sampled spectrum.

The suppression of the harmonics due to the zero-order hold is beneficial, but the main spectrum also suffers a droop which is rather undesirable. So, we need an equalizer near the origin with a spectrum $\frac{\left(\frac{\omega T}{2}\right)}{\sin\left(\frac{\omega T}{2}\right)}$.

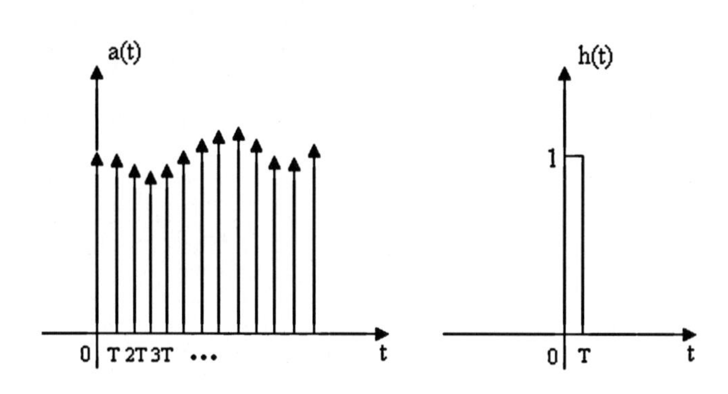


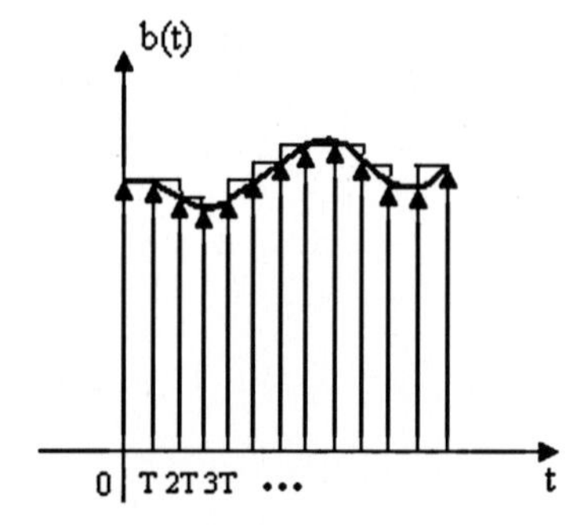


**Fig. 5.54** Graphical description of zero-order hold

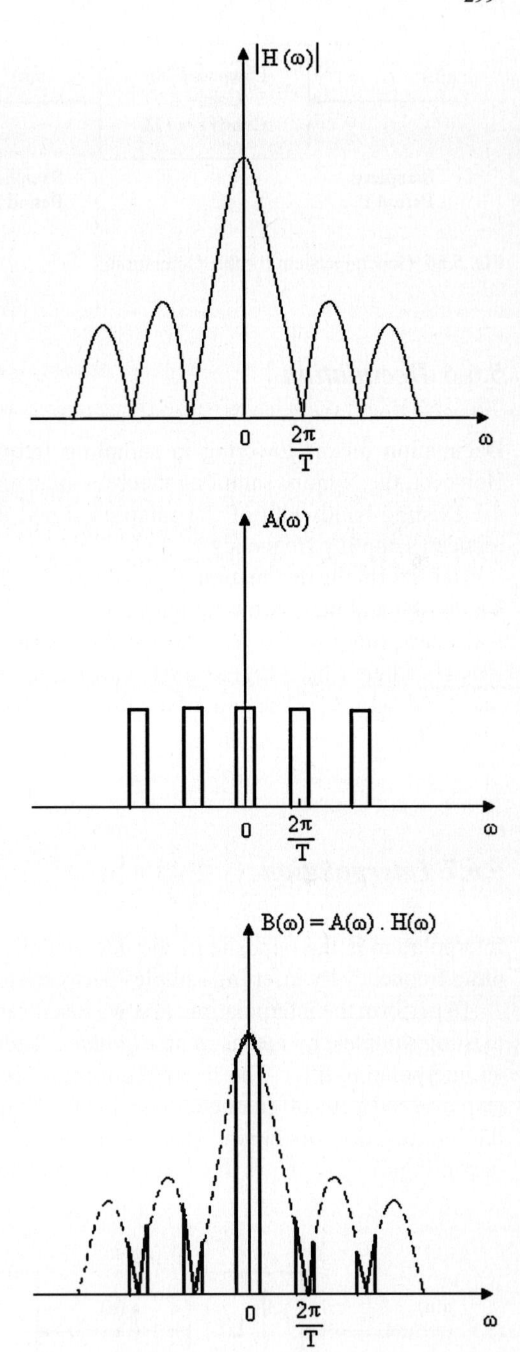


**Fig. 5.55** Effect of zero-order hold

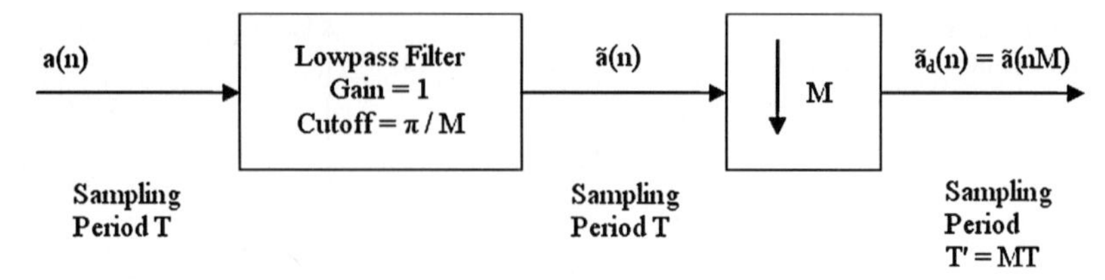


**Fig. 5.56** General system for the decimation

### *5.6.6 Decimation*

Decimation means lowering in sampling frequency by throwing samples away. However, the Nyquist sampling theorem must always be complied with. Therefore, the existing bandwidth of the sampled signal must be less than half of the final reduced sampling frequency.

To perform the decimation, first we lowpass filter the digital signal to the correct bandwidth and then throw samples away by means of a *compressor*. For example, if the sampling rate has to be lowered by a factor $M$, then the signal must first be lowpass filtered by a digital filter with cutoff frequency $\frac{\pi}{M}$ (half of the sampling rate divided by $M$). The gain of the filter is unity. Decimation operation is shown in Fig. 5.56.

### *5.6.7 Interpolation*

Interpolation is the opposite of the decimation operation, i.e., increasing the sampling frequency by inserting samples between the existing samples.

To perform the interpolation, first we insert samples with zero amplitude between existing samples, by means of an *expander*. Then, we lowpass filter the signal with an interpolating filter. The interpolating filter is a digital lowpass filter with a flat response and a cut-off frequency equal to the Nyquist-bandwidth of the signal before the interpolation operation. The gain of the filter is $L$, $L$ being the interpolating factor. Figure 5.57 shows the general interpolation operation.

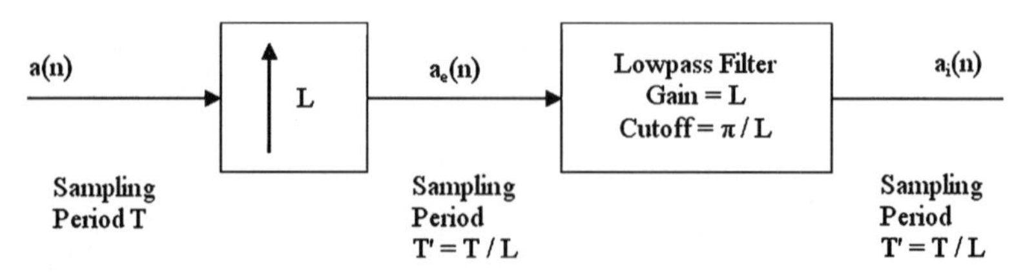


**Fig. 5.57** General system for the interpolation

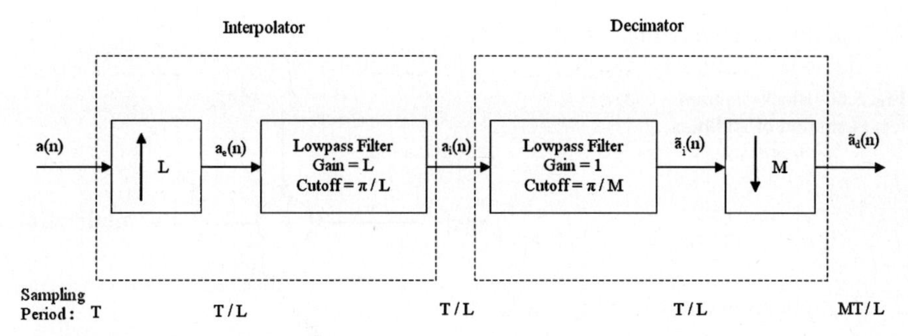


**Fig. 5.58** Changing the sampling rate by a noninteger factor

### *5.6.8 Decimation & Interpolation*

To change the sampling rate by a noninteger factor, we have to use the decimation and the interpolation together. By interpolating with a factor $L$, and thereafter decimating by a factor $M$, the equivalent change in the sampling rate is $L/M$. Figure 5.58 shows the combined operation.

From Fig. 5.58, it is to be noted that we can replace the two filters by one filter with a cutoff frequency equal to min $\left(\frac{\pi}{L}, \frac{\pi}{M}\right)$. This is shown in Fig. 5.59.

## 5.7 Digital Filtering

All the digital signal processing (ultimately) comes to the digital filtering. All the transforms, techniques and so on are utilized in terms of the digital filters. So, we will go through the digital filters in this section.

### *5.7.1 Structures for Digital Filters*

The LSI systems are generally expressed in terms of the difference equations. The block diagram representation for the basic elements (addition, multiplication, delay) of the difference equations are shown below. Using these basic elements, we can, in general, represent any other form.

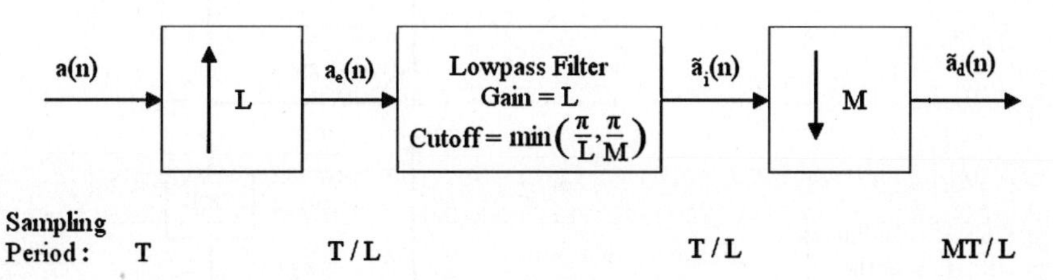


**Fig. 5.59** Combined filter for the decimation and the interpolation

1. Addition (shown in Fig. 5.60):

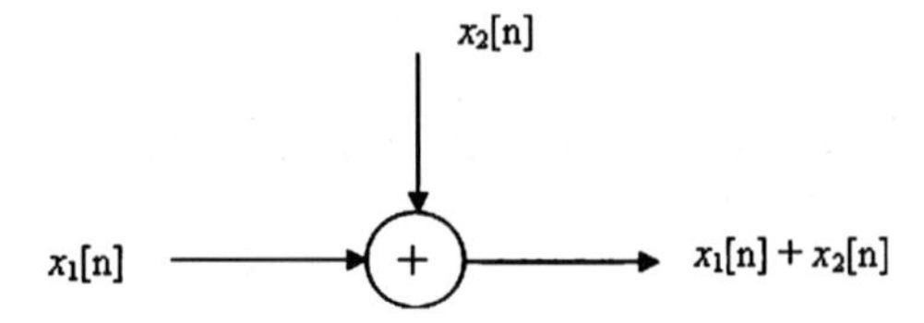


**Fig. 5.60** Block-diagram representation of addition

2. Multiplication (shown in Fig. 5.61):

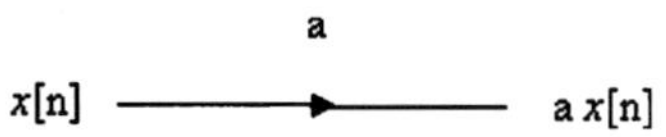

a

x[n] → a x[n]

**Fig. 5.61** Block-diagram representation of multiplication

3. Delay (shown in Fig. 5.62):

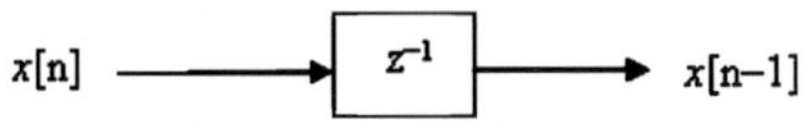

x[n] → $z^{-1}$ → x[n–1]

**Fig. 5.62** Block-diagram representation of delay

So, an example of a typical difference equation representation can be

$$y[n] = a_1\, y[n-1] + a_2\, y[n-2] + b_0\, x[n]\,. \tag{5.169}$$

Equation (5.169) can be represented using the basic block diagrams as shown in Fig. 5.63.
So, we can generalize the difference equation form as

$$y[n] - \sum_{k=1}^{N} a_k y[n-k] = \sum_{k=0}^{M} b_k x[n-k]\,. \tag{5.170}$$

The generalized difference equation could be also represented like

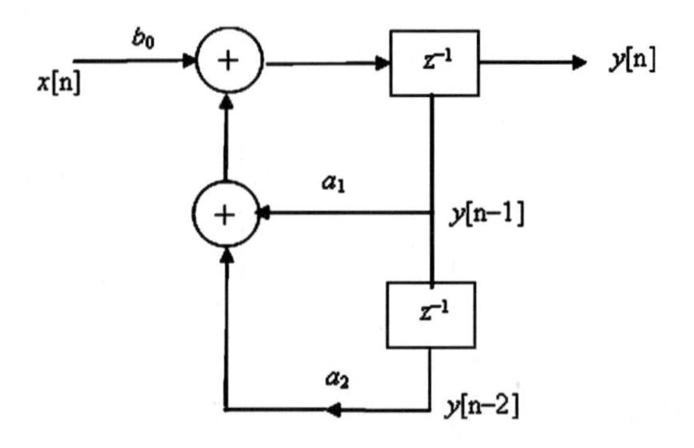


**Fig. 5.63** Block-diagram representation of the difference equation

$$\sum_{k=0}^{N} a_k y[n-k] = \sum_{k=0}^{M} b_k x[n-k]. \tag{5.171}$$

In this case, the coefficient of the term $y[n]$ is normalized to 1. Taking the ZT of (5.170), we get

$$H(z) = \frac{Y(z)}{X(z)} = \frac{\sum_{k=0}^{M} b_k z^{-k}}{1 - \sum_{k=1}^{N} a_k z^{-k}}. \tag{5.172}$$

#### 5.7.1.1 Direct Form-I Structure

Equation (5.170) could also be rearranged in a recursive form,

$$y[n] = \sum_{k=1}^{N} a_k y[n-k] + \sum_{k=0}^{M} b_k x[n-k]. \tag{5.173}$$

The block-diagram representation, shown in Fig. 5.64, is called the Direct Form-I (DF-I) structure.

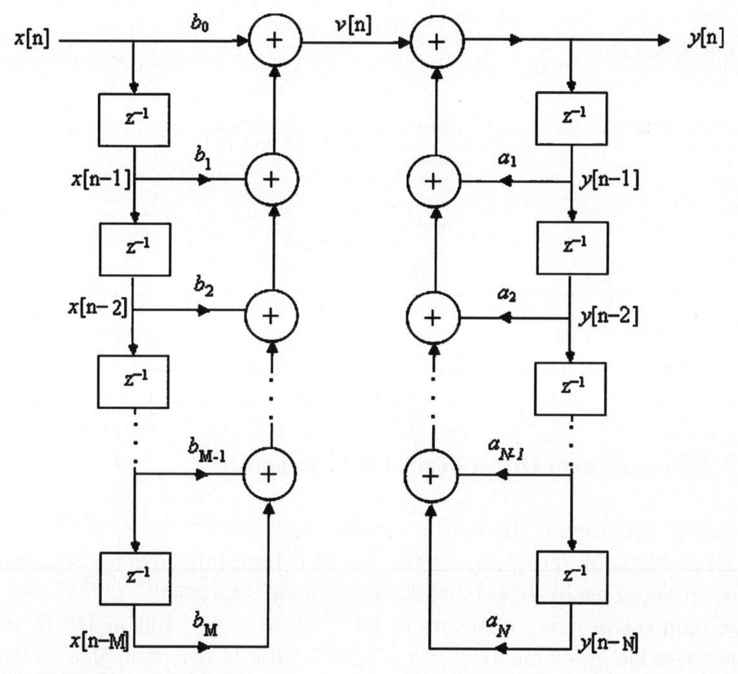


**Fig. 5.64** Direct Form-I structure

#### 5.7.1.2 Direct Form-II Structure

From (5.172), we have

$$H(z) = H_2(z)H_1(z) = \left[\frac{1}{1 - \sum_{k=1}^{N} a_k z^{-k}}\right]\left[\sum_{k=0}^{M} b_k z^{-k}\right]. \tag{5.174}$$

$$Y(z) = H(z)X(z) = H_2(z)H_1(z)X(z) = [H_2(z)X(z)]\, H_1(z). \tag{5.175}$$

Let, $W(z) = H_2(z)X(z)$. So, $Y(z) = W(z)H_1(z)$, where

$$W(z) = \left[\frac{1}{1 - \sum_{k=1}^{N} a_k z^{-k}}\right] X(z), \tag{5.176}$$

$$W(z) - \sum_{k=1}^{N} a_k z^{-k} W(z) = X(z)$$

$$H_1(z) = \left[\sum_{k=0}^{M} b_k z^{-k}\right]. \tag{5.177}$$

In time-domain,

$$w[n] = \sum_{k=1}^{N} a_k w[n-k] + x[n]. \tag{5.178}$$

$$y[n] = \sum_{k=0}^{M} b_k w[n-k] \tag{5.179}$$

From (5.178–5.179), we get the block-diagram representation as shown in Fig. 5.65. This is known as the Direct Form-II (DF-II) structure.

#### 5.7.1.3 Comparison of Direct Form-I & II Structures

- Both the structures are theoretically equivalent.
- In DF-I, zeros of $H(z)$, represented by $H_1(z)$ are implemented first, then, the poles, represented by $H_2(z)$. In DF-II, the order is reversed.
- Total number of delay elements in DF-I is $(N + M)$. But in DF-II, the total number of delay elements is $\max(N, M)$. This is less than that of DF-I, as $\max(N, M) < (N + M)$. So, DF-II structure is an implementation structure

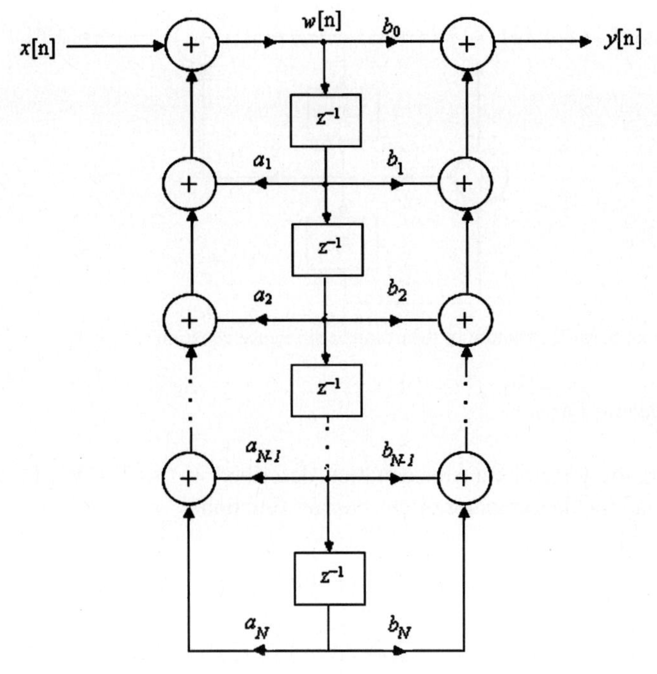


**Fig. 5.65** Direct Form-II structure

with minimum number of delay element. Hence, DF-II structure is also called the *canonic* form.

As an example, we consider the following LSI system function $H(z) = \dfrac{1+2z^{-1}}{1-1.5z^{-1}+0.9z^{-2}}$.

So, here $b_0 = 1,\ b_1 = 2$, and $a_1 = 1.5,\ a_2 = -0.9$.

The DF-I structure is shown in Fig. 5.66 and the DF-II structure in Fig. 5.67.

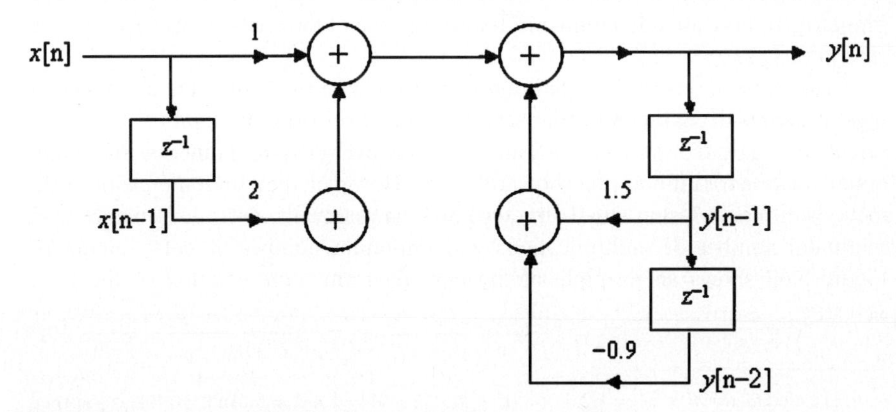


**Fig. 5.66** Direct Form-I structure of the example difference equation

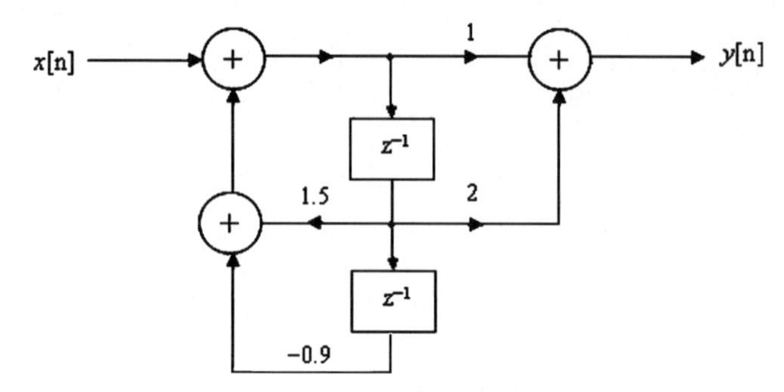


**Fig. 5.67** Direct Form-II structure of the example difference equation

#### 5.7.1.4 Cascade Form

We consider the general transfer equation described in (5.172). We factorize the numerator and the denominator of the transfer function.

$$H(z) = A\frac{\prod_{k=1}^{M_1}(1-g_k z^{-1})\prod_{k=1}^{M_2}(1-h_k z^{-1})\left(1-h_k^* z^{-1}\right)}{\prod_{k=1}^{N_1}(1-c_k z^{-1})\prod_{k=1}^{N_2}(1-d_k z^{-1})\left(1-d_k^* z^{-1}\right)}. \tag{5.180}$$

The general cascade form is obtained by combining complex conjugate and the real roots in the second order polynomial. So, the general cascade form is given as

$$H(z) = \prod_{k=1}^{N_s}\frac{b_{0k}+b_{1k}z^{-1}+b_{2k}z^{-2}}{1-a_{1k}z^{-1}-a_{2k}z^{-2}}. \tag{5.181}$$

This is the case if $M \leq N$ and the real poles and zeros are combined in pairs. Also, $N_s = \left[(N+1)/2\right]$ is the largest integer contained in $(N+1)/2$. If there are odd number of real poles, one of the coefficients ($a_{2k}$) will be zero in (5.181). Similarly, if there are odd number of real zeros, one of the coefficients ($b_{2k}$) will be zero in (5.181).

From above, there are $N!$[23] pairings of poles and zeros. Also, there are $N$ orderings of the resulting second-order sections. So, there would be total of $(N!)^2$ different pairings and orderings. They all have same overall system function and input-output relation for infinite-precision arithmetic. However, their finite-precision arithmetic is different. Using DF-II structure, we can realize the cascade structure with minimum number of multiplications and minimum number of delay elements. Figure 5.68 shows an example of the cascade form representation of the filter structure.

---

[23] $n!$ is called *factorial* $n$. $n! = 1.2.3.\ldots.(n-1).n$. So, $3! = 1.2.3 = 6$. And, $0! = 1$, is a special case.

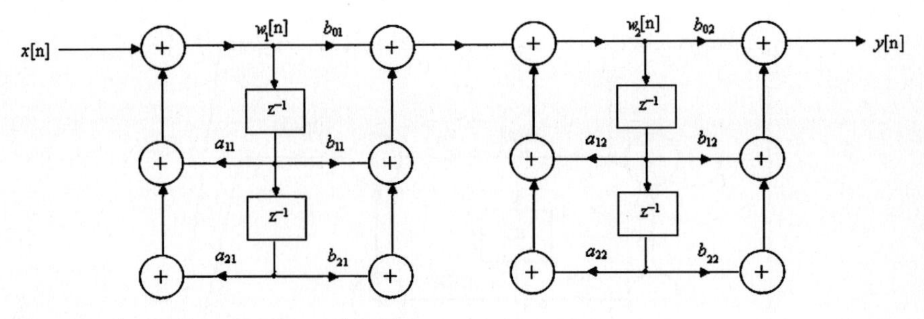


**Fig. 5.68** Cascade structure for a 4th order-system

#### 5.7.1.5 Parallel Form

For the parallel form structure, we express the general transfer function $H(z)$ as a partial function expansion.

$$H(z) = \sum_{k=0}^{N_p} C_k z^{-k} + \sum_{k=1}^{N_s} \frac{e_{0k} + e_{1k} z^{-1}}{1 - a_{1k} z^{-1} - a_{2k} z^{-2}}. \tag{5.182}$$

This is the case with $M \geq N$, $N_p = M - N$ and $N_s = \lceil (N+1)/2 \rceil$. It is assumed that the real poles and zeros are combined in pairs. Parallel form structure for 6-th order system is shown in Fig. 5.69.

#### 5.7.1.6 Transposed Form

Transposition of a flow-graph is accomplished by reversing the directions of all the branches in the network while keeping the branch transmittances as they were. Also, the roles of the input and the output are reversed, so the source nodes become the sink nodes and vice versa. An example is shown in Fig. 5.70.

We can also transform from the DF-II structure to the transposed form. For example, we consider

$$\begin{aligned} w[n] &= a_1 w[n-1] + a_2 w[n-2] + x[n] \\ y[n] &= b_0 w[n] + b_1 w[n-1] + b_2 w[n-2]. \end{aligned} \tag{5.183}$$

The transposed form obtained from the DF-II form is shown in Fig. 5.71.

### 5.7.2 *Filter Types: IIR and FIR*

There are generally two types of digital filters. Those having impulse responses over an infinite range are called the infinite impulse response (IIR) filter. In comparison,

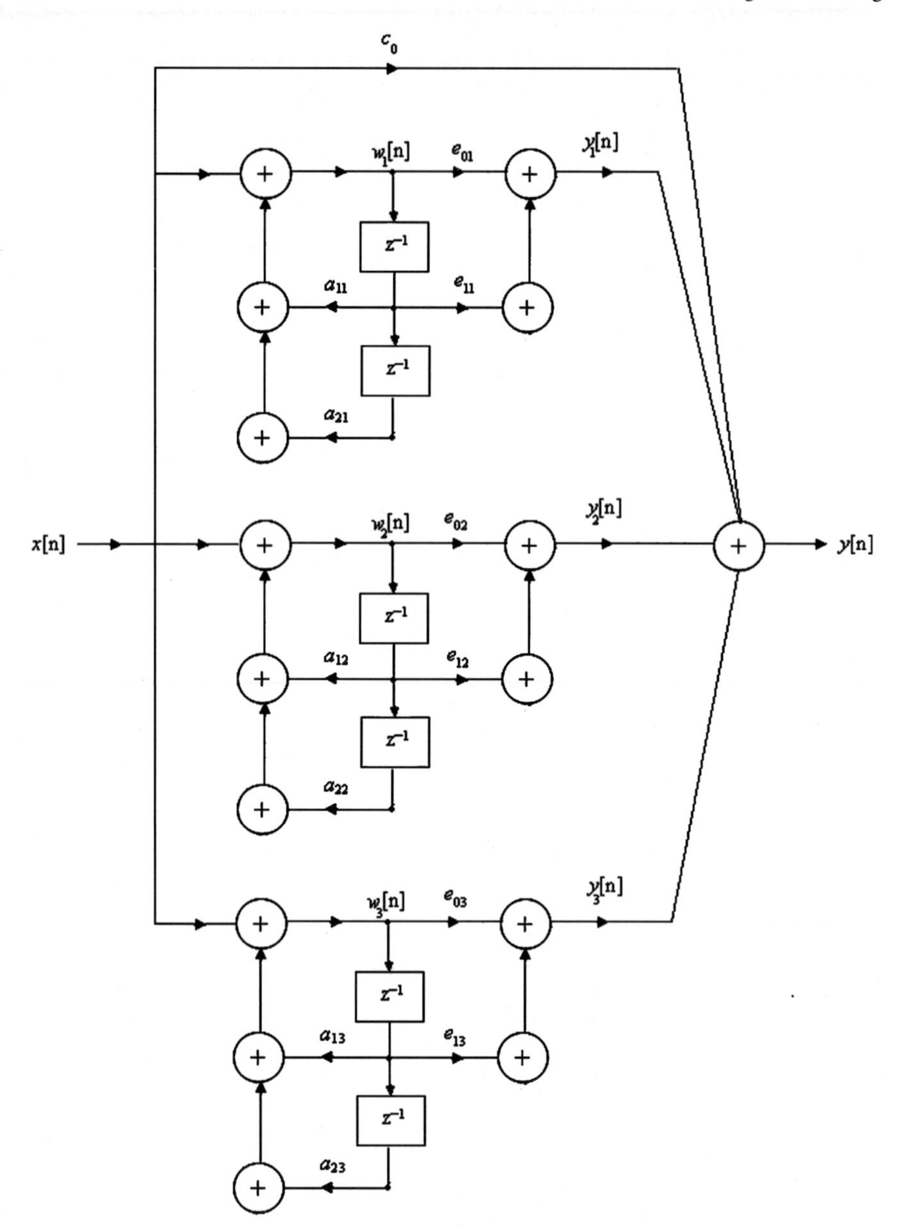


**Fig. 5.69** Parallel form structure for 6th-oder system

**Fig. 5.70** Example of the transposed form

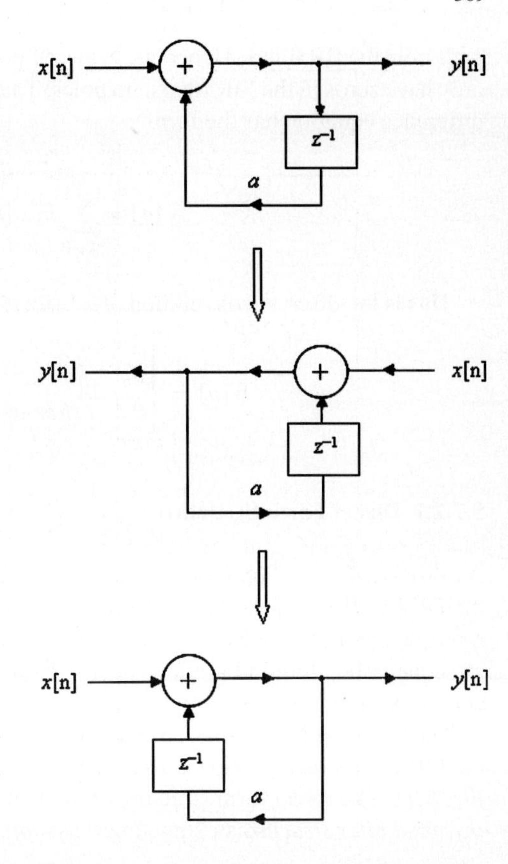


the filters having the impulse response over a finite range are called finite impulse response (FIR) filters.

IIR filters are also called all-pole filters or auto-regressive (AR) or auto-regressive moving average (ARMA) filters. FIR filters, in comparison, are also called all-zero or non-recursive or moving average filters.

From the implementation and structural point of views, an IIR filter has the feedback path. That is, in (5.170), the coefficients $a_k$'s have non-zero values. In other words, from the ZT point of view, the IIR filters have both the poles and zeros (see (5.174)). In comparison, we construct the FIR filters by removing the feedback

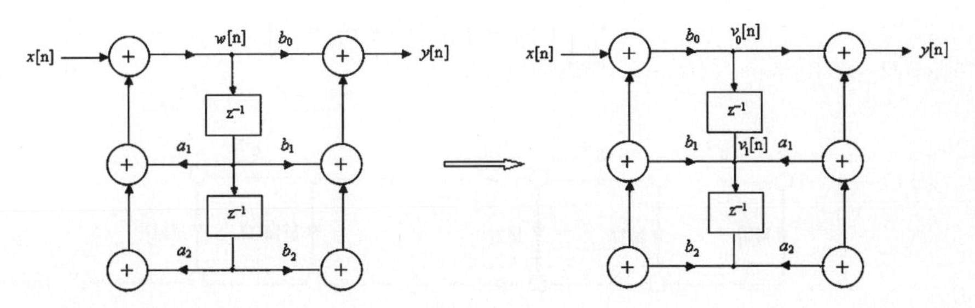


**Fig. 5.71** Transposed form from the DF-II form

path from the IIR filters. Hence, from the ZT point of view, the implication is that we only have zeros in the FIR filters, no poles. That is, for the FIR filter, the generalized difference equation has the form

$$y[n] = \sum_{k=0}^{M} b_k x[n-k]\,. \tag{5.184}$$

This is the discrete convolution of $x[n]$ with the impulse response

$$h[n] = \begin{cases} b_n & for\ n = 0, 1, \ldots, M, \\ 0 & otherwise. \end{cases} \tag{5.185}$$

#### 5.7.2.1 Direct Form Structure

As discussed in the previous section, filters can be effectively represented using the DF-I or II representation. Section 5.7.1.1–5.7.1.2 discuss about the direct form representation of the general filter structure which include the feedback path. Hence, effectively the aforesaid sections deal with the IIR filter type. Hence, we will concentrate on the FIR types in this section.

The FIR-type filters can also be effectively represented using the direct form. For the FIR filter structure shown in (5.184), we represent the direct form realization in Fig. 5.72. The direct form structure of the FIR system in Fig. 5.72 is also known as *traversal filter structure* or sometimes, as *tapped delay line* (TDL).

#### 5.7.2.2 Cascade Form

The cascade form for the FIR system is obtained by factoring the polynomial system function. So, we represent the $H(z)$ as

$$\begin{aligned} H(z) &= \sum_{n=0}^{M} h[n] z^{-n} \\ &= \prod_{k=1}^{M_s} \left(b_{0k} + b_{1k} z^{-1} + b_{2k} z^{-1}\right) \end{aligned} \tag{5.186}$$

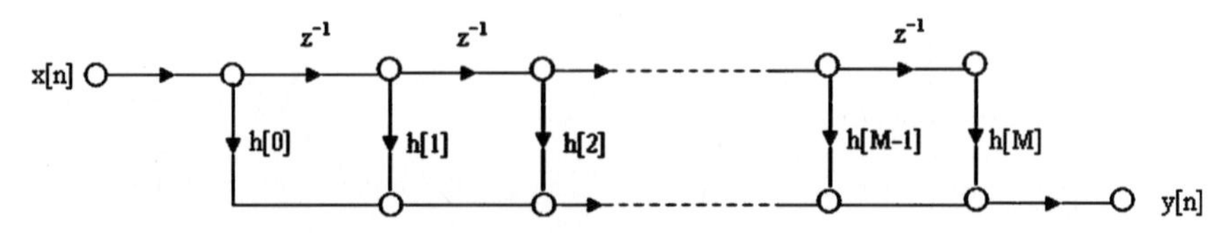


**Fig. 5.72** Direct form realization of FIR system

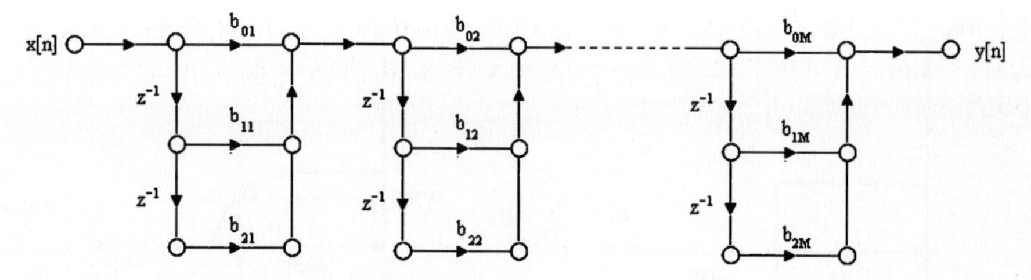


**Fig. 5.73** Cascade form realization of an FIR system

where, $M_s = \left[(M+1)/2\right]$ is the largest integer contained in $(M+1)/2$. If $M$ is odd, one of the coefficients of $b_{2k}$ will be zero, since $H(z)$ in that case would have an odd number of real zeros. Figure 5.73 depicts the cascade form of the FIR filters.

The cascade connection is least sensitive to the effect of parameter quantization. This is an important aspect in the case where use is made of integer arithmetic to realize the filter.

### *5.7.3 Design of Digital Filters*

The following three steps are followed in the design of a digital filter:

1. Specification of the desired properties of the system.
2. Approximation of the specifications using a discrete-time system.
3. Realization of the system.

Step 1 depends on the application, while step 3 depends on the technology used. So, in the filter design we are mostly concerned about the step 2. Figure 5.74 shows the basic structure of the digital filtering in principle.

It is clear from the Fig. 5.74 that if the sampling frequency is high enough and the input is band-limited to avoid the aliasing, then the overall system behaves as an LTI continuous-time system with the following frequency response

$$H_{eff}(j\Omega) = \begin{cases} H(e^{j\Omega T}), & |\Omega| < \frac{\pi}{T} \\ 0, & |\Omega| > \frac{\pi}{T} \end{cases}. \tag{5.187}$$

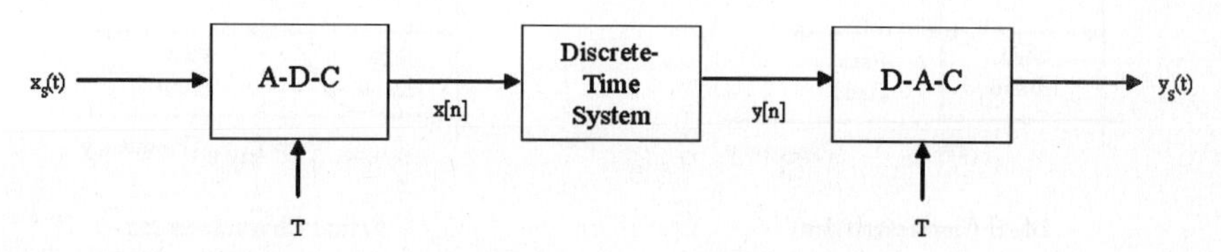


**Fig. 5.74** Basic system for discrete-time filtering of continuous-time signals

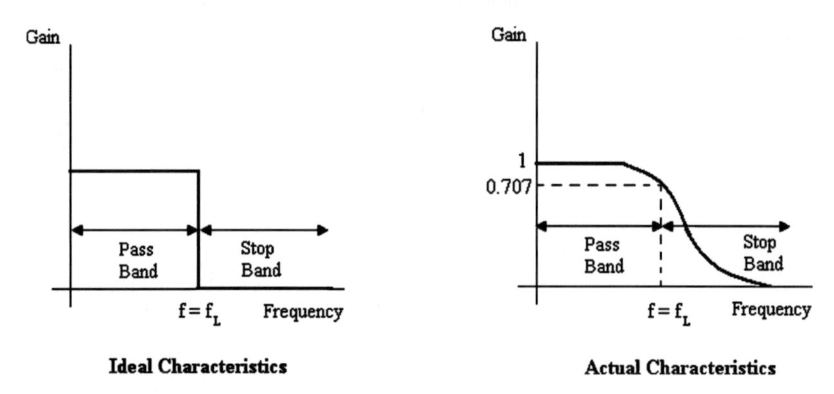


**Fig. 5.75** Lowpass filter

The specification can be converted from continuous-time to a discrete-time filter by using $\omega = \Omega T$. Therefore, $H(e^{j\omega})$ can be specified over one period by

$$H\left(e^{j\omega}\right) = H_{eff}\left(j\frac{\omega}{T}\right), \quad |\omega| < \pi. \tag{5.188}$$

#### 5.7.3.1 Filter Categories

There are many categories of filters. The most common ones are as follows.

1. **Lowpass**. Passes low-frequency signals, stops high-frequency signals. This is shown in Fig. 5.75.
2. **Highpass**. Opposite to lowpass, passes high-frequency, stops low-frequency. This is shown in Fig. 5.76.
3. **Bandpass**. Passes frequencies within a specific band, stops all other. This is shown in Fig. 5.77.

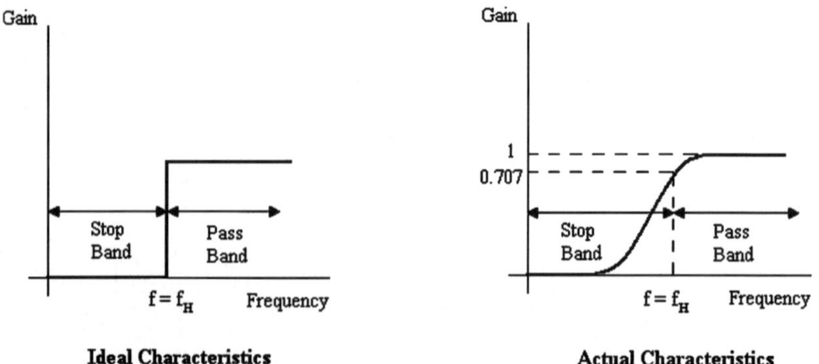


**Fig. 5.76** Highpass filter

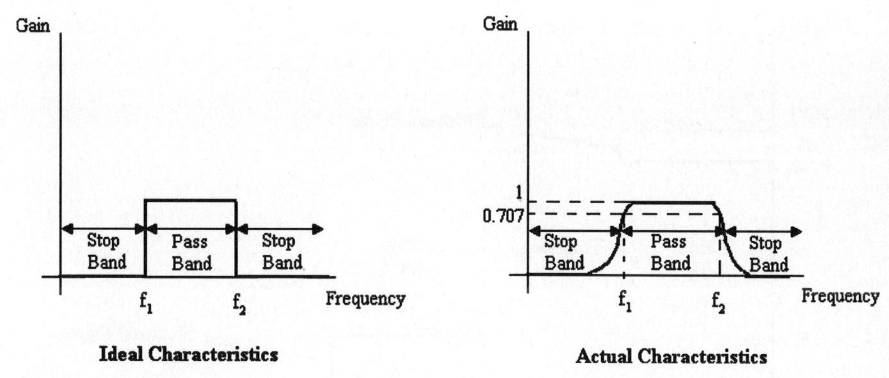


**Fig. 5.77** Bandpass filter

4. **Bandstop**. Opposite to bandpass, stops frequencies within a specified band, passes all other. This is shown in Fig. 5.78.

Depending on the specific application, we choose the different categories of the filters.

### 5.7.3.2 Example

The discrete-time filter, which is to be used as a lowpass filter for a continuous-time signal, must have the following properties when the sampling frequency is $10^4$ samples/second ($T = 10^{-4}s$):

1. Gain $|H_{eff}(j\Omega)|$ has to be within $\pm 0.01 (= 0.086\,\text{dB})$ below unity (zero dB) in the band $0 \leq \Omega \leq 2\pi\ (2000)$.
2. Gain should not be greater than $0.001(-60\,\text{dB})$ in the band $2\pi(3000) \leq \Omega$.

The specifications are depicted in Fig. 5.79 (continuous-time) and Fig. 5.80 (discrete-time).

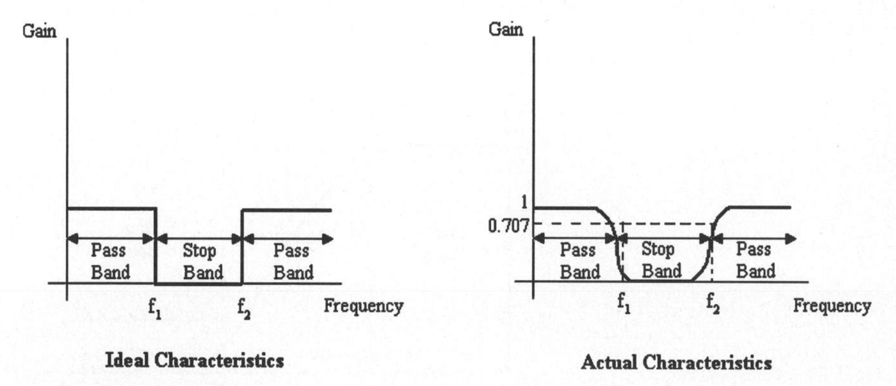


**Fig. 5.78** Bandstop filter

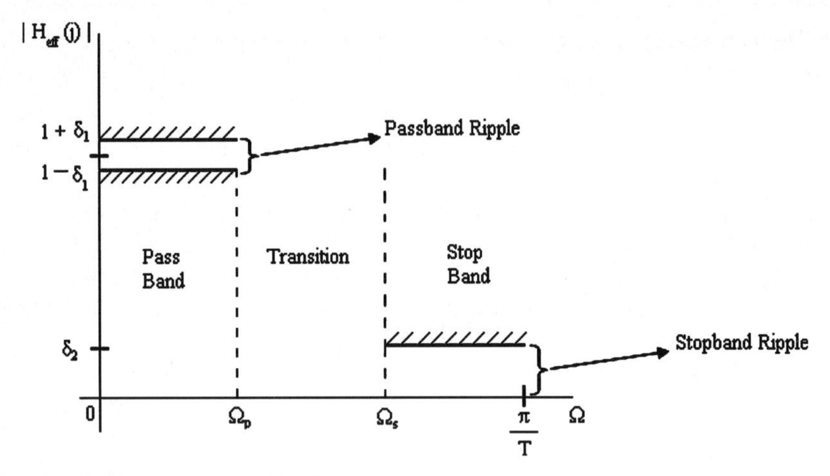


**Fig. 5.79** Continuous-time specification

We have,

Max. Frequency Available (in Hz) $\leq \frac{1}{2}$. Sampling Rate (samples/s).

So, for the sampling rate of $10^4$ samples/s, the maximum frequency available (i.e., gain of the overall system is identically zero above this frequency) is $2\pi(5000)$ Hz.

In Fig. 5.79–5.80, the parameters are as follows:

$$\begin{aligned}
\delta_1 &= 0.01 \quad [20\log_{10}(1+\delta_1) = 0.086\,dB],\\
\delta_2 &= 0.001 \quad [20\log_{10}(\delta_2) = -60\,dB],\\
\Omega_p &= 2\pi(2000),\\
\Omega_s &= 2\pi(3000).
\end{aligned}$$

In Fig. 5.80, the normalized frequency ($\omega = \Omega T$) is used and need only be plotted in the range $0 \leq \omega \leq \pi$.

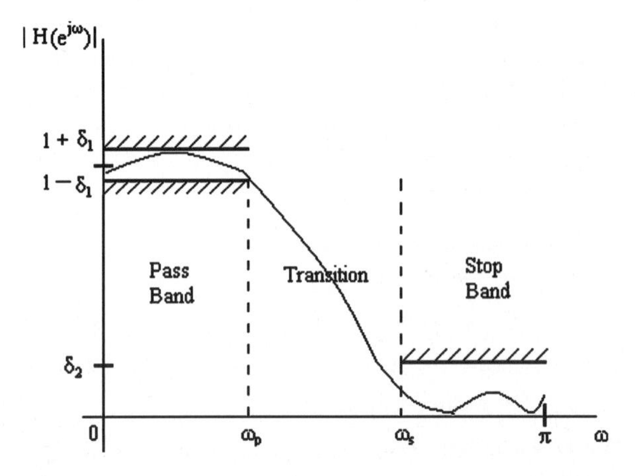


**Fig. 5.80** Corresponding discrete-time specification

The passband, in which the magnitude response must approximate unity with an error of $\pm\delta_1$, is

$$(1-\delta_1) \leq \left|H(e^{j\omega})\right| \leq (1+\delta_1), \quad |\omega| \leq \omega_p. \tag{5.189}$$

The stopband, in which the magnitude response must approximate zero with an error of $\pm\delta_2$, is

$$\left|H(e^{j\omega})\right| \leq \delta_2, \quad \omega_s \leq |\omega| \leq \pi. \tag{5.190}$$

Furthermore,

$$\begin{aligned}
\delta_1 &= 0.01, \\
\omega_p &= 0.4\pi \text{ radians}, \\
\delta_2 &= 0.001, \\
\omega_s &= 0.6\pi \text{ radians}.
\end{aligned}$$

In the transition band, which is nonzero with width $(\omega_s - \omega_p)$, the magnitude response changes smoothly from the passband to the stopband as indicated by the curve in Fig. 5.80.

It is to be noted, when the sampling rate changes with a factor $k$, the cutoff frequencies in the continuous-time domain changes accordingly. In the discrete-time domain the cutoff frequencies stay the same. In most of the filter design techniques, the sampling frequency plays no role in the approximation process of the transfer function.

### 5.7.4 Design of IIR Filters

To design the IIR filters in discrete-time domain, we start with the specification specified in the discrete-time domain. The specification is then transformed to the continuous-time domain and the approximation of the transfer function is then performed in that domain (usually the $s$-domain). Then, the design is transformed back into the $z$-domain (discrete-time) through one of the transformations and the realization of the filter then performed in the discrete-time domain. The main transformation techniques are

- The Impulse invariance method
- Bilinear transformation.

The following conditions should be fulfilled while mapping back the IIR filter specifications from the continuous- to the discrete-time domain.

1. The essential properties of the continuous-time frequency response must be preserved in the frequency response of the resulting discrete-time filter. That is,

we want the imaginary axis of the $s$-plane to be mapped onto the unit circle of the $z$-plane.

2. A stable continuous-time filter should be transformed into a stable discrete-time filter. That is, if the continuous-time system has poles only in the left half of the $s$-plane, then the corresponding discrete-time filter must have poles only inside the unit circle in the $z$-plane.

#### 5.7.4.1 The Impulse Invariance Method

The main principle of the impulse invariance method (Parks and Burrus 1987) is based on the fact that the impulse response of the discrete-time filter is the same as that of the continuous-time filter. This means that the impulse response is invariant of the operating domain, continuous or discrete-time.

The impulse response of the discrete-time filter is a uniformly sampled (sampling period is constant) version of the continuous-time filter's impulse response. An example of the uniform sampling on the continuous-time signal $x(t)$ is shown in Fig. 5.81. The uniform sampler transforms the original continuous-time signal $x(t)$ into the sampled version $x^*(t)$.

From, Fig. 5.81 we get,

$$x^*(t) = \sum_{n=-\infty}^{\infty} x[n\tau]\delta[t - n\tau], \tag{5.191}$$

where $\delta[t - n\tau]$ is the train of scaled impulses. Strength of the impulse is the sampled value at that instant.

$$\begin{aligned} x^*(t) &= \sum_{n=-\infty}^{\infty} x_n\delta[t - n\tau.] \\ x_n &= x[n\tau] = x(t)|_{t=n\tau} \end{aligned} \tag{5.192}$$

Impulse response of the discrete-time filter is chosen as proportional to the equally spaced samples of the impulse response of the discrete-time filter. Therefore,

$$h[n] = t_s h_c(nt_s). \tag{5.193}$$

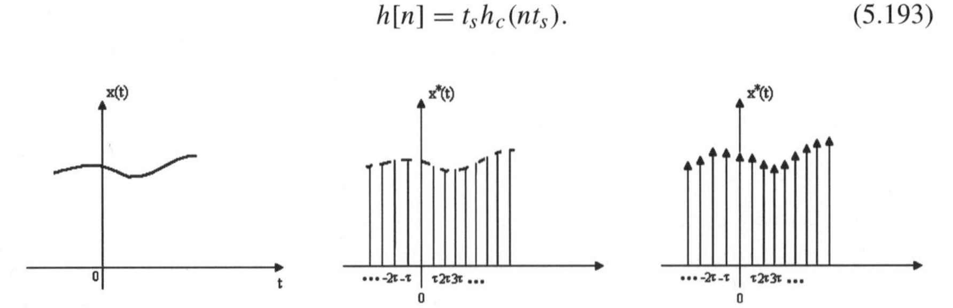


**Fig. 5.81** Uniform sampling

Here, $t_s$ represents the sampling interval.

Following the Nyquist sampling theorem, the continuous-time system must be bandlimited, otherwise aliasing will take place. The extent of the aliasing would depend on the ratio between the sampling frequency and the bandwidth of the system.

The frequency response of the discrete-time filter is related to the frequency response of the continuous-time filter through the sampling theorem.

$$H(e^{j\omega}) = \sum_{k=-\infty}^{\infty} H_c\left[j\frac{\omega}{t_s} + j\frac{2\pi}{t_s}k\right]. \tag{5.194}$$

The frequency response is periodic with period $2\pi$. If the continuous-time filter is bandlimited as follows

$$H_c(j\Omega) = 0, \quad |\Omega| \geq \pi/t_s, \tag{5.195}$$

then,

$$H(e^{j\omega}) = H_c\left(j\frac{\omega}{t_s}\right), \quad |\omega| \leq \pi. \tag{5.196}$$

The discrete-time and continuous-time frequency responses are related by

$$\omega = \Omega t_s. \tag{5.197}$$

The impulse invariance transformation from the continuous-time to the discrete-time is defined in terms of time-domain sampling. Therefore, to carry out the transformation of the system function, we express the transfer function of the continuous-time system as a partial fraction. For simplicity, assuming all single-order poles of the transfer function $H(s)$, we have

$$H_c(s) = \sum_{k=1}^{N} \frac{A_k}{s - s_k}. \tag{5.198}$$

The corresponding time-domain impulse response is

$$h_c(t) = \begin{cases} \sum_{k=1}^{N} A_k e^{s_k t}, & t \geq 0 \\ 0, & t < 0 \end{cases}. \tag{5.199}$$

The sampled impulse response of the digital filter (discrete-time) is obtained by sampling $t_s h_c(t)$,

$$h[n] = t_s h_c(nt_s) = \sum_{k=1}^{N} t_s A_k e^{s_k n t_s} u[n] = \sum_{k=1}^{N} t_s A_k e^{(s_k t_s)n} u[n]. \tag{5.200}$$

The transfer function $H(s)$ of the digital filter is therefore

$$H(z) = \sum_{k=1}^{N} \frac{t_s A_k}{1 - e^{s_k t_s} z^{-1}}. \tag{5.201}$$

Comparing (5.198) and (5.201), it comes of clearly that a pole in the $s$-domain at $s = s_k$, transforms into a pole in the $z$-domain at $e^{s_k t_s}$. Therefore,

$$z = e^{st_s} = e^{(\sigma \pm j\Omega)t_s} = e^{\sigma t_s} e^{\pm j\Omega t_s}, \tag{5.202}$$

and

$$|z| = e^{\sigma t_s}, \quad \angle(z) = \Omega t_s. \tag{5.203}$$

If the continuous-time filter is stable, for the real part of $s_k$ being less than zero, the magnitude of $e^{s_k t_s}$ will be less than unity. Hence, the corresponding pole in the discrete-time filter is inside the unit circle, so the causal discrete-time filter will be stable as well.

#### 5.7.4.2 Bilinear Transformation

The impulse invariance method is used on bandlimited transfer functions. This means that it can only be used in the design of lowpass filters. Therefore, in the case of highpass filter, additional bandlimiting would be required to prevent severe aliasing. Another solution is the bilinear transformation (Parks and Burrus 1987).

The bilinear transformation is a nonlinear transformation to tranform the complete frequency domain in the $s$-plane from 0 to $\infty$, into the $z$-domain from 0 to $\pi$. This will ensure that aliasing cannot take place. There is always a zero at $\pi$ ($\infty$ in the $s$-domain). The relationship between the $s$- and the $z$-domain is

$$s = \frac{2}{t_s}\left(\frac{1 - z^{-1}}{1 + z^{-1}}\right), \tag{5.204}$$

$$z = \frac{1 + (t_s/2)s}{1 - (t_s/2)s}. \tag{5.205}$$

We substitute $s = j\Omega$ and then it can be shown that the continuous-time frequency axis is mapped onto the unit circle by the bilinear transformation as

$$\Omega = \frac{2}{t_s}\tan(\omega/2), \tag{5.206}$$

or

$$\omega = 2\tan^{-1}(\Omega t_s/2). \tag{5.207}$$

To arrive at the above bilinear transformation relationship, we substitute $s = j\Omega$ (continuous-time), and $z = e^{j\omega}$ (discrete-time) in (5.205). Then,

$$e^{j\omega} = \frac{1 + (t_s/2)j\Omega}{1 - (t_s/2)j\Omega}. \tag{5.208}$$

So,

$$\cos(\omega) + j\sin(\omega) = \frac{1 + j(\Omega t_s/2)}{1 - j(\Omega t_s/2)}. \tag{5.209}$$

In polar form,

$$\sqrt{\cos^2(\omega) + \sin^2(\omega)}\angle\left[\tan^{-1}\left(\frac{\sin(\omega)}{\cos(\omega)}\right)\right] = \frac{\sqrt{1 + \frac{\Omega^2 t_s^2}{4}}\angle\left[\tan^{-1}\left(\frac{\Omega t_s/2}{1}\right)\right]}{\sqrt{1 + \frac{\Omega^2 t_s^2}{4}}\angle\left[\tan^{-1}\left(\frac{-\Omega t_s/2}{1}\right)\right]}. \tag{5.210}$$

This gives

$$1\angle\omega = \frac{1\angle\left[\tan^{-1}(\Omega t_s/2)\right]}{1\angle\left[-\tan^{-1}(\Omega t_s/2)\right]} = 1\angle\left[2\tan^{-1}(\Omega t_s/2)\right]. \tag{5.211}$$

Equating the magnitude and the angle parts in (5.211), we arrive at the bilinear transformation condition of (5.207). Figure 5.82 depicts this mapping between the $\Omega$ and $\omega$.

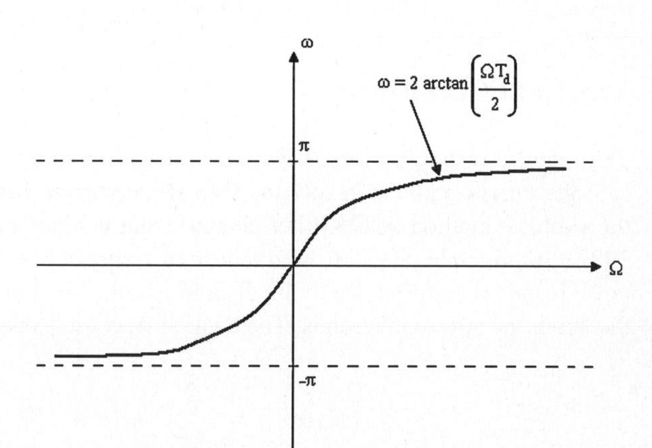


**Fig. 5.82** Mapping of continuous-time frequency axis onto the unit circle of $z$-domain using the bilinear transformation

### 5.7.5 *Design of FIR Filters*

From the discussion of the Sect. 5.7.2, we recall that the finite impulse response (FIR) filters are formed by removing the feedback paths of the output. In order words, in the $z$-transform of the FIR filter difference equation, we have only zeros, no poles.

IIR-filters are usually designed by transforming a continuous-time filter into a discrete-time filter. In comparison, the FIR filters are usually restricted to discrete-time implementations. Therefore, their design techniques are based on the approximation of the desired magnitude of the frequency response of the discrete-time system.

As the FIR-filters do not have a feedback path (i.e., the poles), they usually have a linear phase.

#### 5.7.5.1 Pros and Cons of the FIR Filters

- Pros
  - As the FIR-filters do not have a feedback path (i.e., the poles), they do not have any stability problem. In comparison, for the IIR filters all the poles are very critical for stability. These poles have to lie within the unit circle, otherwise the IIR filter would be unstable.
  - FIR-filters usually have a linear phase.
  - From the implementation point of view, FIR filters are easier to implement using the DSPs utilizing the fast computation capability of the FFT algorithm.
- Cons
  - FIR filters require higher orders than the IIR filters.
  - The group delay is much higher for the FIR filters. For this reason, the response is much slower for the FIR filters.
  - In case of FIR filter the memory requirement is higher.

#### 5.7.5.2 FIR Filter Basics

The impulse response of the IIR filter is of infinite duration. However, if we truncate the infinite impulse response of an IIR filter, we could have a FIR filter. So, the simplest method of FIR filter design is the *window* method (Parks and Burrus 1987). In principle, we start with a desired frequency response, determine the impulse response and truncate to the desired length. The longer the impulse response, the better the approximation is. The desired frequency response is

$$H_d(e^{j\omega}) = \sum_{n=-\infty}^{\infty} h_d(n)e^{-j\omega n}. \tag{5.212}$$

The corresponding impulse response is

$$h_d[n] = \frac{1}{2\pi} \int_{-\pi}^{\pi} H_d(e^{j\omega}) e^{j\omega n} d\omega. \tag{5.213}$$

The infinite impulse response $h_d[n]$ has to be truncated by the *windowing* process. Various types of windows can be used (to be discussed later), the simplest and common one being the rectangular window which is as follows.

We define the finite length of the impulse response (by the truncation method) as

$$h[n] = \begin{cases} h_d[n], & 0 \le n \le M, \\ 0, & otherwise \end{cases}. \tag{5.214}$$

$h[n]$ can be represented as

$$h[n] = h_d[n] w[n], \tag{5.215}$$

where, $w[n]$ is a rectangular window. The rectangular window can be defined as

$$w[n] = \begin{cases} 1, & 0 \le n \le M, \\ 0, & otherwise \end{cases}. \tag{5.216}$$

So, the resultant frequency response of the windowed filter is the periodic convolution between the desired ideal frequency response and the frequency response of the window. It can also be noted from (5.216) that the rectangular window acts like an ideal lowpass filter (see Fig. 5.75). The effect of a rectangular window is shown in terms of the magnitude and log-magnitude response in Fig. 5.83 and 5.84.

#### 5.7.5.3 FIR Filter Types

Let us consider an FIR filter kernel, truncated to $M$ points, defined as $h[M] = [h_0, h_1, \ldots, h_{M-1}]$. Then, from the linear-phase constraint of the FIR filters, we have

$$\angle H(e^{j\omega}) = -\alpha\omega, \quad -\pi < \omega \le \pi. \tag{5.217}$$

where, $\alpha$ is a constant phase delay. Considering (5.217), the FIR filter kernel constitutes two types of symmetries, namely, the even and the odd symmetry. Then, we can have two possibilities for the number of points in the filter kernel, i.e., $M$ might be even or odd. Considering these facts, the FIR filters are generally classified into four types: Type 1 to 4. This is described in Table 5.9.

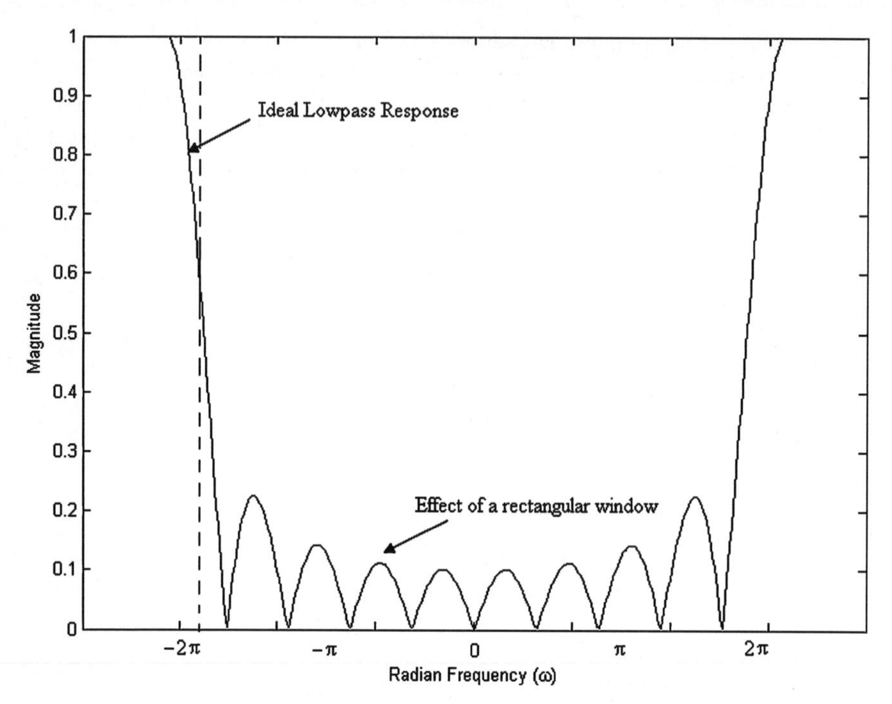


**Fig. 5.83** Magnitude response of an ideal lowpass filter with rectangular windowing

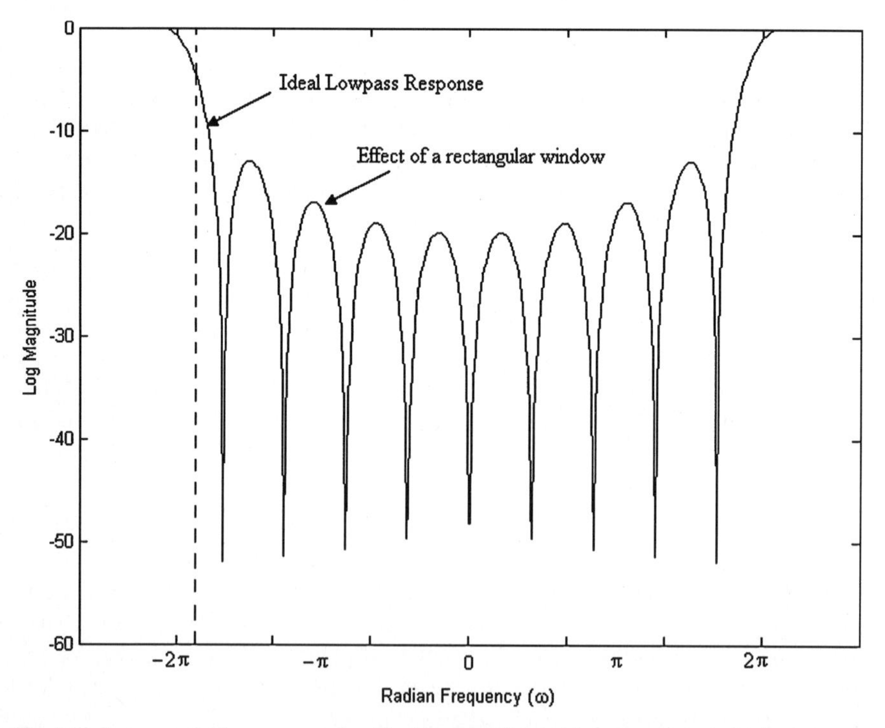


**Fig. 5.84** Log-magnitude response of an ideal lowpass filter with rectangular windowing

**Table 5.9** FIR filter types

| Symmetry Type / No. of Data Points | Symmetry (Even) | Anti-Symmetry (Odd) |
|---|---|---|
| Odd | *Type 1* | *Type 3* |
| Even | *Type 2* | *Type 4* |

### 5.7.5.4 Different Windows

Besides the rectangular window, the following common windows are used in the design of the FIR filter.

1. Barlett (or Triangular) window

$$w[n] = \begin{cases} 2n/M, & 0 \le n \le M/2, \\ 2 - 2n/M, & M/2 < n \le M, \\ 0, & otherwise \end{cases}. \tag{5.218}$$

Figure 5.85 shows a Barlett (triangular) window.

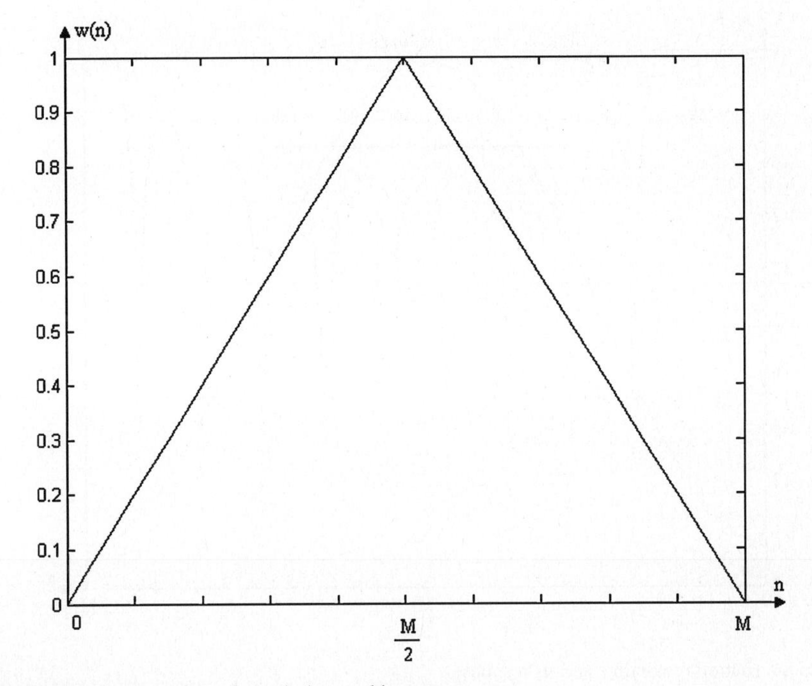


**Fig. 5.85** Barlett (or triangular) window and its response

2. Hamming window

$$w\,[n] = \begin{cases} 0.54 - 0.46\ \cos\left(2\pi n/M\right), & 0 \le n \le M, \\ 0, & otherwise \end{cases}. \tag{5.219}$$

Figure 5.86 shows a hamming window.

3. Hanning window

$$w[n] = \begin{cases} 0.5 - 0.5\cos(2\pi n/M), & 0 \le n \le M, \\ 0, & otherwise \end{cases} \tag{5.220}$$

Figure 5.87 shows a hanning window.

4. Blackman window

$$w[n] = \begin{cases} 0.42 - 0.5\cos(2\pi n/M) + 0.08\cos(4\pi n/M), & 0 \le n \le M, \\ 0, & otherwise \end{cases}. \tag{5.221}$$

Figure 5.88 shows a blackman window.

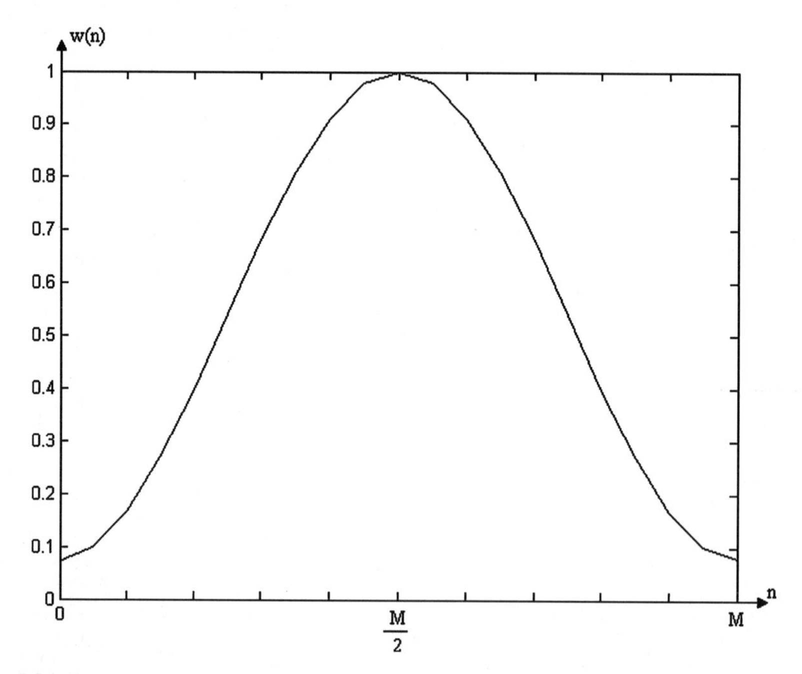


**Fig. 5.86** Hamming window and its response

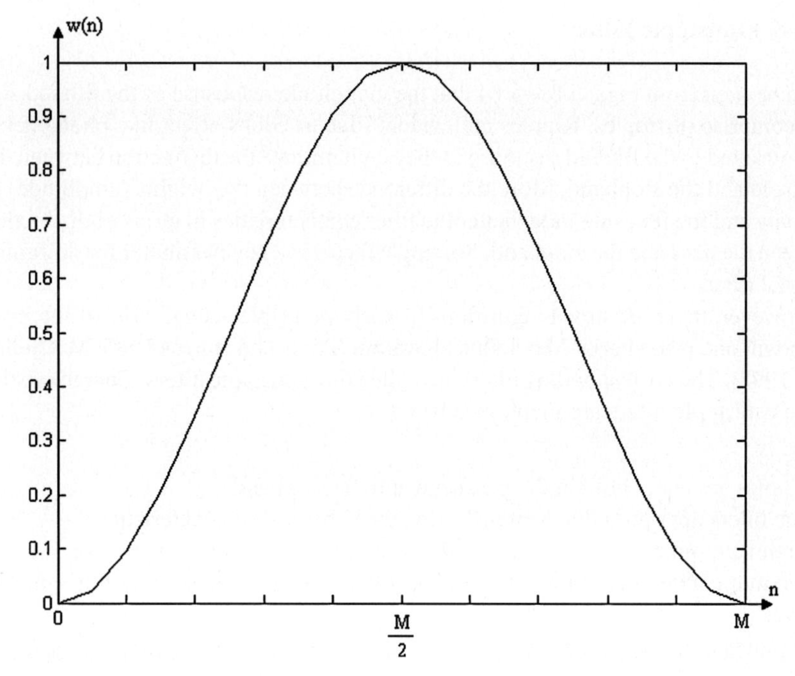


**Fig. 5.87** Hanning window and its response

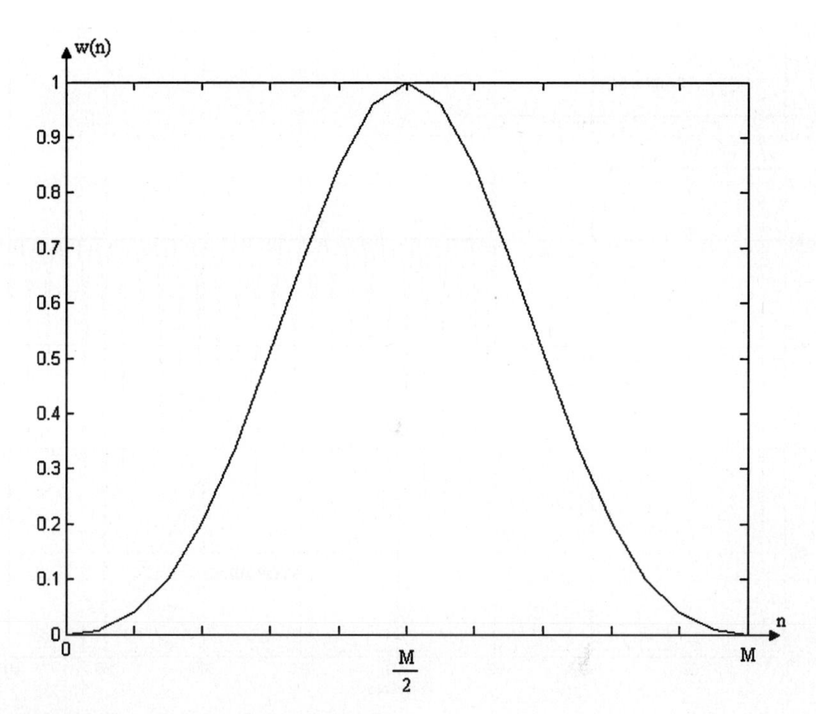


**Fig. 5.88** Blackman window and its response

### 5.7.5.5 Equiripple Filter

It can be seen from Figs. 5.83–5.84 that the magnitude resposnse of the filtered signals comprise of ripples. Ripples are evident (due to Gibbs effect like phenomena) but unwanted in the filtered response as they contaminate the distinction between the passband and the stopband. More the difference between the heights (amplitude) of the main and the first side lobe, better the filter characteristics in terms of distinction between the pass and the stopband. So, ripple factor is a key parameter for designing a digital filter.

However, there are novel algorithms to keep the ripples equal. One of the most important one is the Parks-McClellan algorithm (Parks and Burrus 1987; McClellan et al. 1998). The corresponding filters are called the equiripple filters. Characteristics of the equiripple filters are mentioned below.

- Ripples are equal both in the passband and the stopband.
- The filters are optimally designed using the equiripple characteristics.
- Optimization techniques are applied to minimize the attenuation, i.e., we find the maximum sidelobe, minimize it and all other sidelobes are kept at this minimal level.
- In practice, most equiripple filters are implemented as FIR filters.

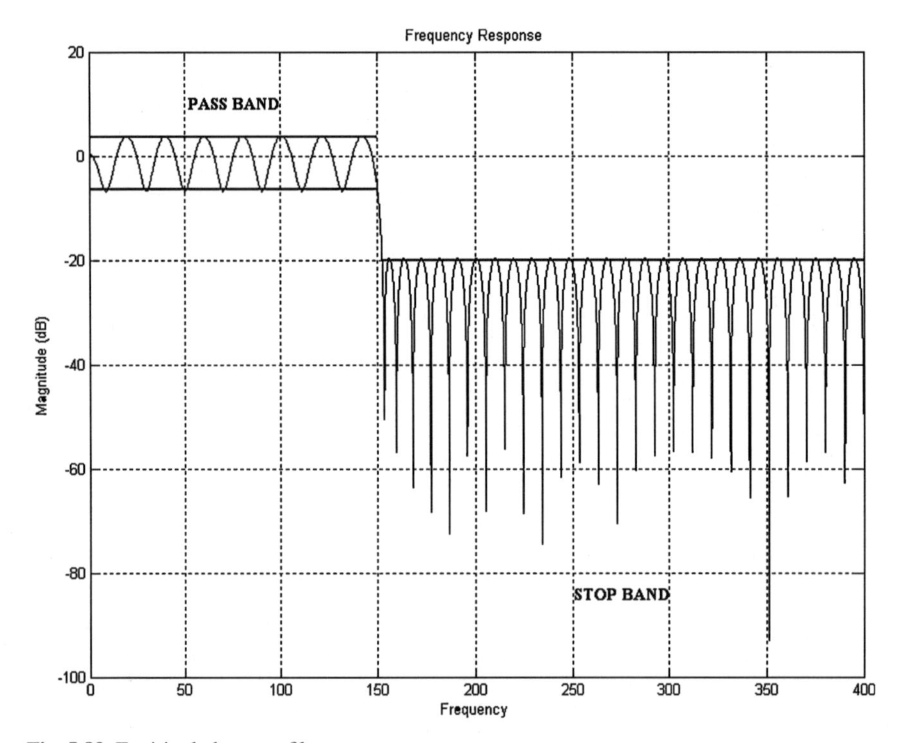


**Fig. 5.89** Equiripple lowpass filter

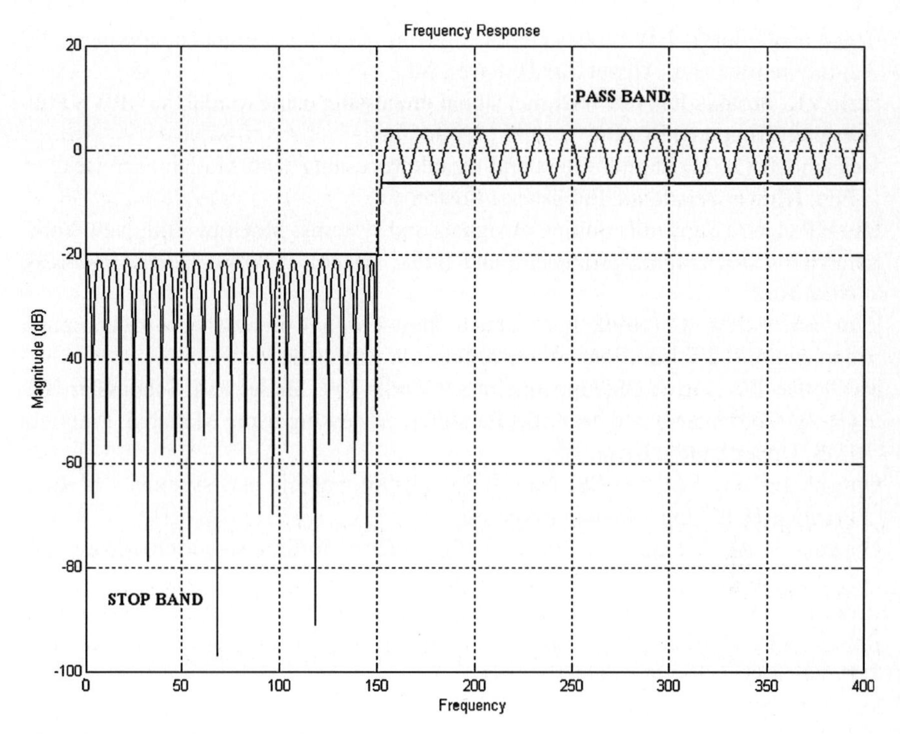


**Fig. 5.90** Equiripple highpass filter

The opposite to the ripple characteristics is the monotonicity. However, equiripple design is the optimal technique to tackle the ripple problem. Figure 5.89 shows the characteristics of an equiripple lowpass filter (cutoff frequency 150 Hz) and Fig. 5.90 of an equiripple highpass filter (cutoff frequency 150 Hz). Table 5.10 lists the important filter types (typically for IIR filters) based on their passband and stopband characteristics: equiripple or monotonic. Details about them can be referred to in (Parks and Burrus 1987; Jackson 2002).

**Table 5.10** Filter categories based on the passband and the stopband characteristics

| Passband Type | Stopband Type | Filter Type |
|---|---|---|
| Monotonic | Monotonic | Butterworth |
| Equiripple | Monotonic | Chebyshev-I |
| Monotonic | Equiripple | Chebyshev-II |
| Equiripple | Equiripple | Elliptic |

# References

Cooley JW, Tukey OW (1965) An algorithm for the machine calculation of complex Fourier series. Mathematics of Computation. 19:297–301

Duhamel P, Vetterli M (1990) Fast Fourier transforms: a tutorial review and a state of the art. Signal Processing 19:259–299

Ifeachor EC, Jervis BW (2001) Digital signal processing, a practical approach, 2nd Edn. Prentice Hall, Upper Saddle River, NJ
Ingle VK, Proakis JG (1997) Digital signal processing using Matlab V.4. PWS Publishing Company, Boston
Jackson LB (2002) Digital filters and signal processing with Matlab exercises, 3rd Edn. Kluwer Academic Publishers, Boston
Hsu HP (1995) Schaum's outline of signals and systems. McGraw-Hill, New York
Lathi BP (2001) Signal processing and linear systems. Oxford University Press, New York
Lynn PA, Fuerst JG (1999) Introductory digital signal processing with computer applications, 2nd Edn. John Wiley & Sons, West Sussex
McClellan JH, Burrus CS, Oppenheim AV, Parks TW, Shafer RW, Schuessler HW (1998) Computer-based exercises for signal processing using Matlab 5. Prentice-Hall, Upper Saddle River, NJ.
Oppenheim AV, Willsky AS, Nawab SH (1996) Signals and systems, 2nd Edn. Prentice-Hall, Upper Saddle River, NJ
Oppenheim AV, Schafer RW, Buck JR (2001) Discrete-time signal processing, 2nd Edn. Prentice-Hall, Upper Saddle River, NJ
Parks TW, Burrus CS (1987) Digital filter design. John Wiley & Sons, New York
Rabiner LR, Gold B (1975) Theory and application of digital signal processing. Prentice-Hall, Englewood Cliffs, NJ
Ramirez RW (1985) The FFT: fundamentals and concepts. Prentice-Hall, Englewood Cliffs, NJ

# Section II: Application Study

## 5.8 Harmonic Filtering

### *5.8.1 Introduction*

Harmonic analysis concentrates on decomposing the signal in question in terms of a fundamental (frequency) signal and signals of integer multiples of that fundamental signals. In this application study, we will encounter analysis of such signal which is a mixture of different frequency. We would like to get rid of other frequencies and concentrate on any particular frequency or frequency band. This can also be termed as harmonic filtering.

In this application study, we consider a signal originating from the 50 Hz system. The signal in question is acquired by a digital instrument operating at a sampling frequency of 1000 samples per second. We want to suppress any frequencies less than the third harmonic component.

#### 5.8.1.1 Specification

The formal specification of the filtering is as follows.

We would like to design a minimum-order IIR filter that will suppress monotonically any frequencies less than the third harmonic component to at least 40 dB. Frequency contents from the third harmonic onwards should have an attenuation of less than 1 dB with an equiripple nature.

### 5.8.2 Specification Analysis

To analyze the above-mentioned specification, first we need to verify the frequency content of the acquired signal shown in Fig. 5.91. So, in Fig. 5.92 we plot the amplitude spectrum of the discrete Fourier transform (DFT), computed using the fast Fourier transform (FFT) algorithm, of the signal in question. The frequency plots in Fig. 5.92 show that the signal in question contains the second and the third harmonic components. This is evident from the two peaks appearing at the 100 Hz and 150 Hz respectively.

Further analyzing the specification and the magnitude plot, we can get the following facts to design the filter.

- Fundamental frequency: 50 Hz
- Third harmonic component: 150 Hz
- Filter type: Highpass IIR filter

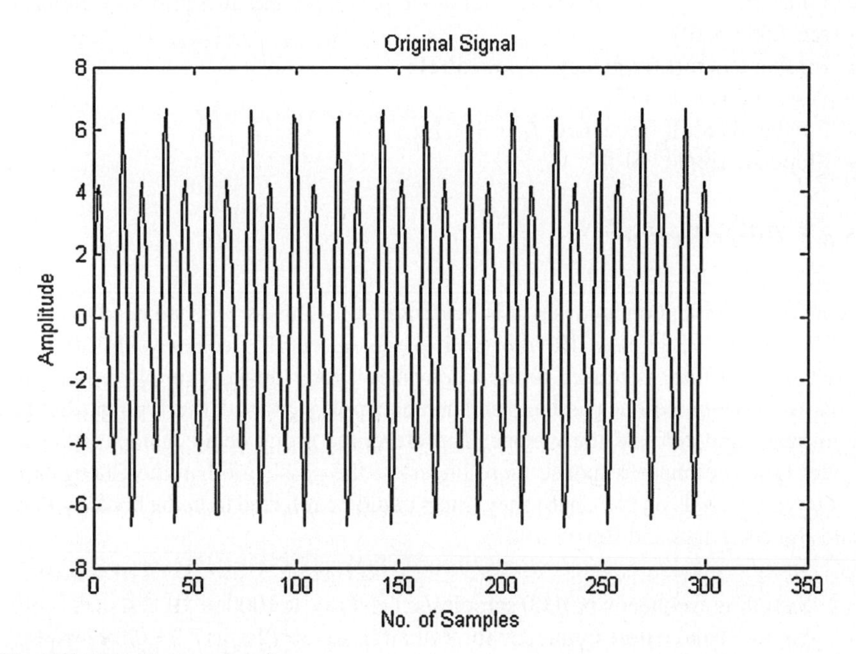


**Fig. 5.91** Signal in time-domain

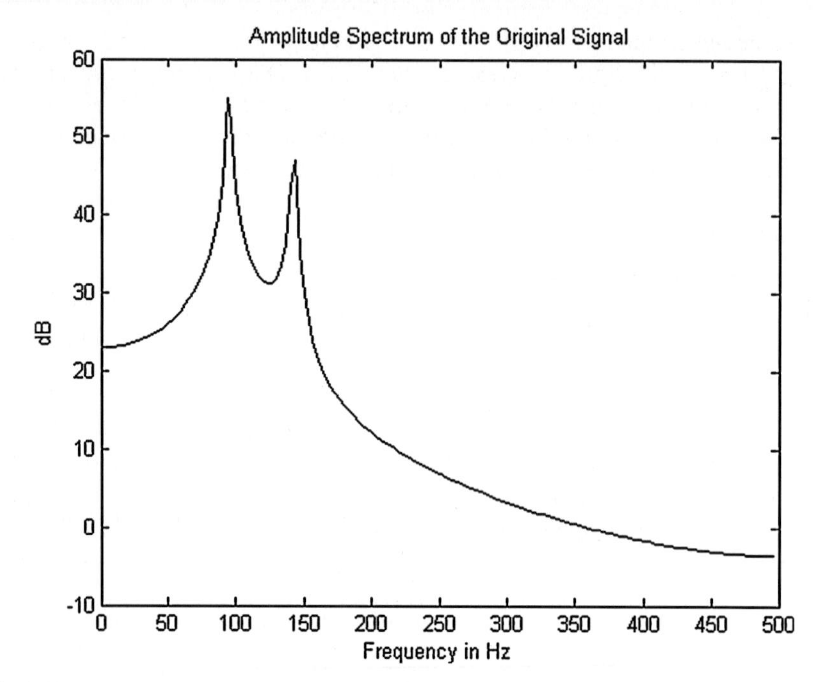


**Fig. 5.92** Magnitude plot of the FFT of the signal

- Filter category: Chebychev-II (monotone passband and an equiripple stopband, see Table 5.10)
- Passband cutoff frequency: $f_p = 150\,\text{Hz}$
- Passband ripple: 1 dB
- Stopband cutoff frequency, $f_s = 100\,\text{Hz}$
- Stopband ripple: 40 dB

### *5.8.3 Filter Design*

From the specification analysis, it is evident that for the purpose we need to design a Chebyshev-II highpass IIR filter with passband cutoff frequency of 150 Hz. A Chebyshev-II filter is related to the Chebyshev-I filter through simple transformation. Due to a monotone passband and an euqiripple stopband, this type of filter has both poles and zeros in the $s$-plane. Therefore, the group delay characteristics are better (and the phase response more linear) in the passband than the Chebyshev-I prototype. Details on the Chebyshev filters could be referred to in the book by Parks and Burrus (Parks and Burrus 1987).

Sampling frequency is 1000 samples/s, i.e. $T_s = 1/1000 = 10^{-3}$ s.
So, passband cutoff frequency (in Radian) is $\omega_p = (2\pi f_p)T_s = 0.3\pi$.
And, stopband cutoff frequency (in Radian) is $\omega_s = (2\pi f_s)T_s = 0.2\pi$.

With these information, we can design the filter using Matlab® signal processing toolbox (Mathworks 2002; Ingle and Proakis 1997). The example script is shown below. However, it is to be noted that some of the functions might be different depending on the version of the Matlab®. Hence, the user is requested to check the compatibility of the functions with his/her Matlab® version.

************************* **FilterDesign.m** **********************

```
% Designing the optimum filter of type Chebyshev II High-Pass

clear;          % Clears Workspace

% Design Specifications
Td = 1/1000;    % Sampling Interval Td=1/1000 sec
Fp = 150;       % Passband frequency
Fs=100;         % Stopband frequency
Rp=1;           % Passband Ripple in dB
As=40;          % Stopband Ripple in dB

Wp = (2*pi*Fp)*Td;  % Passband Edge Frequency
Ws = (2*pi*Fs)*Td;  % Stopband Edge Frequency

% Determining Filter order N, Cut-off frequency Wn for
% Chebysev-II
[N,Wn]=cheb2ord(Wp/pi,Ws/pi,Rp,As);

% Determining polynomials of the digital filter i.e. b & a
% respectively
% Cheby2 function with 'high' determines High-Pass Chebychev II

[b,a]=cheby2(N,As,Ws/pi,'high');

% Coverting from Direct (transfer function) form to Cascade
% (state space) form
% C is Gain Coeffient, M comprise of real coefficients for b's
% and a's

[M,C]=tf2sos(b,a) ;

****************************************************************
```

From the filter design script we get,
Filter order ($N$): 6

$$M = \begin{bmatrix} 1 & -1.9719 & 1 & 1 & -1.0180 & 0.2778 \\ 1 & -1.7994 & 1 & 1 & -1.0388 & 0.4312 \\ 1 & -1.6413 & 1 & 1 & -1.1585 & 0.7545 \end{bmatrix}$$
$$C = [0.3006]$$

$M$ can be splitted into $B$ and $A$ parts to realize the direct form structure.

$B$ is the first three columns (bottom to up) of $M$ from left, and $A$ is the last three columns of $M$.

$$B = \begin{bmatrix} 1 & -1.6413 & 1 \\ 1 & -1.7994 & 1 \\ 1 & -1.9719 & 1 \end{bmatrix}, \quad A = \begin{bmatrix} 1 & -1.0180 & 0.2778 \\ 1 & -1.0388 & 0.4312 \\ 1 & -1.1585 & 0.7545 \end{bmatrix}.$$

So, the filter can be expressed in rational form as shown below.

$$H(z) = \frac{0.3006(1-1.6413z^{-1}+z^{-2})(1-1.7994z^{-1}+z^{-2})(1-1.9719z^{-1}+z^{-2})}{(1-1.018z^{-1}+0.2778z^{-2})(1-1.0388z^{-1}+0.4312z^{-2})(1-1.1585z^{-1}+0.7545z^{-2})}. \tag{5.222}$$

The frequency response of the filter is shown in Fig. 5.93.

The log-magnitude plot (i) in Fig. 5.93 shows the desired filter response as per the specification, i.e., the stopband below 100 Hz as equiripple and the passband

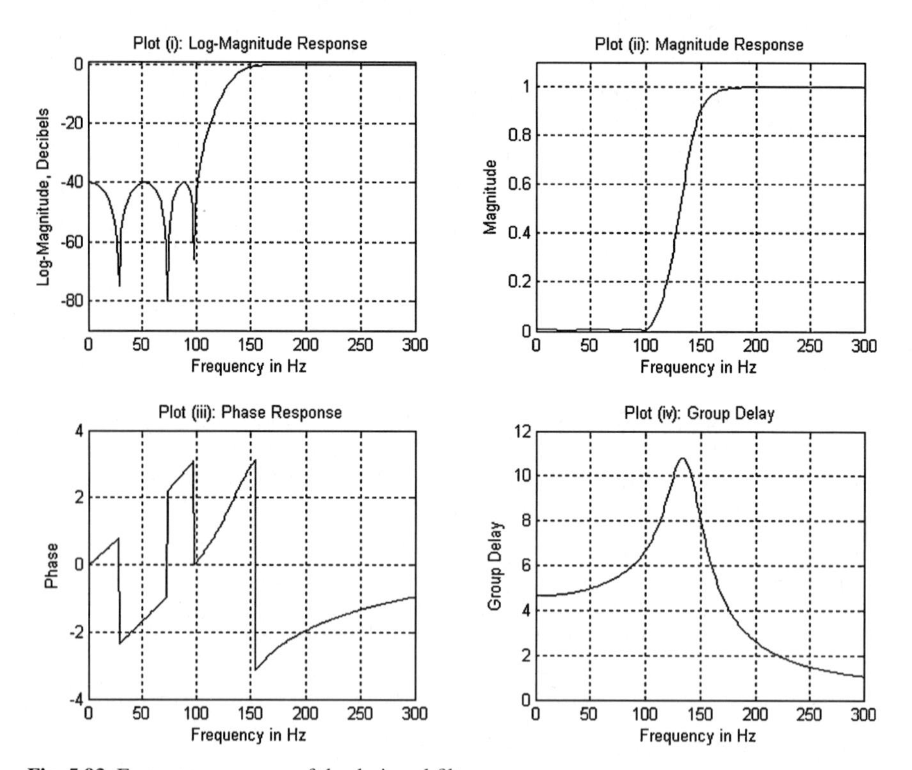


**Fig. 5.93** Frequency response of the designed filter

over 150 Hz as monotonic in nature. It is to be noted that, here the actual cutoff frequency is 150 Hz, i.e., the frequency over which the frequencies are passed (as it is a highpass filter, refer to Fig. 5.76). But in reality, the region between the specified stopband 100 Hz (second harmonic) and the third harmonic (150 Hz) is the rollover frequency region.

### *5.8.4 Harmonic Filtering of the Signal*

The recorded signal (shown in Fig. 5.91) can be passed through the designed filter to get the harmonic filtered signal. The time-domain signals are shown in Fig. 5.94. In Fig. 5.94, the upper plot (i) shows 300 samples of the original time-domain signal, while the bottom plot (ii) shows 300 samples of the filtered signal. Frequency responses of the signals are shown in Fig. 5.95. In Fig. 5.95, the

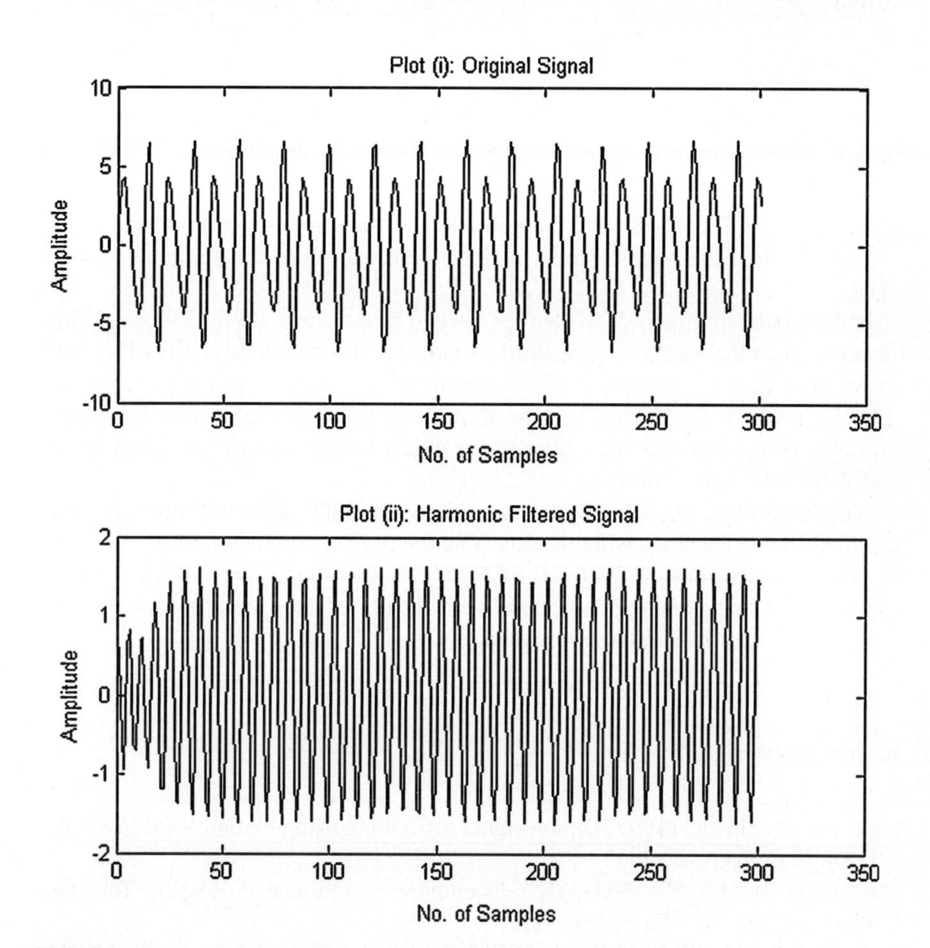


**Fig. 5.94** Time-domain plots of the original and the harmonic filtered signals

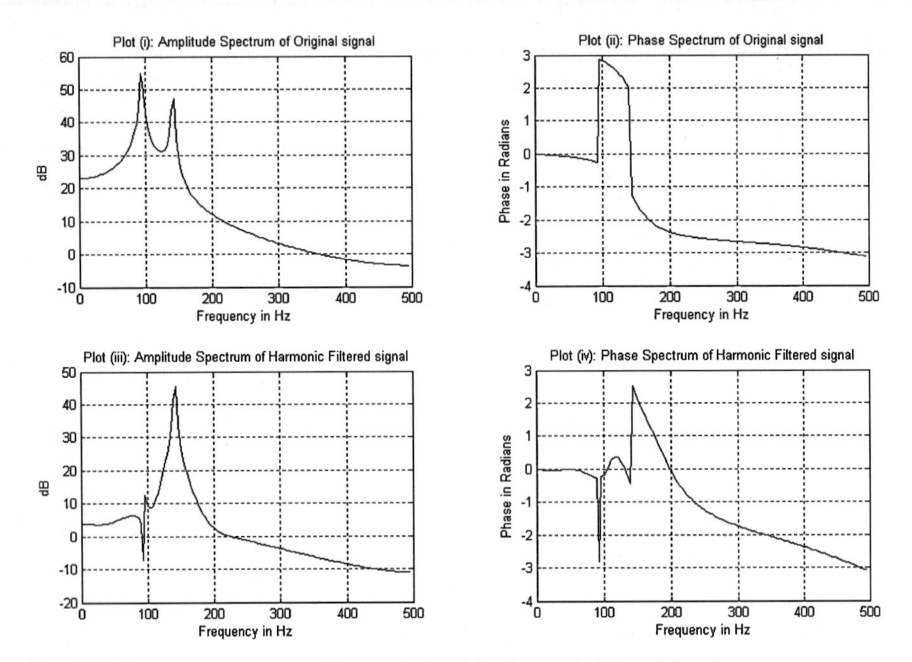


**Fig. 5.95** Frequency responses of the original and the harmonic filtered signals

upper two plots (i & ii) show the amplitude and the phase spectrum of the original signal, while the bottom plots (iii & iv) depict those of the harmonic filtered signal.

In the plot (i) of Fig. 5.95, we can see the two peaks, same as Fig. 5.92, depicting the presence of the second and the third harmonic components. Plot (iii) of Fig. 5.95 shows the frequency content of the harmonic filtered signal. In this plot, we see only one peak at the 150 Hz, i.e., the third harmonic is passed while the second harmonic component has been filtered out which is indicated by the negative dB peak at 100 Hz in the plot (iii).

Another point to be noted is that a properly designed filter only filters out the specified frequency range without disturbing the passing frequency range. This can be noticed in the plots (i) and (iii) of Fig. 5.95, that the dB level of the 150 Hz component (passing range) is kept same while the dB level is different for the 100 Hz component which falls into the stopband.

## References

Ingle VK, Proakis JG (1997) Digital signal processing using Matlab V.4. PWS Publishing Company, Boston

Mathworks Inc. (2002) MATLAB® Documentation–Signal Processing Toolbox, Version 6.5.0.180913a, Release 13

Parks TW, Burrus CS (1987) Digital Filter Design. John Wiley & Sons, New York

## 5.9 Digital Fault Recorder and Disturbance Analysis

### *5.9.1 Introduction*

The analysis of faults and disturbances has always been a fundamental foundation for a secure and reliable electrical power supply. The introduction of digital recording technology opened up a new dimension in the quantity and quality of fault and disturbance data acquisition, resulting in the availability of a huge amount of new information to power systems engineers. Information from the analysis of digital records can provide much-needed insight into the behavior of the power system as well as the performance of protection equipments. Manual analysis of these records, however, is both time-consuming and complex. Today the challenge is to automatically convert data into knowledge, which frees the human resources to implement corrective or preventive action. In this application study we will see some modern trend in disturbance analysis based on digital signal processing techniques and DSP-enabled instruments.

### *5.9.2 Overview of Disturbance Analysis*

The first disturbance recording systems in electrical power systems were based on electromechanical and analog technology (Barth et al. 2003). The measurements were recorded on metallic or photosensitive papers or magnetic tapes. The accuracy, the number of signals, the recording time and the number of data records per paper roll or magnetic tape were highly restricted. Data handling and interpretation were also extremely complex processes.

With the introduction of the digital recording technology, nowadays, engineers in power utilities have more data than can be processed and assimilated in the time available. Users of knowledge extracted from recorded data are engineers and technicians working in operations, maintenance and protection departments.

It is the task of operating personnel to return to service as much of the electric system as practical in the shortest time possible. The following questions concern them: where and what is the problem? Did the line reclose and stay in operation? What equipment was operating? Was everything working correctly? If so, can it be returned to service? If not, what needs to be isolated? A complete analysis should be available for use fairly quickly after the conclusion of the event. Maintenance personnel are charged with repairing and returning outage equipments to service. They require information on what is damaged or operating outside the normal parameters. Typical time requirement to notify the maintenance personnel is in the region of two hours or less. Key questions that need to be answered are: did the right thing respond in the right way? Did the wrong thing respond in the wrong way? Did the right thing respond in the wrong way? The "thing" in these questions might be a relay, relaying system, circuit breaker, or switch (Ukil 2005).

### 5.9.3 Digital Recording Equipments

The following equipments are commonly used for disturbance recording in modern substations and distribution systems.

- **Digital fault recorder (DFR)**. DFRs are highly accurate recording instruments providing sampled waveform and contact data at a relatively high sampling rate (typically above 5 kHz). The DFRs trigger due to the power network fault conditions, protection operations, breaker operation, and the like. Each DFR recording typically consists of binary and analog information in the form of voltages and currents per phase as well as the neutral current. Due to different triggering mechanisms, and the performance and sensitivity of the triggers may vary for different makes of the DFRs. Potential difficulties are associated with the proprietary DFR data formats, not allowing implementation of an open and cross-platform data recording system which can easily be interfaced with the analyzing system.
- **Digital protective relay (DPR)**. It is a complex recording and measurement instrument equipped with decision making control logic. It comes with plenty of monitoring functions and variety of settings (both external and internal). The main problem is the lack of generalized model which usually varies from vendor to vendor. Some of the earlier DPRs have limited frequency representation of the waveforms due to low sampling frequency.
- **Remote terminal unit (RTU)**. The RTU can be a very sophisticated recording instrument that may have a recording performance of a DFR, as well as different monitoring functions like a DPR. As RTUs are part of the supervisory control and data acquisition (SCADA) system, the data is readily available for the analysis at the centralized location. This is also a disadvantage as the user has limited chance to access the data locally before it is fully acquired at the central database.
- **Intelligent electronic device (IED)**. These devices provide internal control and communication through various electrical interfaces. IEDs can be digital fault & data loggers/recorders, power quality analyzers, intelligent switches, breakers, regulators, auto-restoration devices, remote terminal units, substation controllers / gateways, etc. The main issue with IEDs, when used for the analysis, is the open communication architecture and data recording performance. Since the IEDs are not standardized even regarding the functions they perform, it may be very hard to find detailed enough description that will allow generic models to be developed and used for the analysis.
- **Sequence of event recorder (SER)**. These are complex recording instruments implemented using programmable logic controllers (PLCs) and analog waveform data acquisition subsystems. The SERs are capable of monitoring the status changes in the switching equipments with high precision due to a high data sampling rate. Most of the SERs can also be set to provide control function through a number of control outputs. SERs also suffer from the shortcomings that these are usually not designed as open systems. Therefore, interoperability of these devices might be restricted with data analysis devices of other makes.

- **Fault locator (FL).** The stand alone FLs are designed to estimate the fault location very accurately based on the impedance calculation depending on the fact that cables have constant resistance per length. They usually come with fairly advanced built-in data acquisition system. However, as most modern DFRs, DPRs and even RTUs already come up with fault location calculation as an additional functionality, the specific uses of FLs are getting restricted.

### *5.9.4 Digital Fault Recorder*

Digital fault recorders (DFRs) are highly accurate recording instruments providing sampled waveform and contact data using a relatively high sampling rate, typically above 5 kHz (a sample every 0.2 milliseconds). A DFR has all recording functions, as well as some sensing, detection and calculation capabilities. The main characteristics of DFR are as follows:

- the capability of remote interrogation for data analysis and manipulation
- low-cost mass storage.

Because of the different nature and processing techniques of the input signals – analog and binary – two types of data acquisition units (DAU) are designed, namely:

- Analog data acquisition unit (ADAU)
- Binary data acquisition unit (BDAU).

DFRs typically come with 8, 12, 16-bit resolution and maximum scanning frequency more than 5 kHz per channel. The functions of a fault recorder, digital recorder, power and frequency recorder, diagnosis system and message printer are combined in one instrument. Typically when using DFRs in distribution system, every feeder uses three DAUs: 2 $\times$ ADAU (typically for 4 $\times$ voltage and 4 $\times$ current signals), and 1 $\times$ BDAU (typically for about 32 digital signals). One recorder is able to monitor all values of 10 feeders. Per feeder, 8 analog channels (3 phase voltages, 1 phase-to-phase voltage at the busbar, 3 phase currents and the neutral current) and 32 binaries (carrier receive and send, protection operation, status of breakers, etc) can be recorded. The DFR measures continuously, but only a predefined part of the signal will be stored when the DFR is triggered. Usually one second of pre-fault and two seconds after the recorder has been triggered are recorded, that is, the whole record is about three seconds long (Ukil 2005). However, this recording duration and number of DAUs might vary depending on the particular DFR make.

The binary values are either stored as a 0 or a 1 and indicates the status of a contact, for example, breaker auxiliary contact. Binaries are divided into the following four groups:

- Main 1 distance relay contacts
- Main 2 (back up) distance relay contacts

- Other relays
- Circuit breaker and other necessary signals.

Analog values indicate the magnitude of an analog signal (voltage or current) measured at a specific point in time. The analog information consists of

- Voltages per phase
- Current per phase as well as the neutral current.

As discussed before, one of the main problems in DFR is the lack of common framework for data interaction. In past, proprietary DFR data formats did not allow integration of DFRs from different vendors into a particular analysis system, even integration of different analysis system with any particular DFR type. However, the IEEE COMTRADE standard (IEEE 1991) is quite effective is defining a common framework for data exchange.

The IEEE standard for **Com**mon Format for **Tra**nsient **D**ata **E**xchange (COMTRADE) is specified by the IEEE STANDARD C37.111-1991 (IEEE 1991). As per this standard, each event should have three types of files associated with it. Each of the three types carries a different class of information: header (*.HDR), configuration (*.CFG) and data (*.DAT). Each record (row) in the data file (*.DAT) contains information about the disturbance event and recordings. Figure 5.96 shows a sample record from a data file (Ukil 2005).

In the bottom part of Fig. 5.96, we can see 42 data columns, of which first two contains sample number and sampling information as depicted in the top part. The next eight columns are the four analog voltage and four analog current recordings, next 32 columns are the 32 binary points representing the contact changes (Ukil 2005). It is to be noted that Fig. 5.96 shows an example of DFR recording, not a general standard. This might change from DFR to DFR but the structure remains more or less same, only the number of data/channels being variable.

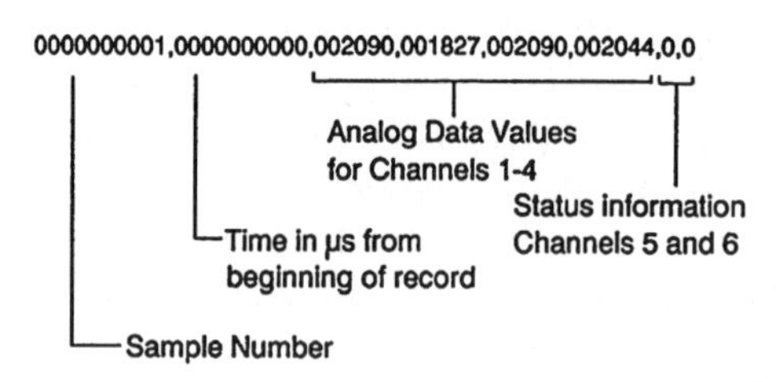


0000000001,0000000000,00856,000947,00050,01030,00056,000054,00006,00005,0,0,0
,0,0,0,0,0,0,0,0,0,0,0,0,0,0,0,0,0,0,0,0,0,0,0,0,0,0,0,0,0

**Fig. 5.96** Example of digital fault recorder data file format

### 5.9.5 Disturbance Analysis Using DFR Data

#### 5.9.5.1 Time-domain Analysis

If we consider the DFR data format shown above, then the sampling frequency can be estimated from the second column data[24].

$$Sampling\ Frequency,\ f_s = \frac{1}{[(2nd\ Col.\ data\ n) - (2nd\ Col.\ data(n-1))] \times 10^{-6}}. \tag{5.223}$$

From the sampling frequency given by (5.223), we can determine how many harmonics[25] can be effectively estimated from the frequency analysis of the fault recordings (voltage and current). To estimate a particular harmonic, we have to maintain the sampling theorem (see Sect. 5.6.2), i.e., the sampling frequency $f_s$ must be equal to or more than double the harmonic.

$$f_s \geq 2f_h, \tag{5.224}$$

$$f_h = nf, \quad n = 1, 2, 3, \ldots. \tag{5.225}$$

$f_h$ is the maximum possible integer ($n$) harmonic of the system fundamental frequency $f$ Hz, at par with the sampling theorem. For example, if we have a DFR with 2.5 kHz sampling frequency, and we consider a 50 Hz system, we have $f = 50$ Hz. Then, using (5.224–5), $n \leq (2500/2)/50 = 25$. That is, theoretically we can estimate accurately upto the 25$^{\text{th}}$ harmonic component, i.e., $f_h = 1250$ Hz.

Data from the column number 3 to 6 can be used to plot the voltage recordings and 7 to 10 for the current recordings. Two examples are shown in Fig. 5.97 and 5.98. Considering the sampling frequency of 2.5 kHz for this example, there are 6463 samples, i.e., about $6463/2500 \approx 2.6$ seconds of recording of the event.

#### 5.9.5.2 Segmentation

It is to be noted from Fig. 5.97 and 5.98 that the voltage and the current plots are matched. That is, the fault section (approximately sample no. 930 to 1130) is represented one-to-one in the voltage (Fig. 5.97) and the current (Fig. 5.98) plot. In the voltage plot, it is associated with the change in amplitude while in the current plot it is associated with increase in the value. This is an example of single-phase-to-ground fault (Ukil 2005). Figure 5.99 shows the matched fault recordings.

---

[24] The factor $10^{-6}$ in (5.223) appears as the sampling interval in Fig. 5.96 is given in μs. This should be adjusted for particular DFR.

[25] In power systems, harmonics are generally the integer multiples of the fundamental frequency. For example, for a 50 Hz system the fundamental is 50 Hz, the second harmonic is 100 Hz, third harmonic is 150 Hz, and so on.

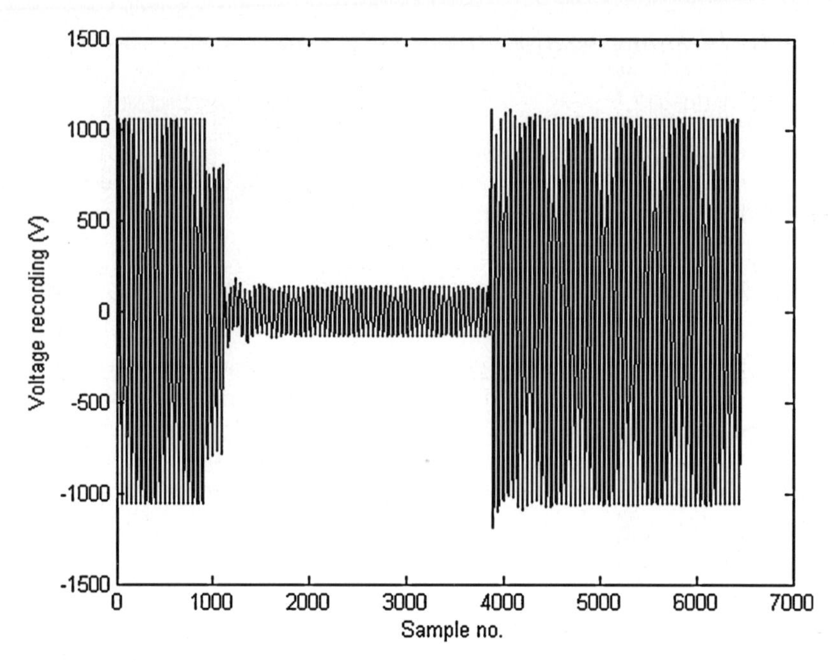


**Fig. 5.97** DFR voltage recording

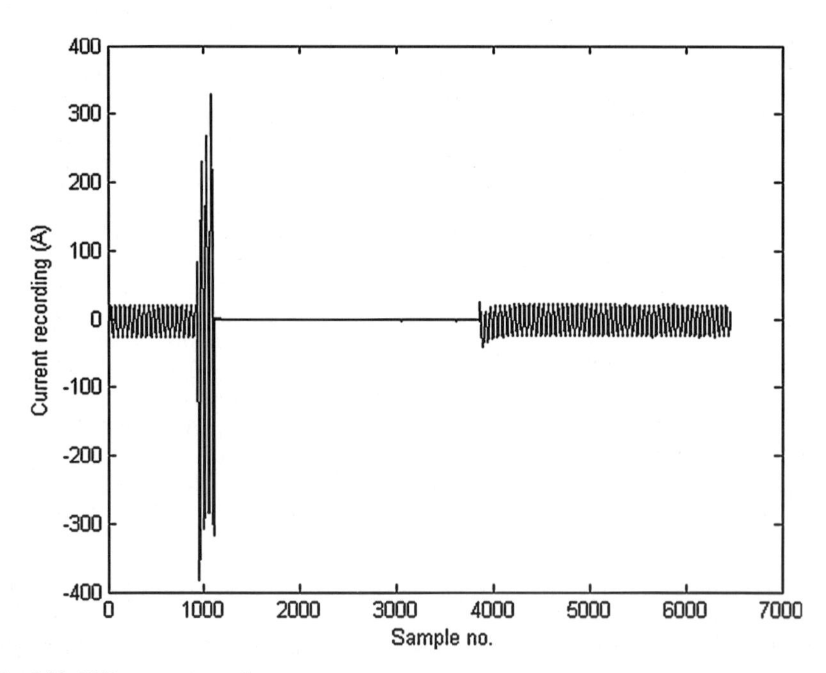


**Fig. 5.98** DFR current recording

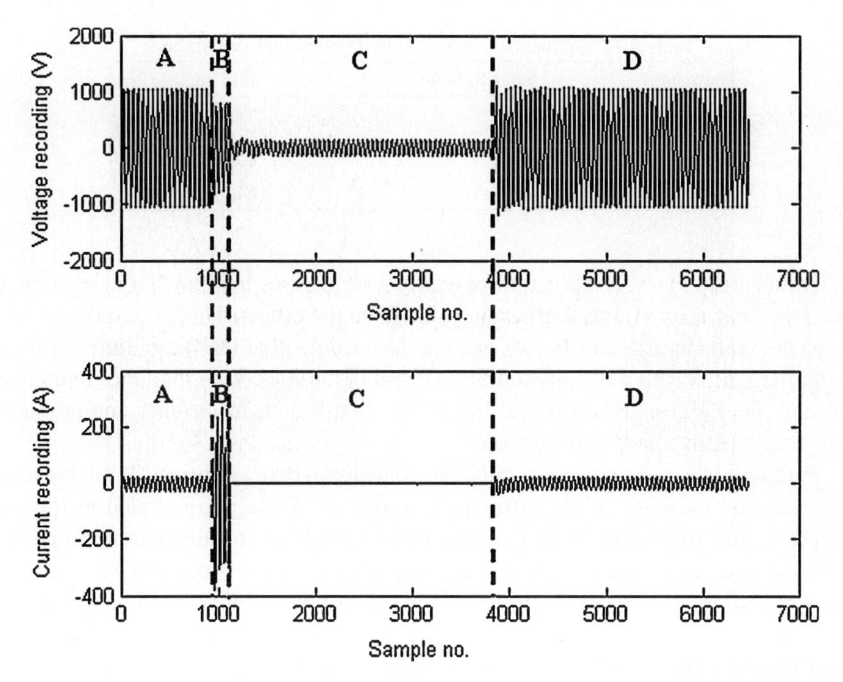


**Fig. 5.99** Matched and segmented DFR recordings

It is to be noted in Fig. 5.99 that we have divided the recordings into four segments, namely, A, B, C, D. Physically these indicate the following series of events during the fault. In Fig. 5.99, segment A represents the pre-fault condition, B shows the inception of the fault and the fault duration, C represents the circuit-breaker opening and D the circuit-breaker auto-reclosing and system restore. Thus, we have segmented the fault recordings into different event-specific segments for further analysis. In this example, we have done this by hand, however it could be done automatically using different algorithms (Ukil and Zivanovic 2006; 2007). The segments in terms of the sample numbers are shown in Table 5.11.

#### 5.9.5.3 Frequency-domain Analysis

In this section we will analyze the segments of the fault recordings in the frequency domain. That is, we will compute the discrete Fourier transform (using the FFT algorithm) for the segments A-D for the voltage and the current recordings. We will check the magnitude response in the frequency domain for the segments of the voltage and the current plots.

Magnitude responses of the segments A-D for the voltage recording are shown in Fig. 5.100.

In Fig. 5.100, we can note that the magnitude responses are plotted against the frequencies which ranges up to the 1250 Hz, which is half of the sampling frequency (2.5 kHz in this case). This conforms to the estimate in Sect. 5.9.5.1. Comparing the

**Table 5.11** Segments in Fig. 5.99 in terms of sample numbers

| Segment | Start Sample No. | End Sample No. |
|---|---|---|
| A | 1 | 929 |
| B | 930 | 1130 |
| C | 1131 | 3870 |
| D | 3871 | 6463 |

magnitude responses for the different sections, we can see that the plot in segment B (i.e., the fault section) has a different value from the others. This is also the section after the fault. In segment B plot, we can also notice that the magnitude response comprises of considerable amount of different harmonics other than the fundamental one (the biggest spike/vertical line). This is different from other conditions as expectable from a post-fault section.

Further harmonic analysis could be concentrated on segment B to estimate correctly the presence of the different harmonics. This is not included in this example as the main aim of this application example is to demonstrate how signal processing techniques can be utilized in power systems disturbance analysis. This functionality along with further diagnostic information estimation are usually

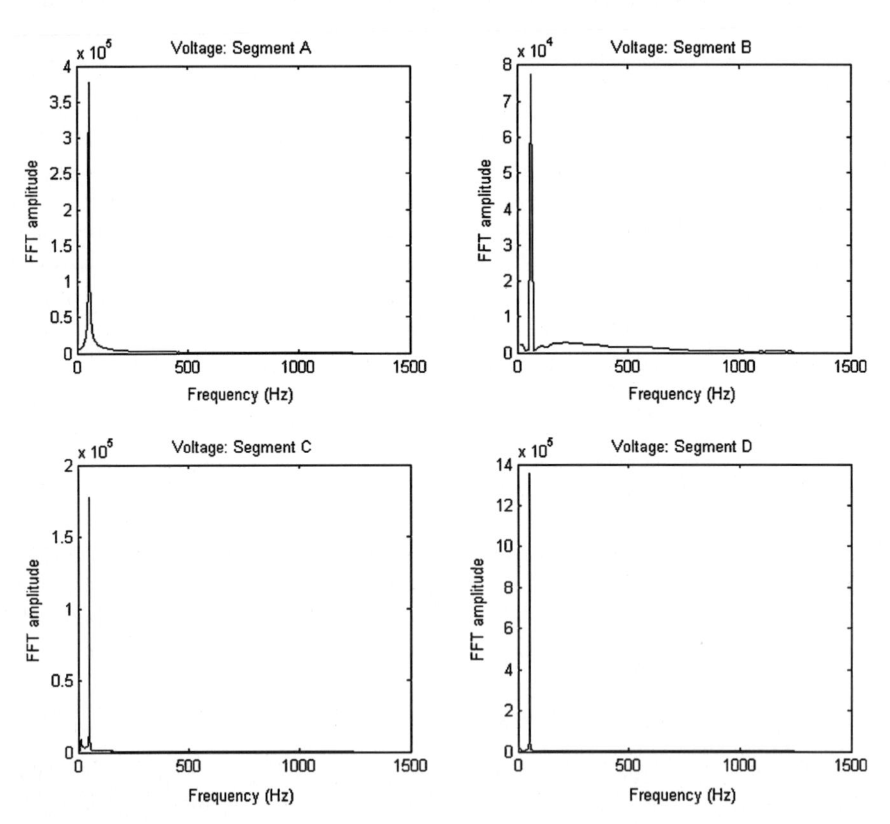


**Fig. 5.100** Magnitude response of the segments of the voltage recording

available/performable in the commercial disturbance analysis programs. Some of these are:

- Faulted phase(s)
- Fault type
- Total fault duration
- Main 1 protection operating time
- Main 2 protection operating time
- Fault location
- Fault resistance
- DC offset
- Breaker operating time
- Auto-reclose time.

Magnitude responses of the segments A-D for the current recording are shown in Fig. 5.101.

In Fig. 5.101, also the magnitude responses are plotted against the frequencies. Comparison of the magnitude responses for the different sections shows that the plots are similar for the segment A and D, which follows Fig. 5.99. It is to be noted that segment D corresponds to circuit breaker auto-reclosing. Therefore,

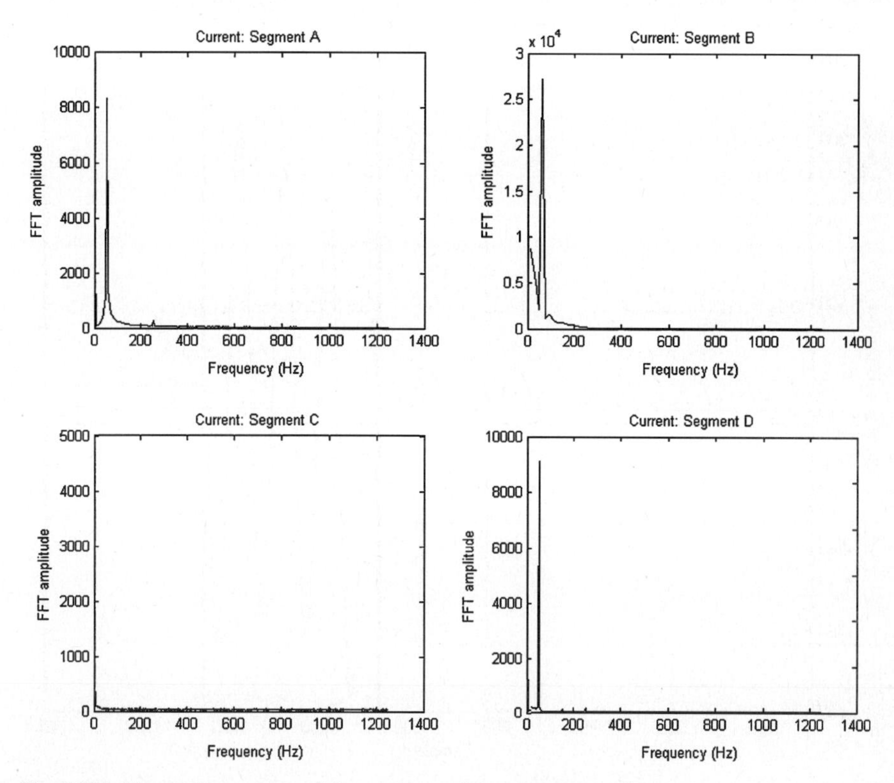


**Fig. 5.101** Magnitude response of the segments of the current recording

equivalence of the section A and D implies successful system restore. Similar to the magnitude response of the voltage recording in Fig. 5.100, we can see that the magnitude plot in segment B (i.e., the fault section) has a different (higher) value from the others, which is expected for a post-fault segment. In segment B plot, we can also notice the presence of different harmonics other than the fundamental one (the biggest spike/vertical line). A difference between the magnitude plots of the voltage (Fig. 5.100) and the current (Fig. 5.101) recording can be noticed in segment C. In Fig. 5.101, segment C plot indicates no current flow which correctly follows from Fig. 5.99. This is correct as segment C is associated with the opening of circuit breaker, implying interruption of the current flow.

Like the voltage recording, further analysis could be oriented towards particular harmonic analysis using different digital filters upto particular harmonics.

#### 5.9.5.4 DC Offset Effect

To visualize the effect of the dc offset in the fault current, we plot the segmented fault current recording in zoomed scale in Fig. 5.102.

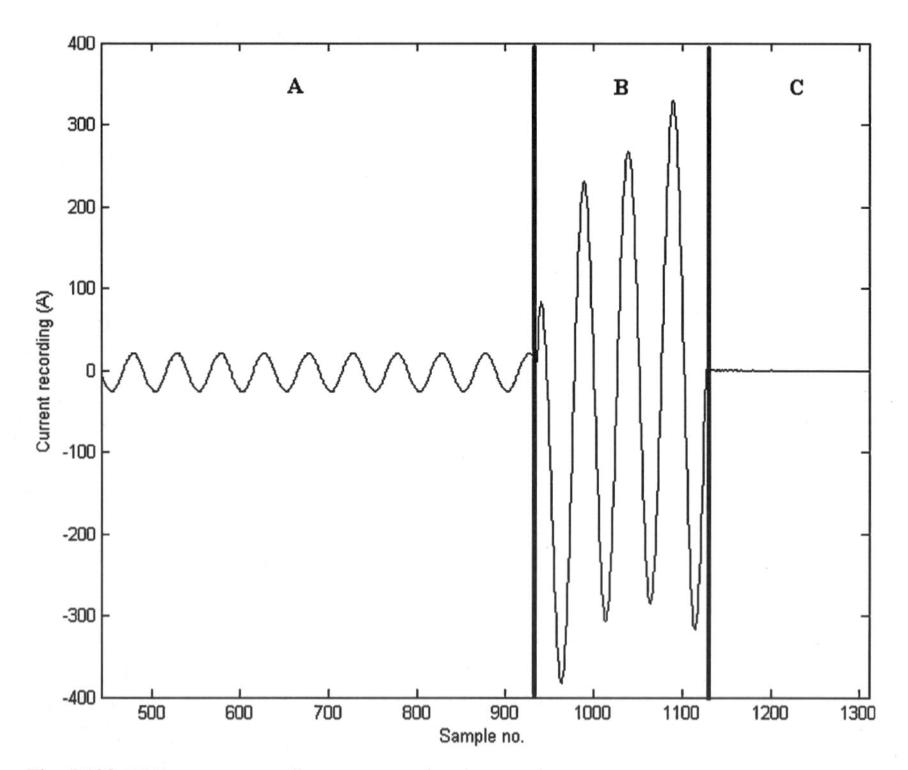


**Fig. 5.102** DFR current recording: segmented and zoomed

In Fig. 5.102, from the segment B, we see that the fault current incorporates some dc offset effect. This can be further clarified in an example plot in Fig. 5.103. In Fig. 5.103, plot (i) shows a pure sinusoidal function, plot (ii) a negative exponential function, plot (iii) sum of the pure sinusoidal function and the negative exponential function.

In Fig. 5.103, plot (iii) depicts the effect of the exponential function when superimposed on the sinusoidal function. Comparing this to the segment B in Fig. 5.102, we can clearly identify the effect of oscillatory exponential drift on the fault current recording (in segment B) due to the dc offset. Although we performed our frequency analysis using the fault current including the dc offset, it is a good practice to remove the dc offset as a pre-processing step to the DFT.

In this pre-processing step, a current is passed through an impedance that is a replica of the power system impedance. The resulting voltage across the replica impedance will not contain any dc offset that may be present in the original fault current.

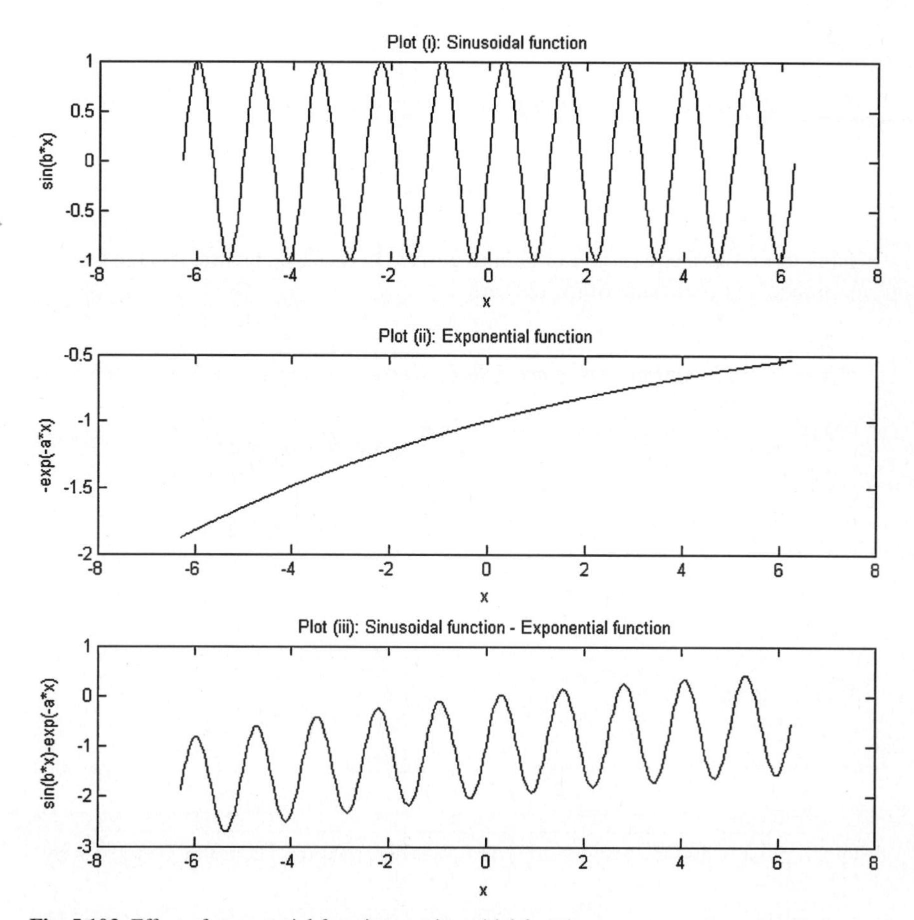


**Fig. 5.103** Effect of exponential function on sinusoidal function

The general equation for a fault current which includes dc offset, as in our application example, is

$$i(t) = I_m \sin(\omega t + \alpha - \phi) - I_m \sin(\alpha - \phi) \exp\left(-\frac{t}{T}\right), \tag{5.226}$$

where,

$I_m$ is the peak value of the current,
$\alpha$ is the angle on voltage wave at which fault occurs,
$\phi = \arctan\left(\frac{\omega L}{R}\right)$, is the impedance angle of power system, $L$ is the inductance and $R$ is the resistance of the system,
$T = \frac{L}{R}$, is the time-constant of the power system.

An impedance $Z_r = R_r + j\omega L_r$ is considered to be a replica impedance if

$$T_r = \frac{L_r}{R_r} = T = \frac{L}{R}. \tag{5.227}$$

The voltage across the replica impedance is given as

$$v_r(t) = R_r i(t) + L_r \frac{di(t)}{dt}. \tag{5.228}$$

Using the general equation for the fault current in (5.226), the first and the second term on the right-hand side of (5.228) are

$$R_r i(t) = R_r I_m \sin(\omega t + \alpha - \phi) - R_r I_m \sin(\alpha - \phi) \exp\left(-\frac{t}{T}\right), \tag{5.229}$$

$$L_r \frac{di(t)}{dt} = \omega L_r I_m \cos(\omega t + \alpha - \phi) - L_r \frac{R}{L} I_m \sin(\alpha - \phi) \exp\left(-\frac{t}{T}\right). \tag{5.230}$$

Summing (5.229) and (5.230), for the exponential part of (5.228) we have

$$v_{\exp}(t) = \left(L_r \frac{R}{L} - R_r\right) I_m \sin(\alpha - \phi) \exp\left(-\frac{t}{T}\right). \tag{5.231}$$

From (5.227), we have

$$\left(L_r \frac{R}{L} - R_r\right) = 0. \tag{5.232}$$

This indicates that the exponential dc offset is removed. The voltage is given by

$$v_r(t) = R_r I_m \sin(\omega t + \alpha - \phi) + \omega L_r I_m \cos(\omega t + \alpha - \phi). \tag{5.233}$$

In general,

$$A\sin(\theta) + B\cos(\theta) = \sqrt{A^2 + B^2}\sin\left[\theta + \arctan\left(\frac{B}{A}\right)\right]. \tag{5.234}$$

Comparing (5.233) and (5.234) we get

$$\begin{aligned} v_r(t) &= \sqrt{R_r{}^2 + (\omega L_r)^2} I_m \sin\left[\omega t + \alpha - \phi + \arctan\left(\frac{\omega L_r}{R_r}\right)\right]. \\ &= Z_r I_m \sin(\omega t + \alpha - \phi) \end{aligned} \tag{5.235}$$

Equation (5.235) shows that the voltage across the replica impedance does not contain any dc offset. It is phase shifted by $\phi$ degrees leading compared to the current.

## References

Barth P, Ludwig A, Schegner P (2003) Development of the fault and disturbance data handling in the German HV-network. In Proc. CIGRE SC-34 Colloquium, Sydney

IEEE (1991) IEEE Standard Common Format for Transient Data Exchange. Standard C37.111–1991, version 1.8

Ukil A (2005) Abrupt change detection in automatic disturbance recognition in electrical power systems. Ph.D. dissertation, Dept. Math. Tech., Tshwane University Technoloy, Pretoria, South Africa

Ukil A, Zivanovic R (2006) Abrupt change detection in power system fault analysis using adaptive whitening filter and wavelet transform. Electric Power Systems Research 76:815–823

Ukil A, Zivanovic R (2007) The detection of abrupt changes using recursive identification for power system fault analysis. Electric Power Systems Research 77: 259–265

## Section III: Objective Projects

## 5.10 Harmonic Analysis & Frequency Estimation

### *5.10.1 Harmonic Analysis*

Harmonic analysis is the study of the representation of functions or signals as the superposition of basic waves. It is not only used in the power systems but many other fields like acoustics, seismology and so on. The analysis concentrates on decomposing the signal in question in terms of a fundamental (frequency) signal and signals of integer multiples of that fundamental signals. These are called harmonics, hence the name harmonic analysis.

In power system, harmonics are detrimental to the normal operation of equipments, like equipment overheating, motor failures, capacitor failure and inaccurate power metering etc (Arrillaga and Watson 2003). Harmonic distortion also has a significant influence upon the power quality (Rosa 2006).

### 5.10.2 FFT-based Harmonic Analysis

Fourier analysis has been the key technology for ages behind the harmonic analysis in many a field. The key technique is to analyze the spectrum of the power system signal based on its FFT. The FFT-based algorithm should be used to determine the fundamental frequency which is 50 or 60 Hz depending on the country of application.

The subsequent harmonic analysis depends on various factors. The main concentration is on decomposing and analyzing the signal in terms of the integer multiples of the determined fundamental frequency. The factors to be considered for further harmonic analysis are as follows.

- Sources of the Harmonics
  - Rectifiers
  - Coverters
  - Thyristor controlled reactors
  - Rotating machines
  - Transformers
  - Arcing devices
- Effects of the Harmonics
  - Effects on rotating machines
  - Resonances
  - Degradation of power quality
  - Interference with protection devices
  - Effects on measuring devices
  - etc.

### 5.10.3 Further Aspects

Further challenging aspects of the harmonic analysis could be the compensation of the harmonic contents towards harmonic elimination. Network-dependent modeling and different filtering techniques, for example, bandpass filters, tuned filters, damped filters (Arrillaga and Watson 2003) etc are particularly important for this.

Another interesting aspect would be to incorporate the effects of the non-integer harmonics. FFT-based approaches are in general very good for analyzing the integer harmonic component, not the non-integer ones. Possible approaches could be to use short-term Fourier transform, using artifical neural networks along with the FFT-based approach and the like.

### *5.10.4 Frequency Estimation*

As evident from the foregoing discussion on the harmonic analysis, reliable estimation of the system frequency is critically important. There have been many approaches towards effective frequency estimation in the power system. Some of these are mentioned in the reference (Sachdev and Giray 1985; Terzija et al. 1994; Yang and Liu 2000).

One particular challenge is the dynamic frequency estimation. In the steady-state operation, the total power generated equals the system losses plus load. When the load and the generation match, the frequency is constant. However, the frequency deviates from its nominal value in case of a mismatch. The frequency increases when the generation exceeds the load and vice versa (Jordaan and Zivanovic 2006).

## References

Arrillaga J, Watson NR (2003) Power system harmonics, 2nd edition. John Wiley & Sons Ltd, Chichester

Jordaan JA, Zivanovic R (2006) Frequency estimation in power systems using dynamic leapfrog method. Measurement 39:451–457

Rosa FCDL (2006) Harmonics and power systems. CRC Press, Boca Raton, FL

Sachdev MS, Giray MM (1985) A least error squares technique for determining power system frequency. IEEE Transactions on Power Apparatus and Systems 104:437–443

Terzija VV, Djuric MB, Kovacevic BD (1994) Voltage phasor and local system frequency estimation using Newton type algorithm. IEEE Transactions on Power Delivery 9:1368–1374

Yang JZ, Liu CW (2000) A precise calculation of power system frequency and phasor. IEEE Transactions on Power Delivery 15:494–499

## 5.11 Phasor Estimation

### *5.11.1 Phasors and PMU*

Phasor is a quantitiy, usually a complex number, with magnitude and phase, measured with respect to a fixed reference. Phasors have a lot of utilizations in the power systems as they simplify computations involving the sinusoids.

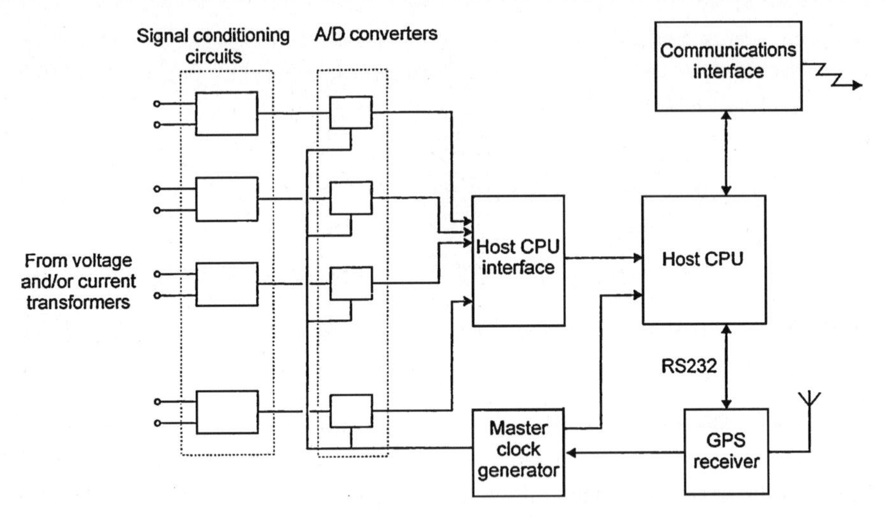


**Fig. 5.104** Block diagram of a PMU

Arising from the computer relaying developments in the 1960s, phasor measurement units, popularly known as PMUs (Grigsby et al. 2007) were developed in the 1980s. Figure 5.104 shows the schematic block diagram representation of a typical PMU.

The signal conditioning circuit interfaces to the voltage or the current transformer. It also isolates the whole system from the live section. The A/D converter then converts the input analog signal into the digital mode and feeds into the host central processing unit. The host CPU is also synchronized with a master clock generator and GPS (global positioning system) to facilitate synchronized phasor measurement. The host CPU unit also has a communication interface.

Figure 5.105 shows a simplified phasor representation.

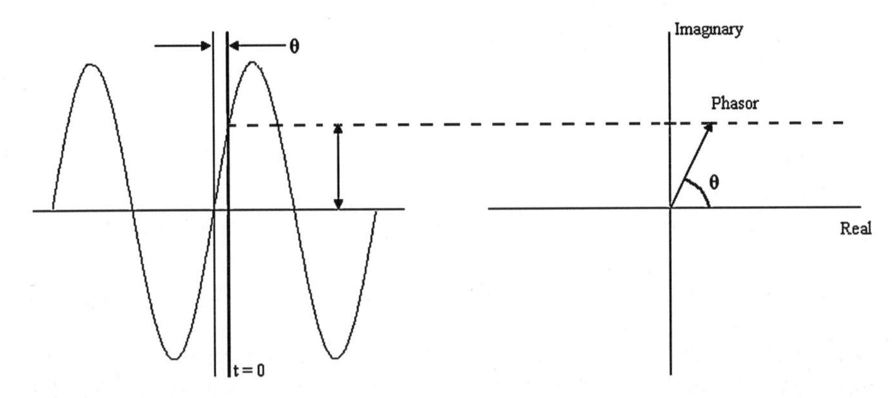


**Fig. 5.105** Simplified phasor

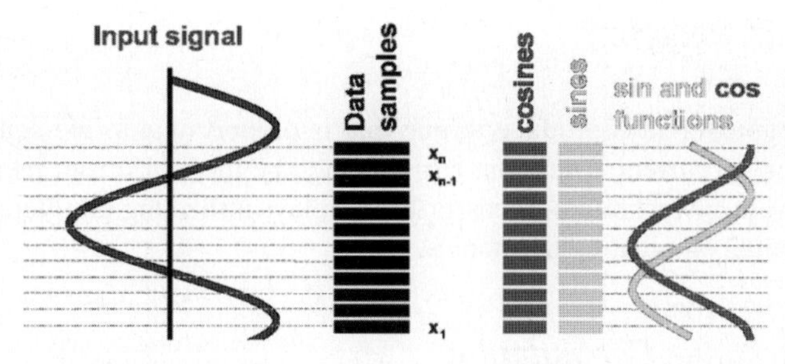


**Fig. 5.106** Fourier filters on the phasors

## *5.11.2 Phasor Estimation*

Figure 5.106 depicts the Fourier filters on the phasors.

Using the Fourier filter, the phasor $X$ can be estimated from the input signal $x_n$ as

$$X = \frac{\sqrt{2}}{N} \sum_k x_k \left(\cos k\theta - j \sin k\theta\right). \tag{5.236}$$

## *5.11.3 Applications of the Phasors*

Typical applications and objective projects involving the phasors are mentioned below.

- Frequency measurement (Grigsby et al. 2007)
- State estimation (Abur and Exposito 2004)
- Phasor-based controller design (Agarwal 2001)
- Protective relaying (Anderson 1999).

## References

Abur A, Exposito AG (2004) Power system state estimation. Marcel Dekker Inc., New York

Agarwal KC (2001) Industrial power engineering and applications handbook. Newnes, Woburn, MA

Anderson PM (1999) Power system protection. McGraw-Hill, New York

Grigsby LL (editor) (2007) The electric power engineering handbook, 2nd Ed., five Vol. CRC Publication, Florida

## 5.12 Digital Relaying

Electromechanical relays produce torque that is proportional to the square of the flux produced by current. The input current could be the RMS (root mean square) or the peak value. Most microprocessor-based relays utilize digital filters to extract the fundamental, eliminating harmonics.

### *5.12.1 Harmonic Computation*

Digital relays usually extract the required computational information from the $N$ samples included in a rectangular window (Phadke and Thorpe 1988). One particular challenge is to compute the harmonics from this fixed sized ($N$) windowed samples. The following methods could be utilized for such computation.

- Short-time DFT (Aravena 1990): using a sliding rectangular window of size $N$.
- DFT based on Walsh transform.
- Moving window FFT (Exposito and Macias 1999)

### *5.12.2 Inrush Currents*

Energizing a transformer often goes together with high magnetizing inrush currents. This high current is due to the remnant flux in the transformer core when it was switched out and it depends where on the sine waves the transformer is switched back in. Transformer protection must be set so that the transformer does not trip for this inrush current (Rahman and Jeyasuya 1988; Yabe 1997).

One interesting application would be the frequency analysis of the magnetizing inrush current in comparison with the fault current to differentiate either.

### *5.12.3 Analyzing Lightning Strike*

A high impulse current is associated with the lightning strike. This is very important from the protective relaying point of view. A possible approach could be to design a proper anti-aliasing filter which effectively attenuates the high frequency inrush current due to the lightning.

## References

Aravena JL (1990) Recursive moving window DFT algorithm. IEEE Transactions on Computers 39:145–148

Exposito AG, Macias JAR (1999) Fast harmonic computation for digital relaying. IEEE Transactions on Power Delivery 14:1263–1268

Phadke AG, Thorpe JS (1988) Computer relaying for power systems. John Wiley & Sons Inc., New York
Rahman MA, Jeyasurya B (1988) A state-of-the-art review of transformer protection algorithms. IEEE Transactions on Power Delivery 3:534–544
Yabe K (1997) Power differential method for discrimination between fault and magnetizing inrush currents in transformers. IEEE Transactions on Power Delivery, 4:1109–1117

# Section IV: Information Section

## 5.13 Research Information

This section is organized towards different applications of signal processing in various power engineering related problems. Power engineering specific signal processing-based research information in terms of relevant books, publications (journal, conference proceedings), reports etc, have been categorized into different sub-sections depending on the applications.

### *5.13.1 General Signal Processing*

Johnson JR (1989) Introduction to digital signal processing. Prentice-Hall, Engle wood Cliffs, NJ
Ifeachor EC, Jervis BW (2001) Digital signal processing, a practical approach, 2nd Edn. Prentice Hall, Upper Saddle River, NJ
Ingle VK, Proakis JG (1997) Digital signal processing using Matlab V.4. PWS Publishing Company, Boston
Lathi BP (2001) Signal processing and linear systems. Oxford University Press, New York
Lynn PA, Fuerst JG (1999) Introductory digital signal processing with computer applications, 2nd Edn.. John Wiley & Sons, West Sussex
Oppenheim AV, Schafer RW, Buck JR (2001) Discrete-time signal processing, 2nd Edn. Prentice-Hall, Upper Saddle River, NJ
Rabiner LR, Gold B (1975) Theory and application of digital signal processing. Prentice-Hall, Englewood Cliffs, NJ

### *5.13.2 Signal Processing and Power Engineering*

Boashash B ed. (1992) Time-frequency signal analysis–methods and applications. Longman Cheshire, Melbourne
Dash PK, Panda DK (1984) Spectral observation of power network signals for digital signal processing. Microprocessors and Microsystems 8:475–480

Heydt GT, Olejniczak KJ, Sparks R, Viscito E (1991) Application of the Hartley transform for the analysis of the propagation of nonsinusoidal waveforms in power systems. IEEE Transactions on Power Delivery 6:1862–1868

Lu ID, Lee P (1994) Use of mixed radix FFT in electric power system studies. IEEE Transactions on Power Delivery 9:1276–1280

### *5.13.3 Disturbance & Fault Analysis*

Din ESTE, Aziz MMA, Ibrahim DK, Gilany M (2006) Fault location scheme for combined overhead line with underground power cable. Electric Power Systems Research 76:928–935

Gan Z, Elangovan S, Liew AC (1996) Microcontroller based overcurrent relay and directional overcurrent relay with ground fault protection. Electric Power Systems Research 38:11–17

Gilany MI, Eldin EMT, Aziz MMA, Ibrahim DK (2005) A PMU-based fault location scheme for combined overhead line with underground power cable. IEE Conference Publication CP 508:255–260

Gu YH, Bollen MHJ (2000) Time-frequency and time-scale domain analysis of voltage disturbances. IEEE Transactions on Power Delivery 15:1279–1284

Kawady T, Stenzel J (2003) A practical fault location approach for double circuit transmission lines using single end data. IEEE Transactions on Power Delivery 18:1166–1173

Lai TM, Snider LA, Lo E, Sutanto D (2005) High-impedance fault detection using discrete wavelet transform and frequency range and RMS conversion. IEEE Transactions on Power Delivery 20:397–407

Lin T, Domijan AJ (2006) Real-time measurement of power disturbances: part 2 implementation and evaluation of the novel complex filter/recursive algorithm. Electric Power Systems Research 76:1033–1039

Lin YH, Liu CW, Chen CS (2004) A new PMU-based fault detection/location technique for transmission lines with consideration of arcing fault discrimination – part I: theory and algorithms. IEEE Trans. Power Deliv. 19:1587–1593

Lin YH, Liu CW, Chen CS (2004) A new PMU-based fault detection/location technique for transmission lines with consideration of arcing fault discrimination - part II: performance evaluation. IEEE Tran. Power Deliv. 19:1594–1601

Neves WLA, Brito NSD, Souza BA, Fontes AV, Dantas KMC, Fernades AB, Silva SSB (2004) Sampling rate of digital fault recorders influence on fault diagnosis. In Proc. IEEE/Power Eng. Soc. Transm., Distrib. Conf., Sao Paulo

Stankovic AM, Aydin T (2000) Analysis of asymmetrical faults in power systems using dynamic phasors. IEEE Transactions on Power Systems 15:1062–1068

Zhang B, Yi S, He X (2001) A novel harmonic current detection technique based on a generalized dqk coordinate transform for active power filter and fault protection of power system. IEE Conference Publication 478 II:543–547

### 5.13.4 Power Systems Protection

Altuve HJF, Diaz IV, Vazquez EM (1996) Fourier and Walsh digital filtering algorithms for distance protection. IEEE Trans. on Power Systems 11:457–462

Fazio G, Lauropoli V, Muzi F, Sacerdoti G (2003) Variable-window algorithm for ultra-high-speed distance protection. IEEE Trans. Power Delivery 18:412–419

García-Gracia M, Osal W, Comech MP (2006) Line protection based on the differential equation algorithm using mutual coupling. Electric Power Systems Research (In press)

IEEE Tutorial Course (1997) Advancements in microprocessor based protection and communication. Chapter 3.4

NengLing T, Yan D (2005) Stator ground fault protection based on phase angle differential of delta third harmonic voltages. Elec. Pow. Sys. Res. 74:203–209

Nguyen TT (2001) Dynamic responses of charge comparison digital differential protection for transmission lines. IEE Conference Publication 478 I:249–253

O'Shea P (2002) A high-resolution spectral analysis algorithm for power-system disturbance monitoring. IEEE Transactions on Power Systems 17:676–680

Wahab MAA, Matsuura K (1994) Modified Fourier transform lightning surge analysis of an overhead line-underground coaxial (CV) cable system part I. modal analysis of underground cables and overhead lines and effect of subterranean water on cable modal propagation characteristics. Electric Power Systems Research 31:31–42

Wahab MAA, Matsuura K (1994) Modified Fourier transform lightning surge analysis of an overhead line-underground coaxial (CV) cable system part II. system lightning surge response. Electric Power Systems Research 31:43–49

Wang YF, Ula AHMS (1991) High speed power system transmission line protection scheme using a 32-bit microprocessor. Elec. Power Sys. Res. 21:195–202

Xia YQ, Li KK (1994) Development and implementation of a variable-window algorithm for high-speed and accurate digital distance protection. IEE Proceedings: Generation, Transmission and Distribution 141:383–389

Zhiyuan C, Li Y, Shaohua M (2004) An intelligent circuit breaker with synchronous closing function based on DSP. IECON Proceedings 2:1536–1539

### 5.13.5 Relaying

Arguelles JAM, Arrieta MAZ, Domínguez JL, Jaurrieta BL, Benito MS (2006) A new method for decaying dc offset removal for digital protective relays. Electric Power Systems Research 76:194–199

Darwish HA, Fikri M (2007) Practical considerations for recursive DFT implementation in numerical relays. IEEE Transactions Power Delivery 22:42–49

Hart D, Novosel D, Calero F, Udren E, Yang LF (1996) Development of a numerical comparator for protective relaying: part II. IEEE Transactions on Power Delivery 11:1274–1284

Kim CH, Heo JY, Aggarwal RK (2005) An enhanced zone 3 algorithm of a distance relay using transient components and state diagram. IEEE Transactions on Power Delivery 20:39–46

McLaren PG, Swift GW, Neufeld A, Zhang Z, Dirks E, Haywood RW (1994) 'Open' systems relaying. IEEE Transactions on Power Delivery 9:1316–1324

Petit M, Le Pivert X, Bastard P, Gal I (2004) Symmetrical components to suppress voltage sensors in directional relays for distribution networks: Effect of distributed generation. IEE Conference Publication 2:575–578

Pinto de Sa JL (1994) Stochastic analysis in the time domain of very high speed digital distance relays part 2: illustrations. IEE Proceedings: Generation, Transmission and Distribution 141:169–176

Rosotowski E, Izykowski J, Kasztenny B (2001) Adaptive measuring algorithm suppressing a decaying DC component for digital protective relays. Electric Power Systems Research 60:99–105

Verma HK, Soundararajan K (1989) Digital differential relaying for generator protection: Development of algorithm and off-line evaluation. Electric Power Systems Research 17:109–117

Yalla MVVS (1992) A digital multifunction protective relay. IEEE Transactions on Power Delivery 7:193–201

Yalla MVVS, Smolinski WJ (1990) Design and implementation of a versatile digital directional overcurrent relay. Elec. Power Syst. Research 18:47–55

Yanquan H, Yong C, Jian X, Lei T, Jin L (2004) Research of implementing least squares in digital distance relaying for A.C. electrified railway. IEE Conference Publication 1:116–118

Youssef OAS (1992) A fundamental digital approach to impedance relays. IEEE Transactions on Power Delivery 7:1861–1870

Zhang Z, Li KK, Yin XG, Chen DS (2001) Adaptive application to impedance estimation algorithms in distance relaying. IEE Conf. Publ. 478 I:269–274

Zhijun G, Elangovan S, Liew AC (1996) Microcontroller based overcurrent relay and directional overcurrent relay with ground fault protection. Electric Power Systems Research 38:11–17

### *5.13.6 Transient Analysis*

Bollen MHJ, Styvaktakis E, Gu IYH (2005) Categorization and analysis of power system transients. IEEE Transactions on Power Delivery 20:2298–2306

Dufour C, Bélanger J (2001) Discrete time compensation of switching events for accurate real-time simulation of power systems. IECON Proc. 2:1533–1538

Gómez P, Moreno P, Naredo JL (2005) Frequency-domain transient analysis of nonuniform lines with incident field excitation. IEEE Transactions on Power Delivery 20:2273–2280

Hauer JF (1991) Application of prony analysis to the determination of modal content and equivalent models for measured power system response. IEEE Transactions on Power Systems 6:1062–1068

Hauer JF, Demeure CJ, Scharf LL (1990) Initial results in prony analysis of power system response signals. IEEE Transactions on Power Systems 5:80–89
Heydt GT, Jun J (1996) Rapid calculation of the periodic steady state for electronically switched, time varying power system loads. IEEE Transactions on Power Delivery 11:1860–1866
Marceau RJ, Rizzi JC, Mailhot R (1995) Frequency-domain transient stability criterion for normal contingencies. IEEE Trans. Power Systems 10:1627–1634
Moo CS, Chang YN, Mok PP (1995) Digital measurement scheme for time-varying transient harmonics. IEEE Transactions Power Delivery 10:588–594
Nguyen TT, Wong KP, Humpage WD (1981) Impulse sampling sequences in time-convolution electromagnetic transient analysis in power systems. Electric Power Systems Research 4:13–20
Semlyen A, Medina A (1995) Computation of the periodic steady state in systems with nonlinear components using a hybrid time and frequency domain methodology. IEEE Transactions on Power Systems 10:1498–1504
Wiot D (2004) A new adaptive transient monitoring scheme for detection of power system events. IEEE Transactions on Power Delivery 19:42–48
Wu CJ, Chen YJ (2006) A novel algorithm for precise voltage flicker calculation by using instantaneous voltage vector. IEEE Transactions on Power Delivery 21:1541–1548

### *5.13.7 Phasor Measurement and Analysis*

Altuve HJ, Diaz I, De la O Serna JA (1998) New digital filter for phasor computation. part II: evaluation. IEEE Transactions on Power Systems 13: 1032–1037
De la O Serna JA (2005) Phasor estimation from phasorlets. IEEE Transactions on Instrumentation and Measurement 54:134–143
De la O Serna JA, Altuve HJ, Diaz I (1998) New digital filter for phasor computation. part I: theory. IEEE Transactions on Power Systems 13:1026–1031
De la O Serna JA, Martin KE (2003) Improving phasor measurements under power system oscillations. IEEE Transactions on Power Systems 18:160–166
Guo Y, Kezunovic M, Chen D (2003) Simplified algorithms for the removal of the effect of exponentially decaying dc-offset on the Fourier algorithms. IEEE Transactions on Power Delivery 18:711–717
Hart D, Novosel D, Hu Y, Smith B, Egolf M (1997) A new frequency tracking and phasor estimation algorithm for generator protection. IEEE Transactions on Power Delivery 12:1064–1070
Lobos T, Leonowicz Z, Rezmer J, Schegner P (2006) High-resolution spectrum-estimation methods for signal analysis in power systems. IEEE Transactions on Instrumentation and Measurement 55:219–225
Messina AR, Vittal V, Ruiz-Vega D, Enriquez-Harper G (2006) Interpretation and visualization of wide-area PMU measurements using Hilbert analysis. IEEE Transactions on Power Systems 21:1763–1771

Rakpenthai C, Premrudeepreechacharn, Uatrongjit S, Watson NR (2007) An optimal PMU placement method against measurement loss and branch outage. IEEE Transactions on Power Delivery 22:101–107

Sidhu TS, Zhang X, Albasri F, Sachdev MS (2003) Discrete-Fourier-transform-based technique for removal of decaying DC offset from phasor estimates. IEE Proc. Generation, Transmission and Distribution 150:745–752

Sidhu TS, Zhang X, Balamourougan V (2005) A new half-cycle phasor estimation algorithm. IEEE Transactions on Power Delivery 20:1299–1305

Wang M, Sun Y (2004) A practical, precise method for frequency tracking and phasor estimation. IEEE Transactions on Power Delivery 19:1547–1552

Working Group H-7 of the Relaying Channels, Subcommittee of the IEEE Power System Relaying Committee (1994) Synchronized Sampling and Phasor Measurements for Relaying and Control. IEEE Transactions on Power Delivery 9:442–552

### *5.13.8 Frequency Measurement & Control*

Backmutsky V, Zmudikov V, Agizim A, Vaisman G (1996) A new DSP method for precise dynamic measurement of the actual power-line frequency and its data acquisition applications. Measurement 18:169–176

Duric M, Durisic ZR (2003) An Algorithm for off-nominal frequency measurements in electric power systems. Electronics 7:11–14

Duric MB, Durisic ZR (2005) Frequency measurement in power networks in the presence of harmonics using Fourier and zero crossing technique. In Proc. PSCC Conference, Liege, Belgium

Eckhardt V, Hippe P, Hosemann G (1989) Dynamic measuring of frequency and frequency oscillations in multiphase power systems. IEEE Transactions on Power Delivery 4:95–102

Jordaan JA, Zivanovic R (2006) Frequency estimation in power systems using dynamic leapfrog method. Measurement 39:451–457

Kezunovic M, Spasojevic P (1992) New digital signal processing algorithms for frequency deviation measurement. IEEE Trans. Power Delivery 7:1563–1573

Lobos T, Rezmer J (1997) Real-time determination of power system frequency. IEEE Transactions on Instrumentation and Measurement 46:877–881

Phadke A, Thorp J, Adamiak M (1983) A new measurement technique for tracking voltage phasors, local systems frequency, and rate of change of frequency. IEEE Trans. Power Applicat. Syst. PAS-102:1025–1038

Sidhu TS, Sachdev MS (1998) An iterative technique for fast and accurate measurement of power system frequency. IEEE Trans. Power Delivery 13:109–115

Thomas DWP, Woolfson MS (2001) Evaluation of frequency tracking methods. IEEE Transactions on Power Delivery 16:367–371

Yang JZ, Liu CW (2001) A precise calculation of power system frequency. IEEE Transactions on Power Delivery 16:361–366

### 5.13.9 Harmonic Analysis

Baghzouz Y, Burch RF, Capasso A, Cavallini A, Emanuel AE, Halpin M, Imece A, Ludbrook A, Montanari G, Olejniczak KJ, Ribeiro P, Rios-Marcuello S, Tang L, Thallam R, Verde P (1998) Time-varying harmonics: part I – characterizing measured data. IEEE Transactions on Power Delivery 13:938–944

Eyad AAAF, Ibrahim EA, Bettayeb M (1994) Power system harmonic estimation: a comparative study. Electric Power Systems Research 29:91–97

Girgis AA, Chang WB, Makram EB (1991) A digital recursive measurement scheme for on-line tracking of power system harmonics. IEEE Transactions on Power Delivery 6:1153–1160

Li C, Xu W, Tayjasanant T (2003) Interharmonics: basic concepts and techniques for their detection and measurement. Elec. Power Systems Research 66: 39–48

Lin HC, Lee CS (2001) Enhanced FFT-based parametric algorithm for simultaneous multiple harmonics analysis. IEE Proceedings: Generation, Transmission and Distribution 148:209–214

Lobos T, Kozina T, Koglin HJ (2001) Power system harmonics estimation using linear least squares method and SVD. IEE Proceedings: Generation, Transmission and Distribution 148:567–572

Lu SL (2005) Application of DFT filter bank to power frequency harmonic measurement. IEE Proc. Generation, Transmission and Distribution 152:132–136

Lu SL, Lin CE, Huang CL (1998) Power frequency harmonic measurement using integer periodic extension method. Elec. Power Syst. Research 44:107–115

Miller AJV, Dewe MB (1993) Application of multi-rate digital signal processing techniques to the measurement of power system harmonic levels. IEEE Transactions on Power Delivery 8:531–539

Moo CS, Chang YN (1995) Group-harmonic identification in power systems with nonstationary waveforms. IEE Proc.: Gener., Transm. Distrib. 142:517–522

Nguyen TT (1997) Parametric harmonic analysis. IEE Proceedings: Generation, Transmission and Distribution 144:21–25

Nunez-Noriega CV, Karady GG (1999) Five step – Low frequency switching active power filter for network harmonic compensation in substations. IEEE Transactions on Power Delivery 14:1298–1303

Radojevic ZM, Shin JR (2007) New digital algorithm for adaptive reclosing based on the calculation of the faulted phase voltage total harmonic distortion factor. IEEE Transactions on Power Delivery 22:37–41

Smith BC, Arrillaga J, Wood AR, Watson NR (1998) A review of iterative harmonic analysis for AC-DC power systems. IEEE Transactions on Power Delivery 13:180–185

Tarasiuk T (2007) Hybrid wavelet-Fourier method for harmonics and harmonic subgroups measurement-case study. IEEE Trans. Power Delivery 22:4–17

Testa A, Gallo D, Langella R (2004) On the processing of harmonics and interharmonics: using Hanning window in standard framework. IEEE Transactions on Power Delivery 19:28–34

Wang G, Dongyang X, Weiming M (2005) Am improved algorithm with high accuracy for non-integer harmonics analysis based on FFT algorithm and neural network. In Proc. PSCC Conference, Liege, Belgium

Watson NR, Arrillaga J (2003) Harmonic assessment using electromagnetic transient simulation and frequency-dependent network equivalents. IEE Proceedings: Generation, Transmission and Distribution 150:641–650

Xue H, Yang R (2002) Precise algorithms for harmonic analysis based on FFT algorithm. Proceedings of the CSEE 22:106–110

Xue H, Yang R (2003) Optimal interpolating windowed discrete Fourier transform algorithms for harmonic analysis in power systems. IEE Proceedings: Generation, Transmission and Distribution 150:583–587

Yang JZ, Yu CS, Liu CW (2005) A new method for power signal harmonic analysis. IEEE Transactions on Power Delivery 20:1235–1239

Zhang F, Geng Z, Yuan W (2001) The algorithm of interpolating windowed FFT for harmonic analysis of electric power system. IEEE Transactions on Power Delivery 16:160–164

### *5.13.10 Power Systems Equipments & Control*

Asnin L, Backmutsky V (2005) Data acquisition in power systems and its digital processing for measurement and automation. Computer Standards & Interfaces 28:176–182

Bohmann LJ, Lasseter RH (1990) Stability and harmonics in thyristor controlled reactors. IEEE Transactions on Power Delivery 5:1175–1181

Bottauscio O, Cardelli E, Chiampi M, Chiarabaglio D, Gimignani M, Raugi M (1994) Comparison between finite element and integral equation modeling of power busbar systems. IEE Conference Publication 384:28–31

Cheng Y, Lataire P (2004) Advanced control methods for the single phase unified power quality conditioner. IEE Conference Publication 3:11–15

Dinavahi VR, Iravani MR, Bonert R (2004) Real-time digital simulation and experimental verification of a D-STATCOM interfaced with a digital controller. Int. Journal of Electrical Power & Energy Systems 26:703–713

El-Naggar KM, Gilany MI (2007) A discrete dynamic filter for detecting and compensating CT saturation. Electric Power Systems Research 77:527–533

Garrido C, Otero AF, Cidra's J (2003) Low-frequency magnetic fields from electrical appliances and power lines. IEEE Trans. Power Delivery 18:1310–1319

Grady WM, Samotyj MJ, Noyola AH (1991) Minimizing network harmonic voltage distortion with an active power line conditioner. IEEE Transactions on Power Delivery 6:1690–1697

Guo J, Zou J, Zhang B, He JL, Guan ZC (2006) An interpolation model to accelerate the frequency-domain response calculation of grounding systems using the method of moments. IEEE Transactions on Power Delivery 21:121–128

Gustavsen B (2005) Validation of frequency-dependent transmission line models. IEEE Transactions on Power Delivery 20:925–933

Han B, Baek S, Kim H (2005) New controller for single-phase PWM converter without AC source voltage sensor. IEEE Trans. Power Deliv. 20:1453–1458

Mattavelli P, Verghese GC, Stankovic AM (1997) Phasor dynamics of thyristor-controlled series capacitor systems. IEEE Trans. Power Syst. 12:1259–1266

Montano A, Juan C, Lopez O, Antonio CI, Manuel GBJ (1990) A TMS320-based reactive power meter. IECON Proceedings 1:267–272

Pal B, Chaudhuri B (2005) Robust control in power systems. Springer, New York

Raghavan N, Taleb T, Ellinger T, Grossmann U, Petzoldt J, Vasudevan K, Swarup KS (2006) A novel lowpass to bandpass transformed PI control strategy for series hybrid active power filter. Elec. Power Systems Research 76:857–864

### *5.13.11 Power Systems Operation*

Ahn BS, Kim BI, Chang TG (2004) A sliding-DFT based power-line phase measurement algorithm and its FPGA implementation. IEE Conf. Publ. 1:44–47

Aiello M, Cataliotti A, Nuccio S (2005) A chirp-z transform-based synchronizer for power system measurements. IEEE Transactions on Instrumentation and Measurement 54:1025–1032

Bakhshai A, Espinoza J, Joos G, Jin H (1996) A combined ANN and DSP approach to the implementation of space vector modulation techniques. In Proc. Conf. Rec. IEEE–IAS Annual Meeting, pp 934–940

Bounou M, Lefebvre S, Malhame RP (1992) A spectral algorithm for extracting power system modes from time recordings. IEEE Transactions on Power Systems 7:665–683

Gu JC, Shen KY, Yu SL, Yu CS (2006) Removal of dc offset and subsynchronous resonance in current signals for series compensated transmission lines using a novel Fourier filter algorithm. Electric Power Systems Research 76:327–335

Guan JL, Gu JC, Wu CJ (2004) Real-time measurement approach for tracking the actual coefficient of ΔV / ΔV10 of electric arc furnace. IEEE Transactions on Power Delivery 19:309–315

Hamed M, Esmail D (1997) Statistical study for switching processes in compensated UHV transmission systems. IEE Proceedings: Generation, Transmission and Distribution 144:237–241

Heydt GT (1993) System analysis using Hartley impedances. IEEE Transactions on Power Delivery 8:518–523

Karimi-Ghartemani M, Iravani MR (2004) Robust and frequency-adaptive measurement of peak value. IEEE Transactions on Power Delivery 19:481–489

Kosterev NV, Yanovsky VP, Kosterev DN (1996) Modeling of out-of-step conditions in power systems. IEEE Transactions on Power Systems 11:839–844

Lev-Ari H, Stankovic AM (2003) Hilbert space techniques for modeling and compensation of reactive power in energy processing systems. IEEE Transactions on Circuits & Systems 50:540–556

Limebeer DJN, Harley RG, Erasmus SJ (1979) An investigation of subsynchronous resonance by Fourier transformation. Elec. Power Sys. Res. 2:133–143

Lu SL, Lin CE, Huang CL, Lu TC (1999) Power substation magnetic field measurement using digital signal processing techniques. IEEE Transactions on Power Delivery 14:1221–1227

Solomou M, Rees D, Chiras N (2004) Frequency domain analysis of nonlinear systems driven by multiharmonic signals. IEEE Transactions on Instrumentation and Measurement 53:243–250

Staroszczyk Z (2005) A method for real-time, wide-band identification of the source impedance in power systems. IEEE Transactions on Instrumentation and Measurement 54:377–385

Svensson J, Bongiorno, Sannino A (2007) Practical implementation of delayed signal cancellation method for phase-sequence separation. IEEE Transactions on Power Delivery 22:18–26

Yu SL, Gu JC (2001) Removal of decaying DC in current and voltage signals using a modified Fourier filter algorithm. IEEE Trans. Power Deliv. 16:372–379

### *5.13.12 Power Quality*

Chilukuri MV, Dash PK (2004) Multiresolution S-transform-based fuzzy recognition system for power quality events. IEEE Trans. Power Deliv. 19:323–330

Dash PK, Jena RK, Salama MMA (1999) Power quality monitoring using an integrated Fourier linear combiner and fuzzy expert system. International Journal of Electrical Power and Energy System 21:497–506

Gu IYH, Styvaktakis E (2003) Bridge the gap: signal processing for power quality applications. Electric Power Systems Research 66:83–96

Heydt GT, Fjeld PS, Liu CC, Pierce D, Tu L, Hensley G (1999) Applications of the windowed FFT to electric power quality assessment. IEEE Transactions on Power Delivery 14:1411–1416

Kandil MS, Farghal SA, Elmitwally A (2001) Refined power quality indices. IEE Proceedings: Generation, Transmission and Distribution 148:590–596

Kezunovic M, Liao Y (2002) A novel software implementation concept for power quality study. IEEE Transactions on Power Delivery 17:544–549

Lev-Ari H, Stankovic AM, Lin S (2000) Application of staggered undersampling to power quality monitoring. IEEE Transactions Power Delivery 15:864–869

Lin T, Domijan Jr. A (2006) Novel complex filter with recursive algorithm for phasor computation in power-quality monitoring. IEE Proceedings: Generation, Transmission and Distribution 153:283–290

Panda G, Dash PK, Pradhan AK, Meher SK (2002) Data compression of power quality events using the slantlet transform. IEEE Trans. Power Delivery 17: 662–667

Poisson O, Rioual P, Meunier M (1999) New signal processing tools applied to power quality analysis. IEEE Transactions on Power Delivery 14:561–566

Ribeiro MV, Romano JMT, Duqe CA (2004) An improved method for signal processing and compression in power quality evaluation. IEEE Transactions on Power Delivery 19:464–471

Ribeiro MV, Park SH, Romano JMT, Mitra SK (2007) A novel MDL-based compression method for power quality applications. IEEE Transactions on Power Delivery 22:27–36

Unsal A, Jouanne ARV, Stonick VL (2002) A DSP controlled resonant active filter for power conditioning in three-phase industrial power systems. Signal Processing 82:1743–1752

### *5.13.13 Load Flow*

Badawy EH, Youssef RD (1987) A method of analyzing sudden loss of loads on power systems. Computers & Electrical Engineering 13:7–15

George TA, Bones D (1991) Harmonic power flow determination using the fast Fourier transform. IEEE Transactions on Power Delivery 6:530–535

Lin WM, Zhan TS, Tsay MT (2004) Multiple-frequency three-phase load flow for harmonic analysis. IEEE Transactions on Power Systems 19:897–904

Saitoh H, Toyoda J, Kobayashi Y (1991) A new index extracted from line flow fluctuation to evaluate power system damping. IEEE Transactions on Power Systems 6:1473–1479

### *5.13.14 Load Forecasting*

Dash PK, Satpathy HP, Liew AC, Rahman S (1997) Real-time short-term load forecasting system using functional link network. IEEE Transactions on Power Systems 12:675–680

### *5.13.15 Power Systems Oscillation*

Chen J, Lie TT, Vilathgamuwa DM (2004) Damping of power system oscillations using SSSC in real-time implementation. International Journal of Electrical Power & Energy Systems 26:357–364

Kakimoto N, Sugumi M, Makino T, Tomiyama K (2006) Monitoring of interarea oscillation mode by synchronized phasor measurement. IEEE Transactions on Power Systems 21:260–268

Lee KC, Poon KP (1990) Analysis of power system dynamic oscillations with beat phenomenon by Fourier transformation. IEEE Transactions on Power Systems 5:148–153

### 5.13.16 State Estimation

Lin SY (1992) A distributed state estimator for electric power systems. IEEE Transactions on Power Systems 7:551–557

Madtharad C, Premrudeepreechacharn S, Watson NR (2003) Power system state estimation using singular value decomposition. Electric Power Systems Research 67:99–107

Monticelli A (1999) State estimation in electric power systems: a generalized approach. Kluwer Academic Publishers, Massachusetts

Rakpenthai C, Premrudeepreechacharn S, Uatrongjit S, Watson NR (2005) Measurement placement for power system state estimation using decomposition technique. Electric Power Systems Research 75:41–49

### 5.13.17 Power Systems Security

Jiang HL, Tang XJ, Dong ZY, Saha TK (2003) An advanced method for eliminating impacts from frequency deviations in power system signal processing. Electric Power Systems Research 67:177–184

Marei MI, El-Saadany EF, Salama MMA (2004) Estimation techniques for voltage flicker envelope tracking. Electric Power Systems Research 70:30–37

Sidhu TS, Cui L (2000) Contingency screening for steady-state security analysis by using FFT and artificial neural networks. IEEE Transactions on Power Systems 15:421–426

### 5.13.18 Power Systems Stability

Clark HK, Gupta RK, Loutan C, Sutphin DR (1992) Experience with dynamic system monitors to enhance system stability analysis. IEEE Transactions on Power Systems 7:693–701

Lavoie M, Qué-Do V, Houle JL, Davidson J (1995) Real-time simulation of power system stability using parallel digital signal processors. Mathematics and Computers in Simulation 38:283–292

Marceau RJ, Galiana FD, Mailhot R, Denomme F, McGillis DT (1994) Fourier methods for estimating power system stability limits. IEEE Transactions on Power Systems 9:764–771

Sharaf AM, Heydeman J, Honderd G (1991) Application of regression analysis in novel power system stabilizer design. Elec. Power Sys. Research 22:181–188

### 5.13.19 Power Management

McGrath DT (1991) Signal processing considerations in power management applications. Digital Signal Processing 1:245–250

### 5.13.20 Renewable Energy

Akkaya R, Kulaksiz AA, Aydo§du O (2006) DSP implementation of a PV system with GA-MLP-NN based MPPT controller supplying BLDC motor drive. Energy Conversion and Management (In press)
Battaiotto PE, Mantz RJ, Puleston PF (1996) A wind turbine emulator based on a dual DSP processor system. Control Engineering Practice 4:1261–1266
Venkatesh B, Rost A, Chang L (2007) Dynamic voltage collapse index – wind generator application. IEEE Transactions on Power Delivery 22:90–94

### 5.13.21 HVDC

Perkins BK, Iravani MR (1997) Novel calculation of HVDC converter harmonics by linearization in the time-domain. IEEE Trans. Power Delivery 12:867–873
Rittiger J, Kulicke B (1995) Calculation of HVDC-converter harmonics in frequency domain with regard to asymmetries and comparison with time domain simulations. IEEE Transactions on Power Delivery 10:1944–1949
Seifossadat G, Shoulaie A (2006) A linearised small-signal model of an HVDC converter for harmonic calculation. Elec. Power Syst. Research 76:567–581

### 5.13.22 Transformers

Bartoletti C, Desiderio M, Di Carlo D, Fazio G, Muzi F, Sacerdoti G, Salvatori F (2004) Vibro-acoustic techniques to diagnose power transformers. IEEE Transactions on Power Delivery 19:221–229
Girgis AA, Makram EB, O'Dell JWNK (1990) Evaluation of temperature rise of distribution transformers in the presence of harmonics and distortion. Electric Power Systems Research 20:15–22
Golshan MEH, Saghaian-nejad M, Saha A, Samet H (2004) A new method for recognizing internal faults from inrush current conditions in digital differential protection of power transformers. Electric Power Systems Research 71:61–71
Guzman A, Zocholl S, Benmouryal G, Altuve HJ (2001) A current-based solution for transformer differential protection-part I: problem statement. IEEE Transactions on Power Delivery 16:485–491

Inagaki K, Higaki M (1998) Digital protection method for power transformers based on an equivalent circuit composed of inverse inductance. IEEE Transactions on Power Delivery 4:1501–1510

Purkait P, Chakravorti S (2003) Investigations on the usefulness of an expert system for impulse fault analysis in distribution transformers. Electric Power Systems Research 65:149–157

Purkait P, Chakravorti S (2004) Can fractal techniques be used for impulse fault pattern discrimination in distribution transformers? Electric Power Systems Research 68:258–267

Sidhu TS, Gill HS, Sachdev MS (2000) A numerical technique based on symmetrical components for protecting three-winding transformers. Electric Power Systems Research 54:19–28

Zhang H, Liu P, Malik OP (2002) A new scheme for inrush identification in transformer protection. Electric Power Systems Research 63:81–86

### *5.13.23 Rotating Machines*

Al-Kazzaz SAS, Singh GK (2003) Experimental investigations on induction machine condition monitoring and fault diagnosis using digital signal processing techniques. Electric Power Systems Research 65:197–221

Calis H, Cakir A (2007) Rotor bar fault diagnosis in three phase induction motors by monitoring fluctuations of motor current zero crossing instants. Electric Power Systems Research 77:385–392

Kang MH; Kim NJ, Yoo JY, Park GT, Yu SY (1994) Variable structure approach for induction motor control-practical implementation of DSP. In Proc. Industrial Electronics, Control and Instrumentation, IECON '94., 1:50–55

Lo KL, Qi ZZ, Xiao D (1995) Identification of coherent generators by spectrum analysis. IEE Proc.: Generation, Transmission and Distribution 142:367–371

Pedra J, Sainz L, Córcoles F (2006) Harmonic modeling of induction motors. Electric Power Systems Research 76:936–944

Poyhonen S, Jover P, Hyotyniemi H (2004) Signal processing of vibrations for condition monitoring of an induction motor. In Proc. 1st Int. Symp. Control, Communications and Signal Processing, pp. 499–502

Yin XG, Malik OP, Chen DS (1990) Adaptive ground fault protection schemes for turbo-generator based on third harmonic voltages. IEEE Transactions on Power Delivery 5:595–601

Zhang J, Makram EB (2000) Analysis of a DC motor drive in the presence of unbalance. Electric Power Systems Research 53:223–230

Zielichowski M, Szlezak T (2006) A new digital ground-fault protection system for generator–transformer unit. Electric Power Systems Research (In press)

### 5.13.24 Power Electronics

De Doncker RW (2003) Twenty years of digital signal processing in power electronics and drives. IECON Proceedings 1:957–960

EL-Kholy EE, EL-Sabbe A, El-Hefnawy A, Mharous HM (2006) Three-phase active power filter based on current controlled voltage source inverter. International Journal of Electrical Power & Energy Systems 28:537–547

Gandhi M, Singh SN, Thakurta A, Kotaiah S (2006) Performance of a dsp based phase-shifted pwm controlled, zero-voltage-switching direct-current regulated magnet power supply. In Proc. IEEE Power India Conf., New Delhi, India

Leonowicz Z, Lobos T, Rezmer J (2003) Advanced spectrum estimation methods for signal analysis in power electronics. IEEE Transactions on Industrial Electronics 50:514–519

Maranzana G, Perry I, Maillet D, Raël S (2004) Design optimization of a spreader heat sink for power electronics. Int. Journal of Thermal Sciences, 43:21–29

Nazarzadeh J, Razzaghi M, Nikravesh KY (1997) Harmonic elimination in pulse-width modulated inverters using piecewise constant orthogonal functions. Electric Power Systems Research 40:45–49

Razzaghi M, Nazarzadeh J, Nikravesh KY (1998) A block-pulse domain technique of harmonics elimination in multilevel pulse-width modulated inverters. Electric Power Systems Research 46:77–81

Warburton WK, Grudberg PM (2006) Current trends in developing digital signal processing electronics for semiconductor detectors. Nuclear Instruments and Methods in Physics Research Section A: Accelerators, Spectrometers, Detectors and Associated Equipment (In press)

# Index